AF708410

About the Cover

The cover picture is a highly stylized rendition of part of the Mandelbrot set. The imaginary space of the Mandelbrot Set is used in Discourse 13: Cosmology of Imaginary Space to provide an idea of how it might be to navigate etheric space. The Set is an excellent analog to represent reality because it is created with a very simple mathematical process based on just a few assumptions. Yet, the resulting imaginary space is infinitely small and contains an infinite number of fractals.

This rendition titled *Fractal Spheres* was produced by Piotr Siedlecki and kindly posted at publicdomainpictures.net. Siedlecki also has an Internet presence at whoami911.deviantart.com.

Acknowledgments

My name is under the title because, admittedly, this book is the world according to me. Should you think I am crazy, it is just me. I might not have become involved in transcommunication had it not been for Lisa, so if you like the book, thank her.

Sarah Estep shined a light on the path which led to this book. Lisa took the first step for me, but the path has been paved by hundreds of past AA-EVP and then ATransC members, other ITC organizations and the pioneers who have been so brave in the light of certain ridicule.

Many people provided important contributions to our understanding of these phenomena. I would especially like to thank Margaret Downey and Cindy Heinen for their experimental support. Of course, members of the Big Circle gave important meaning to communication across the veil with loved ones. Some of the more instrumental participants in the Big Circle include Martha Copeland, Karen Mossey, Vicki Talbott, Teri Daner, Christine Jenkins, Rashelle Eastman Morrison, Erland Babcock and Debbie Caruso for their help in exploring the possibility of EVP.

The paranormalist community provided the background influence which gave me reason to compose this work in the first place. But every time someone made a faith-based comment as if it were fact, I hurried back to my computer and worked on this book. My prayer is that, after a while, there will be no need for a second edition.

Your Immortal Self
Exploring the Mindful Way

by
Tom Butler
With enabling support from Lisa Butler
We are Association TransCommunication Directors

AA-EVP Publishing
Reno, Nevada

First Edition, 2016

Website
EthericStudies.org
ATransC.org

Available from Internet book sellers or by order from local book stores

ISBN 978-0-9727493-9-8

Cover design by Tom and Lisa Butler.
Art by Piotr Siedlecki, publicdomainpictures.net

Printed in the United States by Lightning Source, lightningsource.com

Use License

© 2016 Tom and Alisa Butler

This book is intended to be used as an educational source. The authors reserve their rights under the copyright laws of the United States. Unless you follow the Creative Commons rules noted below, no part of this book may be used or reproduced in any manner whatsoever, including Internet usage, without written permission from the authors, except in the case of brief quotations embodied in critical articles and reviews.

However, the authors offer limited use of this work under the Creative Commons Attribution-Noncommercial-Share Alike 3.0 Unported License. © 2016 by Tom and Lisa Butler.

See creativecommons.org/licenses/by-nc-sa/3.0/

With the Exception of material specifically marked with ©,* you may:

Share: Copy and redistribute the material in any medium or format.

Adapt: Remix, transform and build upon the material with proper attribution.

Under no conditions, may the entire book, sections or chapters be reproduced without written permission from the author.

* I have permission to use some material which may be protected under copyright laws. Please ask if you have questions. (ATransC.org/contact.htm)

A Personal Note

Consider these situations:

- A person cuts another off in traffic or cuts in line at an event.
- A person sits in stopped traffic with music blaring so loud it rattles the lungs of nearby drivers.
- A neighbor takes off from his home on a motorcycle with very loud engine noise or lets a dog bark day and night.
- Target practice near homes or the incessant noise of off-road vehicles near campsites.
- A snide remark on a discussion board or criticism one would not dare make in person.
- Publicly carrying a gun which is only intended to kill.

None of these behaviors are illegal. In some communities, they are not even discouraged, yet individually, people react to being on the receiving end of these behaviors with primal instincts. It is human nature. Loud noise provokes a flight or fight response which is quickly replaced with an aggressive urge to respond in kind.

Cutting someone off in traffic is actually stealing time and space in an already crowded and hurried world. The victim must decide his or her place in the pecking order; a primal reaction follows, based on whether the person submits or responds in kind. It is only human.

Just because we can do something does not mean we should. Mindful living is based on the idea of sharing the world. A good citizen is one who considers the implications of each act before proceeding. How do your actions affect others?

When Lisa and I assumed leadership of the ATransC in 2000, I was very careful to keep my personal assumptions to myself and to remain pragmatically analytical. But as I learned about the implications of survival, I came to realize how important it is for people to understand that everything they do matters. We cannot legislate morality or force people into good citizenship. It must come from a desire to cooperate as part of the collective.

How to Use This Book

What is written here is one thought. In my mind, it is indivisible. If it were otherwise, this would be three books. I talk a lot in this book about the three aspects of a teacher which are acquired understanding, expression of that understanding as a way of living and the results of that way of living as a role model for others. All are aspect of the one thought, which is *teacher*.

This book is written as the three aspects of a teacher. The concepts explained in Section I represents the first aspect as acquired understanding or theory. The community in which those concepts must be developed (the way of living) is discussed in Section II. Learning to work with the phenomena thought to be the result of the concepts (the role model) are discussed in Section III.

While it is possible to read just the theory of Section I, I have taken every opportunity to relate how the theory might apply to the complexities of daily living and the many forms of phenomena. It is my hope that you will treat the book as one thought and do the work to integrate it into your daily living.

Layout

Think of this book as a tool for learning about a relatively complex subject. It is designed to make it as easy as possible for you. Even so, the complexity of some concepts and the way they are interrelated makes it necessary to frequently interrupt discussions to explain related concepts. While confusing at first, I think the confusion will be less as you gain a sense of how the book is designed. So take your time, be patient and keep in mind the potential benefits of understanding your etheric nature.

Internal References

The digital edition has internal hyperlinks to locations within the book which provide further information. Of course, hyperlinks do not work in the hardcopy version, so a bracketed page number has been provided after words or phrases I think require clarification. For instance, clarification of the lucidity concept in the sentence, *"it appears to be more a question of lucidity [151] than one of determination"* as stated on page 224 is available on page 151.

Always look to the Glossary to clarify confusing usage. If that does not help, use the contact tool at the bottom of each ethericstudies.org or ATransC.org web page to ask for further explanation.

Diagrams

Some of the diagrams are used in just about every discourse related to the Implicit Cosmology. To save space, the frequently referenced diagrams are located at the front of the book, just after the Contents and the list of diagrams. I recommend that you place a bookmark there for easy access.

Comments

Footnotes are used when I wish to interject a comment. Please note that, while I attempt to be objective and source-based in the body of this book, these footnotes are usually used to set aside my opinion or personal observations. They are indicated with a superscript number.

Glossary of Terms

As an experienced technical writer, I know to be specific about words I use and consistent in how I use them. Terms are usually used in their common or at least second or third meaning. But the context is novel and you will likely take a while to orient yourself within the context of the Implicit Cosmology.

One good way to adjust to this novel context is to take a little time in the beginning to normalize the terms. For example, you might have a phrase in your mind that means spiritual healing. In this book, that concept is explained as healing intention. Both terms have essentially the same intent, but the way it is worded here aligns with the cosmology.

Always look to the Glossary of Terms to clarify confusing usage. If that does not help, use the contact tool at the bottom of each ethericstudies.org or ATransC.org web page to ask for further explanation. Also check the dedicated pages at ethericstudies.org/immortal_self.htm for updates, corrections and verifications.[1]

---

[1] The Implicit Cosmology is continuing to evolve. It is my intention to maintain notes about its evolution, and other updates, at ethericstudies.org/immortal_self.htm; however, the website is based on old design technology. Soon after this book is published, I will begin updating all of the websites I am responsible for to comply with mobile reader technology. That may mean the dedicated update page for this book will have a new address. Always look to the front page for guidance.

The diagrams and examples do not do well in the grayscale format but they will be on the Immortal Self web page in color. Also, I know there are errors in the book, even after I spent months and used two editors to clean it up. Please consider letting me know via the website contact page if you notice anything that needs to be changed.

References

There is not a lot in this book that is explained in my terms. My only claim is that I have compiled the cosmology, but it is based on evolving thought of the paranormalist community. Most of the references are accessible on the Internet. They may not say exactly what I say, but you will be able to see why I designed the cosmology as it is.

As you seek supporting references, keep in mind that serious consideration of the Survival Hypothesis and transcommunication is relegated to the farthest frontier of popular thought. To be of use, most of the references must be taken with a great deal of reservation. It is the larger picture they suggest that I ask you to consider.

You are directed to the reference list with the appropriate number in parenthesis such as Rupert Sheldrake (9).

Website

Ethericstudies.org/immortal_self.htm will include access to information about this book. This will include corrections, clarifications and updates. The ATransC Idea Exchange discussion board is available if you wish to ask questions, make comments about this book, or discuss concepts.

Content

Diagrams

Some of the diagrams in this book are referred to many times. For simplicity, just one instance of those is located after this list of diagrams. It may be useful if you mark the location of the diagrams with a bookmark.

Pictures, examples and diagrams which are referred to just once are placed with the associated text. The list of diagrams below may be useful for future reference.

List of Diagrams and Examples

Frequently Referenced Diagrams

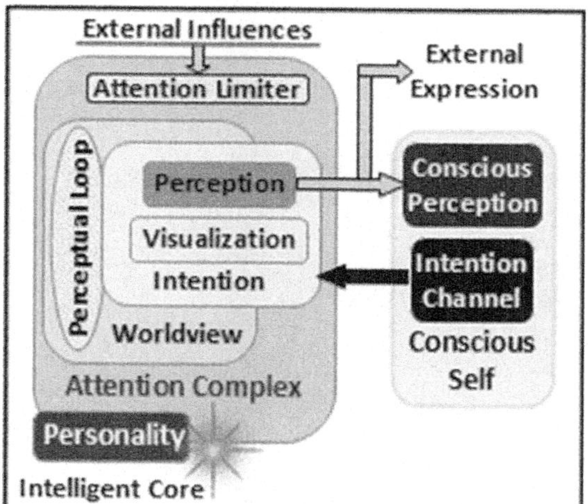

Functional Areas of a Life Field

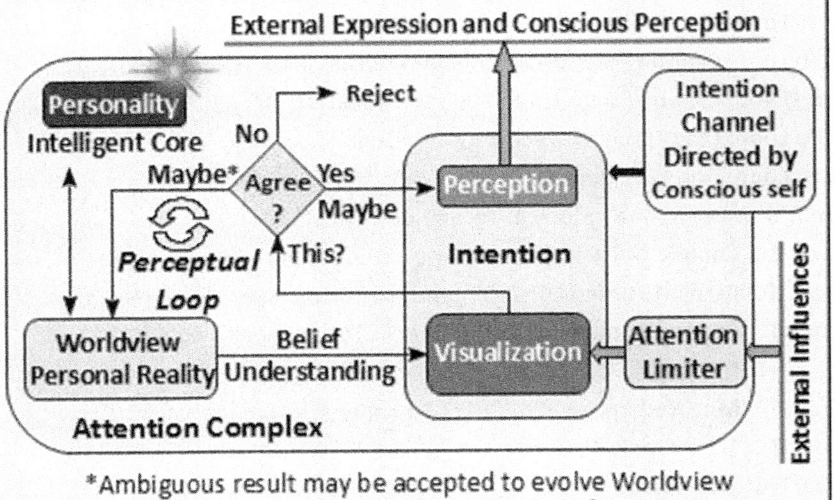

Functional Areas for Perception and Expression

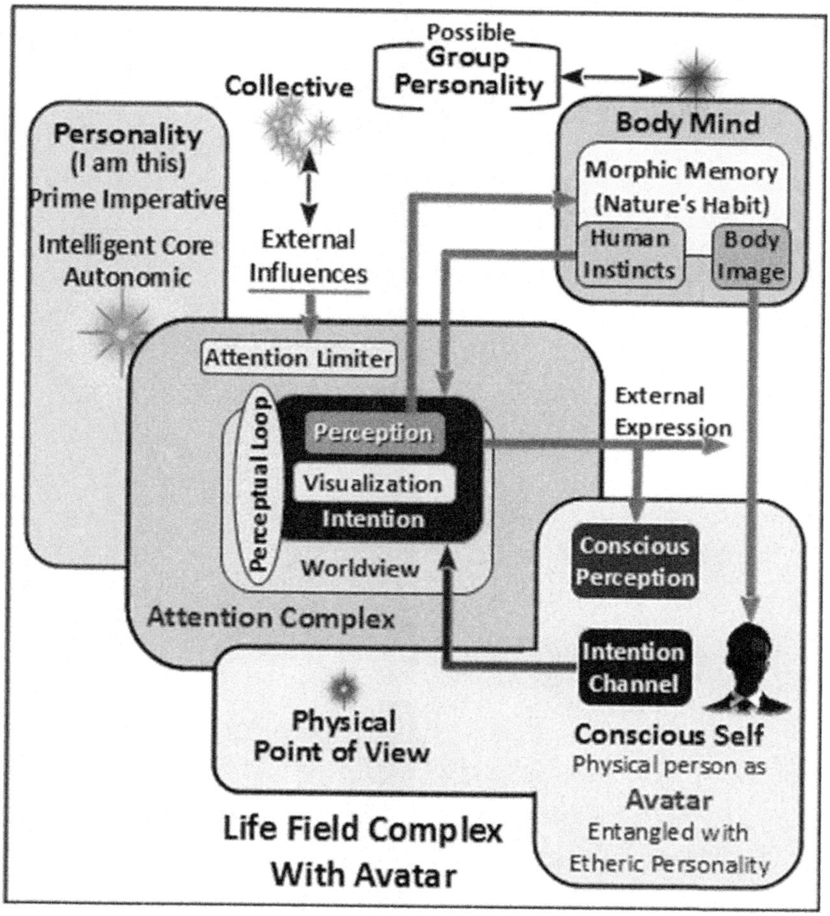
Possible
Group
Personality
Collective
Body Mind
Morphic Memory
(Nature's Habit)
Human
Instincts
Body
Image
Personality
(I am this)
Prime Imperative
Intelligent Core
Autonomic
External
Influences
Attention Limiter
External
Expression
Perceptual Loop
Perception
Visualization
Intention
Worldview
Conscious
Perception
Attention Complex
Intention
Channel
Physical
Point of View
Conscious Self
Physical person as
Avatar
Entangled with
Etheric Personality
Life Field Complex
With Avatar

Your Immortal Self
Exploring the Mindful Way

Introduction to this Book

In the Beginning

The Association TransCommunication (ATransC) was founded by Sarah Estep in 1982 as the American Association of Electronic Voice Phenomena (AA-EVP) to: *"Provide objective evidence that we survive death in an Individual Conscious state."*[2]

Lisa and I assumed leadership of the AA-EVP in 2000 and continued to publish the quarterly newsletter until its last issue, Spring 2014, Number 33-1. That made a total of 129 issues, 57 of which we published.

By 2010, we had come to recognize that transcommunication phenomena are related by a common set of influences. To study one, it is necessary to study all of them. We were also becoming aware of how people influence the message and how understanding the dynamics of transcommunication can facilitate mindful living. To better support the study of these phenomena, and to make the organization more open to people outside of the USA, we changed the name to Association TransCommunication in 2010.

In 2014, we changed the ATransC from a publicly supported membership organization to just publicly supported, meaning that we now depend entirely on donations, income from affiliate programs when website visitors use the Amazon and eBay links to make their purchases, and from the book we wrote, *No Death and There Are No Dead*. (1) As of 2014, the book has earned $37,000 for the Association.

[2] By convention, *Electronic Voice Phenomena* (EVP) is written as a plural. Sarah began with the term as plural and we have continued using it in that way. In fact, there are a number of different forms of EVP and different ways in which they are encountered, and as such, it is a plural expression when considered as a class of phenomena.

Many of the Association members were active EVP experimenters and participated in a number of important studies sponsored by the ATransC. Thanks to those members, during that time we had access to a very large amount of information about how ITC phenomena were collected, what worked, what did not and the problems associated with working with the phenomena of transcommunication.

Prior to becoming involved with the Association, we had been deeply immersed in many forms of human potential and the study of phenomena which included training in several healing modalities and various forms of mediumship. This also included emersion into the National Spiritualist Association of Churches (2) system of thought leading to ordination and certification as healer, medium and teacher.

Lisa writes a *Media Watch* column for the *National Spiritualist Summit* magazine and as of the end of 2015, she has produced 125 issues. She also prepares a Media Watch column for the ASPSI *Searchlight*. (3) I help as I am able, and together we have examined thousands of news articles about things paranormal. This has given me a sense of the various points of view in our paranormalist community and an idea of how the understanding of paranormal phenomena is developing in mainstream society.

The ATransC Collective (4) is a wiki intended to support collaborative development of Best Practices. Related reference material is gradually being added to the Collective, and between the Media Watch archive on ATransC.org and the Collective, we seek to provide easy access to references which may be of value in literature surveys necessary for research.

Since 2006, I have been a Wikipedia editor and have been present for some of the more contentious edit wars between moderate editors and the radical skeptic editors who now control the paranormal-oriented articles in that online encyclopedia. This has given me a sense of how the skeptical community thinks, their talking points and the dynamics of the mainstream-paranormalist interface. (5)

In 1994, I wrote the *Handbook of Metaphysics* (6) in an effort to give sufficient background information about metaphysical concepts to provide a context in which people could understand new concepts. The book was published, but poorly produced, and of course it is now outdated.

A Word About Article Names

It is important to me that you maintain a sense of where you are in this book. It is also important that you have a sense of how to compartmentalize the different parts of this book. The articles in the Implicit Cosmology Section are called *Discourses* because they are reasonably formal expository essays about the various aspect of the cosmology.

By comparison, the articles in the Community Section are called *Commentaries* because they are clearly my opinion. I want you to at least conditionally accept what is said in the discourses, while I would rather you take the commentaries with reservation.

Articles in the Transcommunication Section are called *Studies* because they are based on a lot of formal and informal studies of the subjects. I expect you to accept them at face value just as you would any authoritative report. You may disagree with them, but they should be used as references in your study. They are included in an effort to help you avoid starting from scratch in your study.

Section I: Implicit Cosmology (Theory)

The Trans-Survival Hypothesis Discourse is a rewrite of the Survival Hypothesis with new emphasis on transcommunication. If you accept my contention that Instrumental TransCommunication (ITC) provides important insight into the nature of reality, then you can understand that the Trans-Survival Hypothesis is very different from anything you have previously learned.

It became evident during the process of developing the hypothesis that many of the concepts needed to be addressed individually. For instance, concepts concerning the relationship between a person in a lifetime, the aspect of a person that is an immortal personality and the nature of the greater reality are described in more objective terms than the usual mystical perspective found in most paranormalist literature.

The Implicit Cosmology Discourse was written to explain how all of the concepts fit together. A proper model of reality should hypothesize how reality is arranged, how it works, its purpose for existing, our nature and how we relate to reality. Our purpose, how we can achieve our purpose and why all of this matters should be the bottom line, the *so what?* of the model.[3]

---

[3] It is important to note that the Implicit Cosmology could not exist in its present form without the understanding that has come from the paranormalist community and especially from ATransC members. It has been through the process of ordering my thoughts to write articles for the *ATransC NewsJournal,* and studying to answer questions from members and the public, that I have evolved my present point of view.

Section II: Community (Commentaries)

The paranormalist community consists of people who, in one way or another, accept the reality of paranormal phenomena. To understand what is known, believed and theorized about these phenomena, it is necessary to understand the environment in which they are experienced. The Community Commentaries are intended to describe the paranormalist culture. Of course, they are written from my perspective which has evolved from my study of transcommunication.

The commentaries may seem strident in parts. In actuality, the larger community is fragmented into many subcommunities of interest with little cross-community collaboration. The result is that our community is retreating while our detractors are becoming better organized.

Paranormalists are mild-mannered folk who shy away from contentious discussions. A good example is how hard some of us have tried to get people to help edit paranormal articles in Wikipedia. I was told once that, *"After seeing the abuse you have been subjected to, I would never consider becoming an editor."*

This reluctance to become involved in debate is understandable. Paranormal subjects are not about religion, but they are about abstract concepts such as human potential and connectedness. Most of us seek to understand the actual nature of our world and trying to impose our will on some skeptic is not consistent with that goal.

The Community Commentaries are written as a collaborate with you. Part of that effort is to identify the problem areas so that we can work together to improve. This is not going to happen unless we look behind the curtain and see how messed up the social dynamics really are.[4]

---

[4] I recently felt compelled to aggressively, publically denounce the treatment one of our physical mediums received from supposed scientists for whom he agreed to demonstrate for research. As of this writing, I seem to be pretty much alone in this. Is your silence implicit approval of the treatment?

Our ability to openly study and practice these phenomena is not guaranteed. If we as a community do not police ourselves, that freedom could go away. It has happened before! (7)

Section III: Transcommunication (Studies)

Those of you who know me know that I am always looking for the *so what* meaning of theory and experiences. The transcommunication studies provide the applied theory or *so what* of the Implicit Cosmology.

The section is based on a physical world, hands-on point of view that is informed by the perspective of etheric personality. If you find the theory too abstract, you may more easily relate to the transcommunication studies. I invite you try some of the techniques and go to the ATransC Idea Exchange at ATransC.org/forum/ to ask questions and share results.

Section I: Implicit Cosmology

Introduction to the Implicit Cosmology

> Editor's note: As you read this cosmology, please keep in mind that it is a theoretical model designed to provide a context for my understanding of observed phenomena and the results of considerable study by many researchers. Metaphysics has been studied for centuries; however, modern views, which consider current psi research and instrumental forms of transcommunication, require that contemporary metaphysics be seen in a new light.
>
> Cosmologies are designed to be tested and refined, so what you read here should not in any way be considered final truth. The Implicit Cosmology is only a model with which to study reality. It is the possible understanding to come from that study that one must consider.

Introduction

A common complaint from observers of the paranormalist community is that there is no widely accepted unifying theory to explain these phenomena, especially for the presumed survival phenomena. I have to agree, but I am on the wrong side of the academic-layperson partition to decide if that is true.

I try to be well-informed about the phenomena of interest to our paranormalist community. As discussed in the Community Commentary, the major categories of phenomena include psi functioning, physical and mental mediumship, healing intention, reincarnation, near-death experiences and out-of-body experiences. My observation is that these tend to be studied separately in isolated communities of interest within the paranormalist community.

Studying these phenomena in such isolation seems to cultivate a false sense of authority which has effectively stifled collaboration with other communities of interest. As I discuss in the Science Commentary, the ITC community is mostly ignored by parapsychologists.

That neglect is unfortunate, because from my perspective, ITC phenomena provide convincing, replicable examples that decidedly indicate survival. When I attempt to explain those examples to people, I necessarily include information about psi phenomena and then explain how it might be possible for a person to psychically function in a manner that produces the observed ITC.

You see, that is one of my points. I must include psi influences in order to explain ITC because the most viable explanation for ITC begins with the most viable explanation for psi functioning. I look to the work of parapsychologists so that I can better understand ITC. First Sight Theory [126] is a good example. (8)

As I continue to explain ITC, it is necessary to at least attempt an explanation of why a loved one's voice or likeness appears in some ITC examples. The best explanation I can find is the Survival Hypothesis, so I have added it to the explanation. Somewhere around the twelfth year of my explaining these phenomena, I gave up on science and developed the Trans-Survival Hypothesis.

Few people concerned with transcommunication have the necessary understanding of these complementary subject areas to develop a unifying model. As a director of an international organization concerned with these phenomena, it seems a form of dereliction of duty not to at least propose a model. In my experience, having something for others to change is a lot easier than beginning with a blank page. As such, the Implicit Cosmology is my offering for something to change. I invite you to help evolve it into a form more people in our community can accept as a working hypothesis.

Be mindful of the cosmology's narrow scope. While it probably can be applied to all forms of trans-etheric phenomena, including reincarnation and near-death experiences, it is specifically designed to address survival from the perspective of transcommunication.

Organization of this Book

This is the introduction to a set of discourses written to explain the major concepts which constitute the Implicit Cosmology. The project began as an effort to update the parapsychological version of the Survival Hypothesis while including information related to transcommunication. While the result of that initial effort is the Trans-Survival Hypothesis Discourse, that hypothesis is really just an argument to support the Implicit Cosmology.

After the initial draft of the Trans-Survival Hypothesis, I realized that too many of the foundation concepts were unlikely to be understood without further explanation. Thus, I wrote a series of discourses to explain those foundation concepts. Together the discourses in this section constitute the hypothesis, supporting explanations and a cosmology describing reality based on the hypothesis.

When constructing a cosmology, one must take sides. The point of view in this book is that the evidence sufficiently supports Survival over Super-Psi. Writing from the point of view of the Trans-Survival Hypothesis has resulted in a rather specific vocabulary, or more correctly, a unique choice of word meanings. I attempt to explain key terms as I first use them in each discourse, but remember that the Glossary of Terms is always available.

The cosmology should be considered a closed system. By that I mean the point of view represented by the terminology, internal references and way of describing these concepts are all interdependent. This circular reference has made writing a clear and concise explanation especially challenging. The way I have met the challenge is to repeat portions of explanations several times. So if you think I am repeating myself, just remember that practice is supposed to make perfect.

My Qualifications

I hold a Bachelor of Science degree in electronics engineering. Also, this book is based on over fifty years of study, personal experience and substantial investigation of the many aspects of trans-etheric phenomena. While this background is substantial, you have every reason to question this work and seek alternative explanations. This is especially true of subjects concerning human nature.

Should you have comments, questions or simply decide I do not know what I am talking about, you are invited to contact me via the ethericstudies.org or ATransC.org contact tool and let me know what you are thinking. Alternatively, I am keeping the ATransC Idea Exchange discussion board open at ATransC.org/forum/. I enjoy informed discussion and the Idea Exchange is an ideal forum for collaboration.

Brief Introduction to Important Terms

Each discourse is intended to explain a set of related concepts. The concepts are often explained by referring to other concepts, some of which are not explained until a later discourse. So for context, here are brief explanations of the major concepts.

Energy and Potential

You will probably be familiar with cosmologies that describe reality as planes of existence and that use terms such as higher, lower, finer and vibration. You will also be accustomed to thinking in terms of energy. None of these concepts are useful in this cosmology.

Concepts often spoken of in terms of energy are modeled here in terms of potential which is characterized as the influence of intention. There is no electromagnetic equivalent in the model, although to be technical, electromagnetic energy would be modeled as an expression of intention. See the Etheric Fields Discourse.

Potential has special meaning in that it is used in the sense of emergent form, rather than in the physical sense of a difference of potential between two electrical terminals. In it quiescent state, the etheric can be modeled as a nascent form of potential. More correctly, a nascent process. The idea is that etheric, as the fundamental element of reality, is an undifferentiated process which is defined by intention to produce intended order. This is an etheric or conceptual equivalent of a random physical process such as white noise or the output of a random event generator.[5]

The concepts of *higher* or *lower*, as they are used in New Age conversations to compare an ethereal being with an average physical person, are modeled as having more or less perceptual agreement and understanding. One is not at the same vibration as another. Instead, two people's worldviews are more or less in agreement (perceptual agreement).

Reading the Cosmology of Imaginary Space Discourse will give you a sense of the difference. The mathematical imaginary space is used as an

---

[5] This concept may be easier to understand if you think in terms of how much a process is constrained by design. A very constrained process is not easily changed—say trying to roll a marble out of a deep well. An unconstrained process is relatively easy to change—say trying to roll a marble out of a slight depression.

For instance, a radio is designed to lock onto a radio station and eliminate noise while favoring the broadcasted message. Radio reception is a very constrained process. By comparison, a random event generator is designed to be as free-ranging in frequency as possible. That is a very unconstrained process.

In this model, the initial or quiescent state of the etheric is a nascent process which means it is dynamic but undefined. It is in its most responsive state to intention. One practical result of this is that random noise is a fertile environment for the formation of transform EVP or precipitated form.

analog for reality in that movement from point to point is accomplished by changing original assumptions. Those assumptions are an analog for understanding. In more traditional cosmologies, reality is modeled as a layer cake, so you can see that understanding conceptual (imaginary) space requires an entirely different perspective and terminology.

Conceptual Versus Objective

We are taught to think of the mental processes of our mind as intangible and the physical world we sense with the five senses of our body as objective. The point of view taken in this cosmology is that all of reality is intangible, or more correctly, conceptual. We assign the characteristic of objectivity to parts of our personal reality as we have been taught from birth. For that to work, all of our physical senses must deliver information about what we call the physical world to our mind in a conceptual form. (Careful here: mind is conceptual, brain is objective.)

In this cosmology, the physical world is seen as an expression of life. To use a technical term, these things we call physical are ideoplastic [441] structures, meaning they are formed as a product of thought. The perspective I want you to have while reading this book is that we sense the world in a conceptual way but unconsciously translate perception of that mental information into objective form. This may be easier to understand after you read the Personality-Centric Perspective Discourse. Read that in order of discourses, though. Otherwise it will make little sense without the foundation material.

Life field

In this cosmology, Source represents the first cause for reality and god for the purpose of this cosmology. It is modeled as a life field. Because it is seen as a singularity from which everything has originated, it is also the reality field.

The Functional Areas of a Life Field Diagram [xxiii] shown below illustrates the major functions associated with a life field. All are unconscious processes. The personality functional area represents our immortal presence. When we have aligned our worldview with the actual nature of reality, our conscious self as our *I think I am this* and our personality as our *I am this* come to be in perceptual agreement and are effectively one (lucidity).

The intelligent core represents information required to maintain the life field. It is like a database that contains what Rupert Sheldrake refers to as Nature's habit. (9) It is informed by inherited purpose which serves to bias the influence of Worldview in the Perceptual Loop. See Morphic Fields in the Etheric Fields Discourse.

The attention complex supports perception and expression. You will encounter the functional areas of the Attention Complex many times in the book. An especially important functional area is Worldview which represents our memory, understanding, instincts and beliefs. (Instincts are from Source and from our human avatar.) An important process is the Perceptual Loop which includes the decision to accept or reject information. A third state for that is the *maybe* result which potentially integrates visualized perception with worldview to form new understanding. See the Perception and Expression Discourse.

The external influences include influences from other life fields and the psi signals expressed by ideoplastic structures. When entangled with a human in a lifetime (a person), the human's five senses are transformed into etheric signals in the brain and experienced by our perceptual processes in the Attention Complex. This is an unconscious process, and it is the results of that process we become aware of with our conscious self.

An important input to the Attention Complex is the Intention Channel. That is the only conscious input to the otherwise unconscious part of our mind. Everything begins with intention. It is up to us to intend to pay attention, focus, notice—all of those things that we decide and then expect result from our mind.

Avatar Point of View

The Personality-Centric Perspective Discourse explains the avatar concept. The Life Field Complex Diagram [xxv] shows the relationship between our life field and the life field that is our human body which is our host in this lifetime. This is referred to as a personality-entangled avatar relationship. It is important to note that the human life field is better described as a morphic field with a body image representing *Nature's habit* as described in the Hypothesis of Formative Causation. (9) These terms are explained in the Etheric Fields Discourse.

The diagram will make more sense if you see that our human body shares perceptual processes with us. The relationship is modeled in this way because, while the body maintains a self-image, it also influences our perception with its earth-evolved instincts.

Organizing Principles

You may be familiar with the concept of Natural Law in which reality is thought to be governed by never-changing (immutable) laws. Also in that model, natural laws are part of the expression of Infinite Intelligence (God, Source). As such, learning to understand Natural Law and living in accordance with it is expressed as spiritual maturity. (2) (10)

In the Implicit Cosmology, Source, as the reality field, is modeled as experiencing itself through aspects of itself. All life fields are direct or indirect aspects of Source. As aspects of Source, we gain understanding so that Source may gain understanding. We have a Prime Imperative [61] to do so.

Based on these assumptions, Source is still learning to know itself and is therefore evolving. As Source evolves, so does its expression. In this cosmology, natural laws are referred to as Organizing Principles. The Organizing Principles Discourse provides what I believe is a comprehensive list of these principles.

It is important to understand that the Organizing Principles are different from what you may be familiar with as natural laws. Again, consider them within the context of this cosmology. As a practical matter, for our moment in time, they are immutable, but the model requires that they are able to evolve as Source evolves.

Mindfulness

The implications of this cosmology are addressed in the Mindfulness Discourse. The idea of mindfulness is not unique to me, but as I have studied these phenomena and spent so much of my life contemplating their implications, the idea of mindfulness has become my most important understanding.

Our perception is filtered by our unconscious in the Attention Complex. As I understand current trends in psychology, the fact that our decision making is based on processes thought to be beyond our conscious control has led to the idea that we cannot have free will. (11) However, mindfulness includes a technique by which it is possible to moderate worldview. So in any moment, we may not have free will (self-determination), but if a person has practiced mindful living for some time, then self-determination can come from how that way of living has produced a more *correct* worldview.

Let me make that point another way. Think of it: our conscious self, that which we consciously assume is who we really are, is the source of intention for all of our unconscious processes. As we intend, so will we live. The problem is that there is considerable lag between the time we intend and the time our worldview allows us to change. Worldview, representing how we have learned to think, is very slow to change. Mindful living is a way to consciously begin the process of aligning worldview with the true nature of reality.

In a nutshell, that is the *so what* of this book.

Developing this Cosmology

Before I present this hypothesis, it is important for you to understand how it was developed. As an engineer, I am trained to look for how things work. For instance, it is not enough to say that a sound was made and we heard it. The question that tells us about who we are, the metaphysical view, is how we heard the sound. The physiological explanation is that bone and flesh structures in our head converted the mechanical sound into a small bioelectric stream of information to be presented to the brain.

Of course, the mainstream view for how a bioelectric signal is translated into awareness is that nerve cells and such work together to compare it with another bioelectric signal called memory. The result of that comparison is presented to the conscious mind. All of this is supposed to be biological body function. That is, the mainstream view is that mind is a product of brain.

In my fifty-plus years of seeking understanding about my spiritual nature and the world we live in, I have come to see that mind precedes brain and continues to function as a living entity after brain is dust. I understand current research to also say this, and so, a base assumption in this study is that brain is just a transmitter/receiver for mind. Any model that results from this study must address the reasons that have led me to that acceptance.

Mind is modeled here as a conceptual thing, meaning that it is a field of influence I refer to as personality. Personality is the person's intelligent core as it acts as the organizing influence for its life field. A person's conscious self as the *I think I am this* is the physical expression of personality while personality is the true *I am this.* In this view, the bioelectric sound signal representing objective sound must have an etheric component which can be sensed by the etheric personality. This component must be included in the model.

Black Box Analysis

Development of the hypothesis was a gradual process as I attempted to integrate each new concept into a unified model. The technique I used is sometimes referred to as black box analysis. The idea is that any process can be reduced to functions that respond to certain inputs to produce specific outputs. It is possible to design functional areas operating in the box that will respond to known inputs to produce known outputs.

When this process is used to reverse engineer products, it is important that the resulting functions produce the desired output, but it is helpful if the new design does not replicate the example design, else there may be patent infringement problems. This is usually not a problem because, until hardware design begins, black box analysis is entirely a thought exercise.

It must be understood then that black box analysis can produce meaningful models that are only representative of reality but not necessarily the same as reality. I am satisfied that this modeling technique is useful to explain the cosmology, but I make no claim to say that it is exactly the same as reality. A person with a different background might imagine very different solutions to the same problem.

The first step in the analysis was to locate the physical-etheric interface. (From the metaphysical view, it is the same to say *objective-conceptual interface*.) Mainstream scientists (11) (12) are beginning to agree with parapsychologists such as James Carpenter, author of *First Sight Theory* (13), in saying that we first unconsciously process environmental information and that we only become aware of what we unconsciously think of that information. Paradoxically, we experience only a version of reality based on our cultural training. The interface, then, is where the conscious self generally stops and the unconscious mind begins. As it turns out, this is at the objective boundary of the physical body. That is, the biological brain is objective, mind is conceptual.

Signals from the body's five senses must pass through the objective-conceptual interface, but otherwise, all of the inputs to the black box are conceptual and originate in etheric space. Outputs are expressions resulting from the perception and expression process in the Attention Complex and Understanding, which is shared with personality.

A second factor in black box analysis is a set of rules and implicit processes which moderate the functions in the box. This is important when there are conditional outputs such as expression.

It may help for you to understand that expression means everything that comes from the perceptual process. This is based on the understanding that perception is an unconscious mental process, while conscious awareness of the results of that process is in the form of an expression. This includes such signals as understanding, decision to speak, and selection of specific words, emotions, vision, hearing, feeling, smelling, tasting and influence which produce an ideoplastic structure. In effect, the physical body is just an input device for the perception into the Attention Complex and an output device for expression into physical space. The self only becomes aware of what is to be output from the unconscious as expression and what emerges from the unconscious perception process about the physical.

Consider mental mediumship as an example. Information previously consciously unknown to the medium is the input. A second input is intention to access information. The output is also that information, but it is colored by the medium's worldview. The implicit process is that there is a database function in the box that contains information which is used to color the medium's expression. This process is described as the Perceptual Loop and Worldview functional area in the Perception and Expression Discourse.

In practice, for the complex system represented by reported paranormal phenomena and human nature, many single-threaded design solutions such as that for mediumship could be proposed for the black box functions. Fortunately, there are similarities amongst paranormal phenomena, and shared functions emerge when they are considered together.

The emergent characteristics are identified in the Perception and Expression Discourse, and their interaction is illustrated in the Functional Areas of Perception and Expression Diagram [xxiv]. Unconscious information processing is accomplished in the Attention Complex, and the output of that is referred to as expression. The world around us which we think of as real is modeled as an expression of intended order.

It is important to note that, when black box modeling is used in a metaphysical cosmology, a plausible explanation may be developed that does not correctly describe reality. While these explanations are useful to develop the context of this cosmology, it is for you to decide to what degree the model makes sense.

Six Pillars

It might be helpful as a memory aid to summarize the Implicit Cosmology in terms of foundation concepts. Because it is about reality, there are many, but these six sum up the cosmology rather well.

Pillar 1, Reality: Reality consists of life fields and expressions of life fields (Reality Discourse).

Pillar 2, Person: A person is an immortal etheric personality entangled with a human in an avatar relationship (Life Fields Discourse).

Pillar 3, Purpose: We inherited the urge to gain understanding about the nature of reality from our source personality (Prime Imperative Organizing Principle).

Pillar 4, Mind: Unconscious psi sensing precedes conscious perception with unconscious worldview as the arbiter and conscious intention as the motivator (Perception and Expression Discourse).

Pillar 5, Creation: The creative process is attention on an imagined outcome to produce an intended order (Perception and Expression Discourse).

Pillar 6, Progression: Integrating acquired understanding of the nature of reality into worldview constitutes progression and expands opportunities for greater understanding (Perceptual Agreement Discourse).

Implicit Cosmology

Discourse 1:
Trans-Survival Hypothesis

Introduction

The Survival Hypothesis you may be accustomed to is little more than a proposal that some of the phenomena people experience can be better explained if mind is separate from body. In that view, the brain is only a transducer between the physical and some other aspect of reality in which mind exists before and after a lifetime. Of the contending hypotheses, the Super-Psi Hypothesis and the Physical Hypothesis are the most important.

None of the versions of the Survival Hypotheses I have seen explicitly consider the phenomena of Instrumental TransCommunication (ITC). Being familiar with ITC, and being both unable to account for examples of ITC with known physical principles as well as by applying Super-psi, I have compiled the Trans-Survival Hypothesis as an exercise to model reality as if Survival is literally true.

In the Trans-Survival Hypothesis, a person is modeled as a physical body and an etheric personality entangled in a symbiotic relationship, with the physical body functioning as an avatar for personality's presence in the physical. The core personality is modeled from an etheric or nonphysical personality-centric perspective, while the conscious self is modeled from a human avatar-centric perspective. From the perspective of the conscious self, personality is modeled as a normally unconscious aspect representing a person's mental musing and dream world.

The model includes speculation that the purpose of the avatar-personality relationship is to permit personality to gain understanding about the operation of reality given the constraints of physical principles. Speculation about a purpose is not necessary for the hypothesis but does provide useful context in which mindful living makes more sense. It also offers a possible mechanism by which a person might experience a more self-directed lifetime.

The etheric personality is seen as having evolved separately from the physical body, while the human avatar is modeled as an organism that has evolved in the physical. In this view, the conscious self is seen as an aspect of personality which is entangled with the avatar during a lifetime.

About Hypotheses

The objective of this hypothesis is to provide a context in which relevant phenomena, both mundane and anomalous, can be considered as a system for critical evaluation. If properly developed, this should include itemization of what is considered to be given as initial considerations. As a unifying model of reality, initial considerations would include observed phenomena, the circumstances under which they are experienced, and the scope or limits to which the hypothesis applies.

Next, this hypothesis should propose a way in which the givens combine to interact as a system. This includes physical properties such as technology and human interaction. Known or proposed properties should be substantiated with some measure of rational, and if available, empirical or objective research.

It is important that this hypothesis provides predictive implications of the proposed theory. Ideally, these should be testable. Any hypotheses should be able to be proven wrong; however, in a practical sense, verification might be partial or delayed, awaiting new technology or funding for research. Considering that reality, this hypothesis might remain *open for discussion* for many years.

Perspective

Most hypotheses are designed from the perspective of a physical person surrounded by the universe. In the first two hypotheses discussed below, the physical perspective constitutes all of reality. Be mindful that the Trans-Survival Hypothesis is designed from the perspective of a person standing at the hypothesized physical-etheric interface so that both etheric and physical characteristics are considered as a system. You should also note that perspective in terms of a physical person and an etheric person might be discussed in the same paragraph.

Dominant Views of Reality

The Physical and Super-Psi Hypotheses are briefly addressed here to provide perspective. There is little sense in spending a lot of time with the Physical Hypothesis but Super-psi will be addressed from time to time throughout the book. Three dominant views of reality considered in this book are described along with their primary features.

The Physical Universe Hypothesis

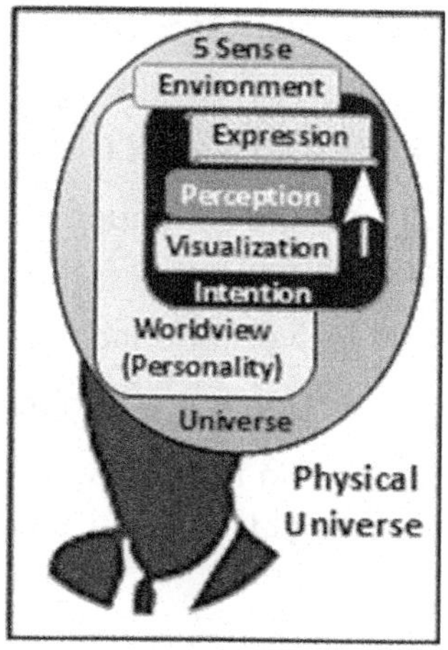

- All that exists is the physical universe.
- The universe has evolved from a singularity into what it is today.
- Life has evolved on earth into what it is today.
- Mind has evolved as a product of brain which is a product of evolution.
- Memory is an artifact of mind.
- When the brain dies, mind and memory cease to exist.
- People have five senses: smell, sight, hearing, touch and taste.

The Super-Psi Hypothesis

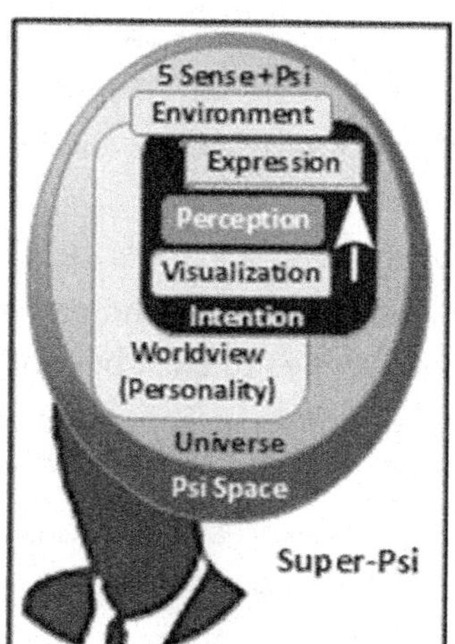

- All that exists is the physical universe.

- The universe may have evolved from a singularity into what it is today.
- An as yet unidentified form of space called psi (psi field) permeates all of physical reality.
- Life has evolved on earth into what it is today.
- Mind exists in the psi field and continues beyond death of the brain as differentiated, residual energy.
- Brain is a transmitter/receiver for mind.
- Thought, memory and emotions are retained in the psi field.
- People experience reality via five bodily senses that are informed by impressions from the psi field.

The Survival Hypothesis

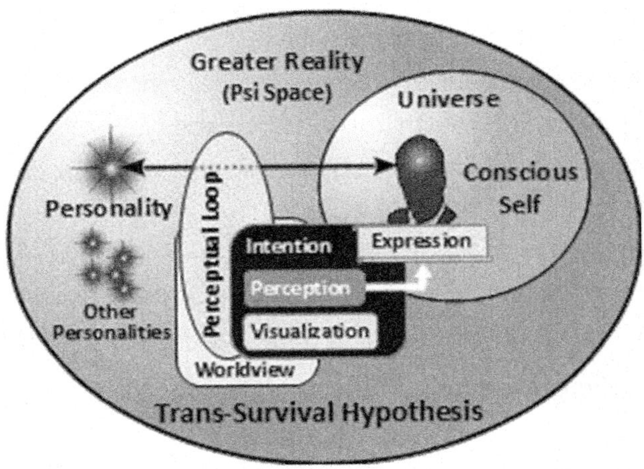

- There is a greater reality of which the physical universe is an aspect.
- An as yet unidentified form of space called *psi* permeates all of reality.
- The psi field is an aspect of the greater reality.
- Mind, with its thoughts, memories and emotions, has evolved in the greater reality and continues to exist beyond death of the brain.
- For a lifetime, mind and brain are entangled to produce a physical-etheric link: a person.
- During a lifetime, mind is expressed as consciousness (*I think I am this*) and a mostly unconscious etheric personality (*I am this*).
- Unconscious mind is informed by the person's five physical senses and psi signatures from the environment.

- Unconscious mind expresses to conscious self an understanding of the environment as it is informed by worldview (memory, experience, human and personality instincts).

Scope of the Three Hypotheses

Of course, the Physical Universe Hypothesis represents the mainstream view. It does not address any of the observed anomalous phenomena and is only listed here for comparison. The Super-Psi Hypothesis is largely supported by the parapsychological community and the Survival Hypothesis is primarily championed by the community studying transcommunication. The Super-Psi and Survival Hypotheses will be considered here.

The Super-Psi and Survival Hypotheses are described here as sharing the characteristic of psi functioning, but they are shown to differ in how this is accomplished. The contention of the Survival Hypothesis that anomalously accessed information may be from a still living personality is a contradiction of the Super-Psi Hypothesis which holds that such information must be from a subtle energy field containing memories of the past. The Survival Hypothesis will be described using many of the Super-Psi assumptions and proofs since the validity of Trans-Survival depends to a large extent on the validity of many aspects of Super-Psi.[6]

Characteristics That Must Be Addressed

Any theory intended to explain reality should address characteristics of experienced phenomena. The major categories of these experiences and the communities of interest studying them are listed here.

Physical World

This is the entire suite of physical-world experiences, including such mundane characteristics as gravity, heat, light, sound, space and evolution. Known physical principles should be accounted for so that magic is not required. Physical process might be modeled as characteristics of ideoplastic structures [104], but then they need to be related to how a nonphysical intelligence might interact with them. How gravity is related to levitation is an example.

---

[6] Note that the above description of the Survival Hypothesis is fairly common usage in the paranormalist community. The last item describing how conscious self becomes aware of its environment is a relatively new concept just now becoming part of parapsychology thanks to First Sight Theory. (13)

This category is addressed by mainstream science, at least within bounds of the Physical World Hypothesis.

Psi Phenomena

Information accessed by means other than with the five senses or noncontact influence is referred to as *anomalous information access* in parapsychology. This includes human abilities referred to as psychic, psi sensing or psi functioning. Examples include remote viewing, healing intention, psychometry, environmental sensing and psychokinesis. Most theories explaining these experiences are based on the Super-Psi Hypothesis and seldom consider survived personality as a source of information.

Transcommunication

The term *Transcommunication* applies to any form of information passing across the physical-etheric interface. Instrumental TransCommunication (ITC) is the name we use for the primary forms of influence in this category. ITC phenomena include the audible subset known as Electronic Voice Phenomena (EVP) and visual forms such as images found in visual noise. ITC may be spontaneous but is usually studied as an induced form of trans-influence.

After-Death Communication (ADC) is well known as a spontaneous form of transcommunication. A good study of this has been conducted by Bill and Judy Guggenheim who reported their work in *Hello from Heaven.* (14)

Other forms of transcommunication include apports, precipitation, materialization, transfiguration and mediumship. I often repeat in this book that parapsychologists have little interest in transcommunication if it is not seen as psi functioning. Mediumship is an important exception. A few researchers are working very hard to establish that information acquired via mental mediumship is at least sometimes from discarnate personalities.

As I will explain in this book, the model I use in the Implicit Cosmology for the functional areas of a person suggests that all anomalous information access is the same and the only distinction is whether it is from a discarnate personality or from one still entangled with a human avatar.

Reincarnation, OBE and NDE

Reincarnation, Out-of-Body Experiences (OBE) and Near-Death Experiences (NDE) provide important insight into the relationship between mind and body. While they appear to show that mind is separate from body, there remains considerable discussion as to whether or not this separation means survival of mind. These three experiences, along with mediumship, are often referred to as the best evidence for Survival. (15)

From the perspective of transcommunication and survival, the phenomena of reincarnation, OBE and NDE are seen as providing very important evidence (16), but there has been little support from the parapsychological community to relate that evidence to transcommunication. For this hypothesis, these experiences are considered natural consequences of etheric personality-physical body entanglement.

The Trans-Survival Hypothesis

You might consider reading these cosmology discourses and then refer back to them while examining the references. This hypothesis is offered here with the understanding that it will have little impact on the community if it is not understood and recommended by community opinion setters.

Statement of the Trans-Survival Hypothesis

It is useful to remember that this hypothesis is developed and explained from an etheric personality-centric perspective, rather than a physical body-centric perspective.

Top View

As a general statement, reality exists as life and expressions of life. Characteristics necessary for the operation of reality are detailed in the Organizing Principles Discourse.

Initial Conditions

For completeness, the singularity called Source is used as the initial condition for this hypothesis. Since this hypothesis is concerned with the progression of personality, if just personality is considered, the cause and nature of reality may be ignored. Initial Condition include:

1 There is a Source as the top life field and the top organizing field. As a singularity, Source is the reality field. See the Etheric Fields and Life Fields Discourses.

2 For simplicity, Source is assumed to be (at least) self-aware and curious about its nature. This curiosity is expressed as a creative process to explore its nature. See the Perception and Expression Discourse.

The Anticipation Corollary of First Sight Theory, [126] which essentially states that mind seeks to anticipate events, provides reasonably good support for this idea of a personality having a natural tendency to understand its environment in order to anticipate changes. If this is a valid argument, then all else follows.

3 Seeking to understand itself, Source has imagined aspects of its personal reality and aspects of itself to experience those sub-realities.

4 The process of imagining with the intention of understanding has produced a hierarchy of personalities (aspects of Source) as a Hierarchy of nested sub-subfields [98].

5 In response to a prime imperative [61], aspects of Source express aspects of themselves with the intention to understand the nature of reality (which is the nature of Source). This translates in an individual as an inherited urge to gain understanding through experience (progression [73]).

A Person

Personalities, in this hypothesis, are the experiential aspect of reality. One of the reasons for proposing Source as part of the initial condition is to establish a purpose which might explain the nature of the conscious self. According to the model, personalities are directly or indirectly aspects of, and inherit purpose from, Source. As such, inherited purpose is characterized as a prime imperative to gain understanding about the nature of reality (the source field or reality field). This is comparable to the Spiritualist's effort to understand and live in accordance with the principles of Natural Law.

To facilitate that search for understanding, personality acts as a creator to express an aspect of itself into an imagined venue for learning. In this model,

1 A person is an etheric personality entangled with a human in an avatar relationship.

2 From a person-centric perspective, etheric personality is a mostly unconscious influence on the person's conscious self. The relationship is explained in the Life Fields Discourse.

3 When the human's lifetime ends, personality's perspective returns to some other aspect of reality in a process referred to as transition.

4 Trans-etheric influences, which are the expression of personality's intention, are facilitated by the avatar relationship.

5 Lucidity between conscious self and mostly unconscious personality is regulated by a perception and expression process representing worldview.

Purpose of a Lifetime

As it was shown in the initial conditions, individual personalities (aspects of Source) respond to the Prime Imperative by seeking to understand the nature of reality. As it is modeled in the Implicit Cosmology (which is based on this hypothesis), many rounds of aspectation, resulting from individual efforts to understand the nature of reality, will produce a hierarchy of nested fields.

In the same way, it is argued that a *local* source functions as an attractor for a collective of entangled aspect personalities (aspects of that local source personality). In this model, personalities in the same collective will share the same inherited expression of the Prime Imperative. They will express that urge to learn differently, depending on such factors as circumstance and ability of the avatar.[7]

Scope of Survival

The question of survival necessarily addresses that which survives. In this model, a person is a personality entangled with an avatar. Both the etheric personality and the human organism are modeled as fields. The Life Fields Discourse explains the characteristics of life fields and the Life Field Complex Diagram [xxv] illustrates a person's functional relationship between personality and avatar.

With this perspective, it is clear that survival is really a misnomer. The evidence seems to suggest that personality exists prior to its entanglement with a human for the purpose of a lifetime. As such, the more correct term might be something like *cycle of experience*.

During a lifetime, the aspect of personality, which is actually entangled with the human, is modeled as conscious self. In this view, the self shares the perception and expression process with the human's body consciousness. When the human's lifetime is finished, the conscious self is free to return its

---

[7] The phrase in John 14:2, In my Father's house are many mansions, can be construed as a reference to the idea that there are many different venues for learning. These venues would be the local aspects of reality I am speaking of here.

focus of attention to nonphysical aspects of realty. A process of transition from one venue for experience to another is expected.

In this view, it is probably more descriptive to refer to the Survival Hypothesis as the *Transition Hypothesis*. Of course, people are more familiar with the survival issue, and I expect the term will remain for now.[8]

Supporting Material in the Form of Direct Evidence

Transcommunication is an umbrella term used for many types of trans-etheric influences. Visual and audible forms of ITC produce important forms of objective evidence. The examples shown here are considered anomalous in that they represent areas of apparent intended order that have emerged out of noise.

Reflected Light Phenomena

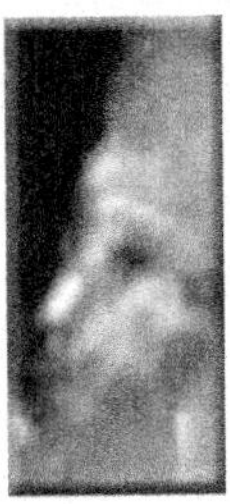

Moving Water ITC Example: Features formed in chaotic light are considered an induced form of ITC. In this example, you may see a man facing toward your left. The image was found in light reflected from moving water.

Anomalous features sometimes found in neutral-density optical noise are known as reflected light phenomena. An example of this is provided in the articles "Faces in Reflected Light" (17) and "Faces in Light Reflected from Crystals." (18)

[8] This is a brief description of how I see reality and our relationship with it and our body. Standing back for a while and looking at the above explanation with fresh eyes, I see that it is complex and very condensed. I have been working with these concepts for years and have learned to compress a lot of information into a few concepts.

As I suggested earlier, perhaps it would be best if you did not worry about whether or not you understand my point. Read on instead! There are twelve more discourses designed to explain the details. By the time you have read this section and Section III on transcommunication, you will be an expert on Trans-Survival and the Implicit Cosmology.

Moving water is a useful medium for generating medium-density optical noise for feature formation. A good example is in the article "ITC Experiments Using Light Reflected from Water." (19) See the Techniques – Visual ITC Study.

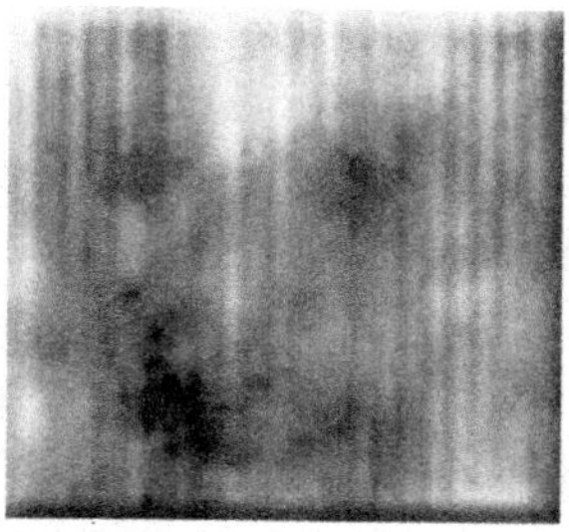

Video-Loop ITC Feature: Video loop ITC Example 1 of the "Perception of Visual ITC Images" study. Website visitors correctly described the feature 81% of the time. See Video Loop ITC Example 2 in the Techniques-Visual ITC Study [325].

An important form of induced ITC is the video loop method for generating optical energy suitable for feature formation. A study was conducted to determine how well website visitors are able to see features in video loop examples. The study, *Perception of Visual ITC Images* (20), indicated an average of 61% of respondents correctly identified the paranormal feature. A comprehensive discussion of visual ITC is in the Techniques-Visual ITC Study.

Audible ITC Phenomena

Transform EVP are formed out of available audio-frequency noise. A series of studies was conducted to determine how well website visitors were able to correctly understand EVP examples. On average, visitors correctly understood 25% of the examples. Similar studies by other researchers have indicated an average of 20% to 25% correct understanding. (21) A comprehensive discussion of audio ITC is in the Audio ITC a.k.a. EVP Study.

Opportunistic EVP (random selection) are thought to be formed by psi influence on a random process to select preexisting voice fragments (sometimes from a library of words). The computer software known as EVPmaker (22) is an example. Another example is randomly processed speech synthesis EVP. (23)[9]

---

[9] As directors of the ATransC, our policy is to discourage use of this form of EVP except for research. A number of studies based on this technique failed to reproduce the expected quality of EVP. (24) Also see the Psi Research Study in the Transcommunication section.

Implications of ITC

The fact of anomalous voices and images is well established, and mundane explanations have not accounted for many of their characteristics. In some instances, ITC images and voices have been identified as clearly indicating a known discarnate person. (19) Examples of ITC are relatively easy to produce by an experienced practitioner, making the phenomena one of the most replicated form of induced transcommunication. Because of the objective nature of the examples, we have recommended that EVP should be used as a kind of *lab rat* for the study of other forms of psi functioning.

The nature of EVP is better understood than visual forms of ITC, and it provides most of the supporting evidence for Survival. Common characteristics of EVP that indicate intelligent interaction with an aware personality include direct responses to questions, comments about local activity and reference to prior activity. See Unique Information in this discourse. Also see the Audio ITC a.k.a. EVP Study.

The implication of ITC is that the communicating personality is the same as that of the person while still in a lifetime, but who is now operating under different circumstances. While practitioners have learned a little about the nature of those *different circumstances*, more has been learned from EVP about the nature of the Attention Complex.

One of the reasons I designed the mostly unconscious perception and expression functional areas of the Attention Complex as I did is to accommodate the apparent influence of the practitioner on the message. We see the same sort of coloring in mental mediumship. (10) See the Perception and Expression and Life Fields Discourses.

The Super-psi perspective of ITC is that the information is accessed from a person still in the physical or from residual energy in the form of memory. I agree that some examples are clearly from current memory (from a still physical person). It is common for me to access current memory, which is a point of confusion when I am giving spirit greetings in Spiritualism. Testing for this possibility is one of the challenges for testing the Tran-Survival Hypothesis. See the *"DoJa, no!"* example [38] for a discussion comparing survived memory and living personality as the source of information in ITC.

Mediumship

In mental Mediumship, a person who is sensitive to the influence of discarnate personalities is able to convey information from them to people still in the physical. The study of mental mediumship began in the late 1800s with convincing evidence of anomalous access to information. However, whether or not that access has been via psi sensing of residual information and telepathy between living agents (Super-Psi), or from discarnate people (Survival), remains an open question. (25) A comprehensive discussion of mediumship is in the Mediumship Study.[10]

Since there is no known way to shield from psi, the Survival verses Super-Psi question is not easily answered via mediumship alone. However, taken in the larger context of transcommunication, mental mediumship can be seen as one of many forms of transcommunication that can be used to study the question. While there are examples of mental mediumship that arguably support Super-Psi, it is usually the indirect evidence in this discourse that supports survived personality as the source of information.

An important instance of mental mediumship is the contact between Jane Roberts and an entity known as Seth. Analysis of the Seth Material to determine the likelihood that Roberts was capable of producing the over 21 years of consistent, sophisticated and metaphysically profound material is discussed in *The Content-Source Problem in Modern Mediumship Research.* (26) Because of clear differences in style between the abundant work of Jane Roberts, her husband and the presumed Seth, the conclusion of the study was that there is a high probability that a different personality initiated the Seth Material.

---

[10] **On the veracity of *Revealed* Information via trance-channel:** Amongst the indirect evidence for survival is the formulation of information received via transcommunication. Anyone who has witnessed someone learning how to trance-channel will have noted the difference between what is said when a person is probably trying but not actually channeling and when a person is possibly channeling a different personality.

As a general observation, a person who is behaving *as if* in contact with another personality typically speaks obvious phrases intended to seem profound, often hesitates while searching for words and frequently contradicts prior comments. When this *as if* performance turns into interestingly profound, smoothly delivered and consistent material, it becomes clear that a more lucid person is speaking.

The Seth Material study cannot be applied to all ostensibly channeled material, and in fact should not be; however, it does establish that at least one instance of channeled information appears to originate via psi sensing from an agency outside of Roberts' physical environment.

Exoplasmic Hand: Ectoplasmic hand can be seen rising from a pile of ectoplasm at the feet of the Felix Experimental Group's medium. (184) ©

Physical Mediumship

Physical mediumship is represented by a suite of phenomena which includes materialization of sound (knocks, direct voice) and light (*spirit lights*), reunions between discarnate and physical people, precipitation (apports) and ectoplasmic structures which are sometimes in the form of a discarnate person. Information delivered from discarnate personalities via the physical medium is considered a form of channeled or deep trance mental mediumship.

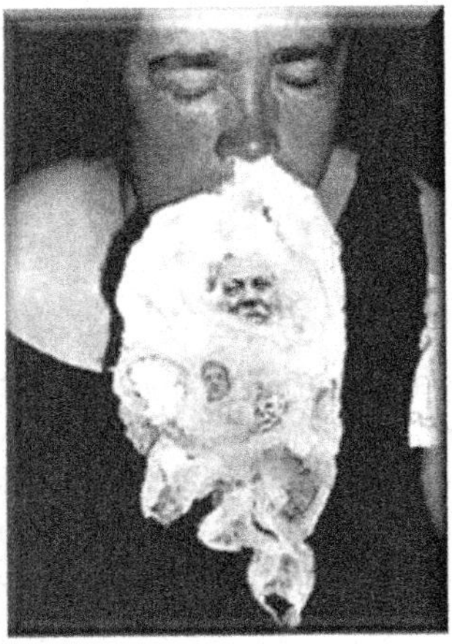

Faces in Ectoplasm: Photograph titled *Conan Doyle's Return*. An enlargement from the Dr. Thomas Glendenning Hamilton collection of physical medium, Mary Marshall. (196) ©

The FEG medium has produced similar cameo-like faces and some have been recognized by sitters as the likeness a loved one. (184)

Articles providing first-hand accounts of physical mediumship are under the Mediumship tab of ATransC.org. The mediumship of Stewart Alexander (27), Kai Mügge of the Felix Experimental Group (FEG) (28), David Thompson

(29), Fay Wright (30) and Hoyt Robinette (31) have been carefully examined by skilled witnesses.

Most séance room phenomena provide evidence for the existence of the psi field, the influence of intentionality on the psi field (psychokinesis) and the contact field [100] concept. This primarily supports Super-Psi. However, reunions between discarnate and still physical people provide a most profound form of evidence for survival.

In a Class A example, a loved one (discarnate communicator) might materialize as an ethereal form cloaked in ectoplasm to resemble the person's body in physical life. The materialized form might walk about and feel warm to the touch just as in physical life. The materialized person might speak with an appropriate voice and say things unique to that person, sometimes conveying information unknown to anyone in the room. This form of demonstration is typically carried out in total darkness; however, there have been instances of the phenomena in reasonably good light.

Class A reunions are not common, but they are documented experiences. During a David Thompson séance (29), which I attended, sitters were treated to the physical touch of Timothy, David Thompson's playful, discarnate child who is said to be responsible for physical phenomena during séances. It was totally dark in the room, but he came to each sitter without hesitation or fumbling and held a finger with his tiny hands.

While the appearance of Timothy would be considered a Class B example in terms of ITC because the child could not be seen, the contact was very evidential in that all sitters were holding hands to account for their whereabouts and Timothy's hands were warm, soft and way too small to have been faked by an adult. It is interesting to note that the tiny hands came to us from a position some few feet off of the floor, just as one would expect when being touched by a child. Every sitter in the room had this experience. (32)

Such phenomena are relatively common amongst the small community of people privileged to sit with a publicly demonstrating, evidential physical medium. As of this writing, Kai Mügge is probably the most studied contemporary medium. (28) The phenomena produced by him are well documented, often in good red light. There is little doubt as to the paranormality of the effects he produces. While it is conceivable that an elaborate explanation for the phenomena depending solely on psi and residual energy might be concocted, at some level of complexity, such explanations are no longer tenable. See the Mediumship Study and the Science Commentary.

Precipitation Mediumship

Precipitation mediumship is a form of physical mediumship in which a substance is psychokinetically deposited into the physical as intended form. Precipitation art is thought to be facilitated by the medium but under the direction of a communicating personality.

Precipitated Art on Cloth: Image on cloth thought to be Madam Blavatsky. Hoyt Robinette was the medium. *(31)*

Hoyt Robinette is often described as a precipitation medium because of the precipitated images he produces during some of his public demonstrations. In practice, pictures of people and/or animals appear on blank paper cards or pieces of cloth. (31) The pictures often appear to be cut out of a magazine, and often, are identified as loved ones or people important to the sitter. Two important concepts associated with Robinette's phenomena are intended order and apports.

In transcommunication, the common factor for both visual and audio ITC is a source of physical energy which is modulated with a chaotic process that can be influenced by the etheric communicator to produce an intended form. More or less random regions of order naturally emerge in chaotic processes. In ITC, that emergent order is transformed into the voice of EVP or the image of visual ITC, apparently under the influence of the communicating entity. A common characteristic of visual ITC intended order is the presence of both primary and secondary features (usually faces).

Images precipitated onto paper by Hoyt Robinette typically have just one feature; however, at least in the example shown here which was received by me, a secondary feature was visible but faded within a day or so after formation. This secondary feature was much as is seen in video loop ITC.

Precipitated Art on Paper: Image precipitated onto a three-by-five-inch paper card by Hoyt Robinette. (31) It is not known who the man is, but he bears a strong resemblance to the Butler family

The images produced by Robinette often appear to have been taken out of existing media. Of course, this draws accusations of fraud, which might be difficult to counter were it not for the lack of means for executing the fraud, as witnesses will attest. Also, there are other forms of trans-etheric influence that appear to be taken out of the media. The most common example of this is the apport, in which physical objects are taken from one part of the world and deposited in another in an apparent process of dematerialization, *movement* in the etheric, and materialization in the new location. Some examples of EVP and visual ITC also seem to have been taken from existing media, seemingly supporting the apport theory. See Apports in the Nature of Transcommunication Study.

Precipitation phenomena represent a possible energy-saving approach to transcommunication because they typically produce objects (images, thoughts, words, things) that have been previously embodied in the physical. In the Implicit Cosmology, I argue that the physical expression of intended information requires the avatar relationship. This objective expression of trans-etheric influence is also apparently constrained by physical principles and appears to have physical energy limitations. In the transform form of objective influence, the physical principle of stochastic amplification [276] is an example. Apports may represent an alternative energy saving technique, but expression of the apported object would appear to depend on the influence of intended order.

Healing Intention

Healing Intention provides an important tool for the study of intentionality. In the context of this discourse, the question of whether or not healing occurs is not as important as are the distant, objective changes thought to be caused by the influence of intention on the psi field. The verifiable existence of noncontact influence related to intention is necessary for both the Super-Psi and Survival Hypotheses to be true. In the Trans-Survival Hypothesis, it is proposed that noncontact healing intention involves a person's expression of the intention to benefit the wellbeing of a sitter, which is sensed by the sitter's perceptual process. See the Healing Intention Study.

Indirect Evidence

The rationale for even considering the possibility of the Trans-Survival Hypothesis is not based on the primary phenomena so much as it is on the implications of the often startling ways and circumstances under which individual instances occur. An excellent example is how a response to the practitioner's question in EVP will sometimes be found in the sound file before the question is spoken.

Yes, one can resort to quantum theory showing that information may exist independent of time and space, but if that were true, then it should be common for all sorts of events to occur out of sequence. In fact, the vast majority of experiences are appropriately sequential.

Trans-Channel Communication

The characteristic of some EVP containing answers before questions are asked supports the model described in the Life Fields Discourse and illustrated in the Perception and Expression Diagram [xxiv]. The characteristic is predicted by the contention in the Trans-Survival Hypothesis that transcommunication occurs via a mind-to-mind exchange of information between etheric personalities.

The hypothesized communicator is an etheric personality with no physical body entanglement. The practitioner is an etheric personality entangled with a human body. The communication channel between the practitioner's conscious self and its mostly unconscious personality is filtered by the perceptual and expression function in the Attention Complex. It is in the Attention Complex that the practitioner's question is formulated. As it is modeled in this cosmology, that functional area is also monitored by the practitioner's etheric personality. Thus, communication between the communicating personality and the practitioner is seen as an exchange via the practitioner's unconscious perceptual process.

It is known about normal human behavior that a person first imagines what will be asked before actually expressing the question. (13) Based on the Trans-Channel Communication circuit described here, our mental processes routinely provide information to the etheric community of personalities as a purely mental signal prior to our vocal expression.

A second point of support for this hypothesis is that the time interval between answer and question in EVP is consistent with the time usually required to mentally formulate a question. Of course, the rest of the story is the intelligent response which indicates a sentient entity and not residual energy.

Unique Information

To establish whether information a person senses is from the physical environment or from the psi field, it is necessary to somehow blind a person from access to physical information. This can be accomplished via various sensory deprivation techniques such as blindfolds, flotation chambers and acoustical suppression.

It is also necessary to design a target for psi sensing that is not known to the research subject who is supposed to be functioning as a psychic. Since witnesses are always alert for lucky guesses, cheating and unexpected prior knowledge, computers have been very successfully used by researchers for picking targets. In that practice, targets are selected by the computer out of a database of possible targets. The researchers only know the target after the experiment is finished.

Testing for mediumship poses an even greater challenge. It must be possible to show that the target information could not have previously existed in the physical. Since there is no known way to shield from psi influence or hypothesized residual psi fields, it is necessary to design an event which can be initiated by the researcher without knowing the exact nature of the event.

Finally, time and distance are thought not to exist in psi space (nonlocal and nontemporal). The *everywhere is here* characteristic of psi space (the etheric) has been demonstrated in EVP, but from my study, the *all-time is now* characteristic is questionable. I expect presentiment is explainable as the trans-channel communication. Longer-term futures are probably explainable with the concept of potential thoughtforms. In that, the expression of intention produces the potential for many possible futures. These are represented by potential thoughtforms which may be psychically sensed. This might also help explain why some predictions fail to occur.

Nonlocality means that having a person in another room or another country makes no difference when it comes to psi sensing. The nontemporal characteristic would mean that, if anyone ever becomes aware of the target for psi sensing, that information would become available to the psychic during the experiment. This means that, given a nonlocal and nontemporal nature of psi space, it is probably impossible to design a laboratory test that would assure the information came from a discarnate personality.

A Possible White Crow

EVP can be induced in that often questions can be asked by the practitioner or an interested observer and answered by the possibly discarnate personality. Sometimes EVP are clearly a spontaneous response to an unexpected situation in the physical. An example of intelligent interaction involving previously unknown information is the *"Doja, no!"* EVP recorded by Martha Copeland. (33)

Martha Copeland's daughter, Cathy, transitioned as a young woman and Martha assumed care for her dog, Doja. Martha had accidentally left Doja in the house when she left to go shopping. It was her habit at the time to always leave her voice-activated digital recorder turned on, and so it was left on in the house with the dog. No one was in the house except for the animals. Doja tore up a potted plant and made a great mess. You can hear an EVP from Cathy as she scolds the dog saying, *"Doja, no!"*

Discarnate Cathy scolding her still physical dog is an example of an EVP in which no physical person was aware of the dog's behavior. Even if Martha had been aware of the dog's actions, she would not have had the technological wherewithal to manufacture Cathy's voice saying the dog's name in a scolding tone. It is possible to conceive of other alternative explanations such as psychic access to information. For instance, a person might have snuck into Martha's home and seen the dog tearing up the plant. Martha could then have psychically sensed what the person witnessed, and thinking how Cathy would have scolded her dog, unconsciously created the EVP. All of this is possible within the Super-Psi Hypothesis.

A single instance of the *"Doja, no!"* type of EVP could be set aside as an anomaly amongst anomalies—an outlier to be discarded. However, it is a good example of a not so uncommon form of EVP. So, while it is tempting to contrive exotic alternative explanations, at some level of complexity, it becomes simpler to include this example amongst the collection of experiences indicating survival. The example can be heard in the ATransC.org under the Examples tab. (34)

As it is modeled in the Implicit Cosmology, the etheric communicator is able to witness physical events via the senses of a physical person or animal. In the Doja example, Cathy would have witnessed the activity via her dog.

In a similar way, we have had a number of reported EVP that included a request for the practitioner to have a mirror in the room. Thinking the communicators were seeing them from behind the mirror, practitioners have even tried writing a question and holding the writing against a mirror. It is more likely that the communicators are able to see what we look like when we see ourselves in a mirror.

Intended Order

A common factor in ITC is the transformation of broad-spectrum, chaotic energy into apparently intelligently intended features, such as the voice in EVP (21) and faces in visual ITC. (20) Current research is showing that output of random event generators increase in order when in the presence of strongly expressed intention. (35) (36) There is evidence in precipitation phenomena, such as spirit art, that the precipitation material undergoes a transformation from chaotic noise to intended order as raw material is apparently apported to the scene and organized as the intended image. (31) These examples suggest that the basic element of the expression is the emergence of intended order.

If there is a physical mechanism responsive to intention, and if that interaction can be seen to operate across many forms of trans-etheric influence, then it can be said that psi functioning is a common experience operating in more than just parapsychological phenomena (psychokinesis, distant healing intention, anomalous information access). If this influence can be demonstrated to affect formative fields, then the scope of psi functioning might be extended to include trans-etheric influences, collective consciousness and evolutionary influences.

Lucidity

In the context of the Trans-Survival Hypothesis, *lucidity* is used to describe how consciously aware a person is of unconscious personality. This is further discussed in the Personality-Centric Perspective Discourse.

A form of false lucidity is frequently encountered amongst new ITC practitioners. This is referred to here as hyperlucidity [152]. In this, a person claims the ability to produce astounding communication as if completely lucid. While there are practitioners who do produce impressive examples of transcommunication, these exaggerated claims are characterized by a suite of behaviors which indicates a person who is sincere, but generally mistaken about this ability. Hyperlucidity is assumed to resolve itself as the practitioner becomes better informed.

The demonstrable difference between imagined (hyperlucidity) and real ability (lucidity) makes it clear that transcommunication is an ability that can be distinguished and developed. As such, it can be quantified.

Animal Contact via EVP

EVP have been reported containing animal sounds, mostly dog barks or cat meows. A few EVP have been reported containing English-language phrases that appear to be initiated by dogs. Etheric communicators sometimes refer to family pets being with them on the other side. A relatively common form of apparition is a small animal, often thought to be a pet that has transitioned. The implication is that animals, certainly pets, experience transition in some form and interact with the physical much as discarnate people are thought to interact.

The Trans-Survival Hypothesis supports the idea that a discarnate animal's interaction with the physical would necessarily be via the practitioner's or interested observer's etheric-to-physical entanglement with an avatar. However, the Super-Psi Hypothesis might explain the impression of animal communication as either illusion or psi access to residual energy—a sort of *energetic essence of animal*. There is also the possibility that a person might remember how the animal sounded and inadvertently inject those expected sounds into the media.

In fact, the Implicit Cosmology uses the same life field model for all physical life forms. The difference between a tree and a human, or even a cell of a human and the entire body, is the extent to which functional areas are implemented, content of worldview and threads of entanglement. There is nothing in the cosmology that prevents animals from being avatars.

In this view, the etheric aspect of a pet is virtually the same as that for a person and only perception of reality distinguished the two. Also in this model, all information accessed by a person is sensed via the mostly unconscious perceptual process. In that, the information may be sensed as a gestalt thoughtform which must be translated via the perceptual Loop which is informed by worldview. Finally, the cosmology holds that EVP are formed as intended order expressed via the practitioner's or interested observer's perception and expression processes. As such, a message from a dog could very easily produce a human-sounding utterance in the recording.

As with all of these comparisons, the question that must be asked is how complex theoreticians are willing to make the Super-Psi Hypothesis to avoid considering survival. If the animal contact is unexpected, persistent and complex while retaining the animal's personality, then it is unreasonable to say that its presence is illusion created by the witness?

Cultural Influence

An important aspect of ITC is the extent to which a person's upbringing influences how trans-etheric influences are experienced. This is evident in mental mediumship in which the medium must always be mindful not to allow personal impressions to color the sensed message.

In EVP, there is a clear relationship between religious upbringing and the nature of EVP recorded. For instance, a very religious person might record demon-like utterances, while under the same circumstance, a more agnostic person might record neutral to friendly messages. Trans-etheric influences are thought to begin as conceptual information. They must be *embodied* in some way via etheric-physical entanglement. It is the person's worldview (or that of an interested observer) which is thought to color the message during translation from conceptual to objective form.

This effect is applicable to both hypotheses, but it especially makes sense if the avatar model is applied. It is also a factor moderating how literally trans-etheric information is understood. See Normalize in the Glossary [450].

Personality Type

It is proposed in the Trans-Survival Hypothesis that a person's life field is formed by a (creator) personality from which the person inherits an urge to gain specific understanding, and that a physical lifetime is an opportunity to satisfy that urge. As such, the physical aspect of reality is seen as a venue for learning.

In this model, at least for the physical, a person begins at birth with a worldview that is blank except for the inherited urge to gain specific understanding and a baseline of comprehension based on the progression of the person's entangled etheric personality. Possibly by design, this inherited understanding is experienced by a person as a nonspecific knowing to assure unhindered participation in this lifetime (thus the term *comprehension* rather than *knowledge*). A common term for this comprehension is *spiritual maturity*.

People think and behave in different ways that seem to be established at birth. People also experience fundamental changes in point of view as they gain understanding about the world they live in. Understanding gained through experience is a function of the person's innate ability to comprehend the nature of experiences. Thus, a person who was born with greater prior comprehension (understanding) may tend to gain more understanding from experience than a person with relatively less comprehension.

For instance, in the ancient wisdoms, the Major Arcana of the Tarot are designed to guide the student's progression. The Hermetic relationship of comprehension with progression in the Tarot is closely related to the personality types described in astrology. In more modern times, it has been noted that people tend to fit into personal styles [120] (37). This knowledge has been an important tool for training salespeople. On close examination, the Tarot, astrology and modern personal styles, like the concept of intelligence quotient, are generally in agreement, suggesting the concept of increasing comprehension is a real experience for which the hypothesis should account.

The comprehension a person is born with and subsequent ability to learn and understand is not well explained today; however, it can be explained as comprehension gained by a *survived* personality prior to embodiment in the current lifetime. Intelligence Quotient (IQ) and personal style are two different concepts. The commonality I am addressing here is the inherent difference people exhibit that is not necessarily related to physical ability, body type, education or culture.

Language and Cultural Conditioning

In this cosmology, the natural form of information is the conceptual thoughtform, but to be compatible with the physical world, it is expressed in objective terms. Also in this cosmology, the translation of nonphysical information into objective expression is modeled as occurring in the

Attention Complex, which is an unconscious part of mind dominated by worldview. See the Perception and Expression Discourse.

A person is a personality entangled with a human avatar. That entanglement produces a channel, via the shared Attention Complex, through which conceptual information is translated into objective expression. See the Life Field Complex Diagram [xxv]. Worldview can be thought of as a database populated with a person's beliefs, understanding and inclinations. Because Worldview is the moderator of the perception and expression process, information that emerges into consciousness is colored by the person's cultural training.

One of the more important indications that this is true is the way in which information is consistently colored by the person. For instance, EVP are virtually always recorded in a language the practitioner (or an interested observer) understands. As an example, with the practitioner speaking and understanding only English, and while recording for a Korean television crew, a Korean-language EVP was recorded. Otherwise, the practitioner only records English-language EVP.

The only apparent constraints on who acts as the channel for the transformation of information from etheric to physical are that the person is in the physical and is aware of the communication. For instance, knowing that a practitioner in New York is conducting an EVP experiment to contact a loved one on his or her behalf, a person in Sacramento could provide the channel by which the response occurs in the practitioner's recorder in New York.

The Implicit Cosmology, which is based on the Trans-Survival Hypothesis, provides a mechanism that predicts the coloring of messages one sees in mental mediumship and the obvious cultural bias of EVP. It also provides a mechanism by which a person may modify worldview and improve lucidity with mindfulness.

This mechanism can be applied to Super-Psi as well, but existence of an Attention Complex which is shared by the human body consciousness makes considerably more sense if the source of information is a living intelligence, whether or not it is embodied in a lifetime.

Implications of Direct Evidence for Survival

The existence of anomalous voices, images and influences is well established. To be clear, I refer to these phenomena as anomalous because mainstream science does not explain their existence. In fact, these phenomena can be replicated by qualified practitioners on request. The requirement of science that experiments must be replicable is satisfied.

A second complaint about hypotheses concerned with survival is that they cannot be proven false (falsifiability). I think this complaint is primarily directed toward the god concept. That is, how do we prove God does not exist? Well, in fact, the existence of a Source (god) can be considered a logical initial condition, but it is not critical to the main point that people have a nonphysical aspect which is independent of a lifetime.

The idea of an intelligent influence from outside of the physical is the essence of the god concept. We routinely see such intelligent influence in ITC and other forms of transcommunication, and that is falsifiable if ITC can be accounted for by mundane means.

People skeptical of paranormal phenomena are mostly arguing from a very out-of-date point of view. The mainstream seems to be still clinging to the idea that mind is a product of brain, but as emergent understanding becomes widely accepted, mind will be seen as a psi field characteristic of consciousness. (11) (38) (39) This is not survival. The mind visualized by this emergent understanding might cease to exist as the body dies. The point is that the falsifiability of survival is in the same boat as current understanding about the nature of mind.

The anatomy of an entangled personality-avatar relationship is largely accepted in parapsychology. First Sight Theory (40) is an example of growing understanding that people have an unconscious nature which moderates conscious perception.

Discussion

The Super-Psi Hypothesis depends on the retention of information in the psi field and some means to access it with the mind. The following items address anomalous access to information.

Retention of Information in Psi Space

The retention of information in psi space is supported with a number of concepts in quantum theory, but the applicability of those concepts is not clearly established.

Psi phenomena seem *like* some quantum phenomena, but alternative theories may be more appropriate. For instance, research concerning the influence of intentionality on quantum-level phenomena in split-beam studies (41) clearly shows the influence of intentionality on photon entanglement, but our understanding of the influence of intentionality on chaotic systems might also explain the observed effects.

The expression of intention can produce physical effects, but that does not imply that those physical effects explain the influence of intention.

Retention of Information in Psi Space

Conservation of energy has been proposed as a mechanism for retention of information in the psi field. However, it has not been shown that physical principles apply to the psi field (etheric). In fact, there has been selective application of physical principles as needed to explain observed effects in an *"It seems like"* kind of logic.

Based on observed phenomena involving a conceptual-to-physical production of objective material, physical principles are arguably derivatives of the expression of intention as an ideoplastic structure. Just as the influence of intention is an etheric equivalent to physical energy, so is attention an etheric equivalent to conservation of energy. In this hypothetical view, without personality's attention on information, it will undergo what appears to be a form of attenuation in which it eventually dissipates.

Apports are an excellent test for the question of if and how physical principles apply to psi space. As the concept is understood, an apport such as a stone is a physical object that has been removed from one place in physical space and returned to physical space at a different location. In effect, the object is dissolved into purely nonphysical information, translated from its original perceptual orientation to a new and intended orientation, and then precipitated into the intended location as its original physical form.

Here, *perceptual orientation* is modeled in the Implicit Cosmology as that of the personality expressing the intention necessary to cause the apportation. This is an aspect of perceptual agreement as discussed in the Organizing Principles Discourse.

For apportation to occur, it is necessary that physical principles governing the object's behavior are suspended and then reinstituted in the new location. If the evidence of apports is substantiated, the process appears to represent an instance in which physical principles do not apply to psi space.

Access of Residual Information

There is a need for Super-Psi theorists to better explain organization of information in psi space. As I understand First Sight Theory (13), as information is composed in a person's mind, it is even then a thoughtform that is made specific by a person' unconscious perceptual process. In my experience, First Sight Theory is an excellent model for how information in psi space is translated into objective form.

I am defining information as a gestalt-like thoughtform with specific characteristics of subject, relationship and quality. As a mental process conceptual element of thought, information is conceptual in its etheric form, but its expression is objective. The conceptual-to-objective translation is under the influence of a person's intention. See the Perception and Expression Discourse for further explanation of this concept.

Assuming the information does remain in the psi field via some as yet unidentified mechanism, the best speculation I have seen is that the information would be a gestalt-like thoughtform of conceptual information that requires the source personality's worldview to correctly translate it into objective form. To understand this point, go back to First Sight Theory. You may also want to look ahead to the Perception and Expression Discourse.

The information only exists as a concept until it has been processed in a person's mostly unconscious perceptual mechanism. That mechanism is informed (actually filtered and translated) by the person's worldview. No two people have the same worldview, and so, any two people will express the same source information in slightly different ways. The objective component of information retained in psi space as a memory will look different to different people.

The point is that in Super-Psi the information is in the form of memory with no intelligent component to help someone accessing it understand the original meaning. In transcommunication between a discarnate person and a physical medium, the information is also presumed to come to the medium as a gestalt-like thoughtform. But in a real-time interactive exchange, the source personality is able to modify the message to account for how the medium's perceptual process might color it.

This unconscious refinement of information exchange is indirectly supported by First Sight Theory. The conceptual nature of information exchange in the psi field is generally better addressed by the Trans-Survival Hypothesis in which a still living personality is proposed as the source of anomalous information accessed by the mental medium.

Of course, this perspective also needs to be tested, but at least the hypothesis provides a context for some of the more astoundingly accurate mediumistic messages.

Comment

It should be obvious by now that I have decided we survive physical death and then gathered information that seems to support that belief. It has been a life-long process involving many conceptual dead-ends and many more *aha* moments in which concepts complemented other concepts, eventually forming my understanding that we are at least conditionally immortal. There is the likelihood that some of my conclusions just sound reasonable but are otherwise poorly conceived. Be mindful of this.

The remaining discourses in this section are designed to explain a model of reality based on survival. It is for you to decide if the cosmology makes sense. I will say that the model has worked very well for me to explain many paranormal experiences. So in keeping with the idea of a black-box analysis, the cosmology is a pretty good simulation of reality.

Discourse 2:
Introduction to the Implicit Cosmology

Objective

The Implicit Cosmology describes the relationship amongst the various aspects of reality with a focus on our nature, our survival and the reality in which we survive. It is based on the Trans-Survival Hypothesis described in the preceding discourse.

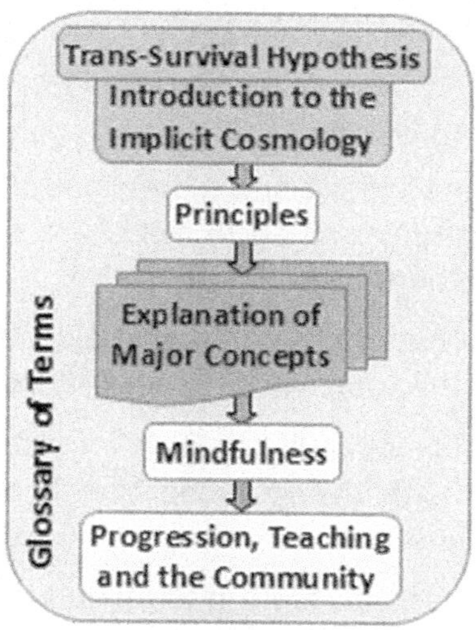

Organization of the Cosmology Series

This book is designed to help you understand the concept of survival and Organizing Principles described in the Implicit Cosmology. As you can see in the Organization of the Cosmology Series Diagram, this discourse introduces the other discourses. It also includes a three-point outline intended to give you an easy guide for talking about the cosmology with others.[11]

[11] This is not about religion; however, the subject matter involves concepts and terms that are also involved in religious thought. References to Spiritualism, references to the science and philosophy aspects of the Spiritualism, not the religious aspect.

The Cosmology

Our purpose is to follow a path toward progression, and that progression depends on understanding our nature. This understanding can begin with comprehension of this cosmology, but it will take work, and I cannot do the work for you. It is something you must decide for yourself.[12]

Basic Assumptions

This book constitutes the full definition of the Implicit Cosmology. With the Trans-Survival Hypothesis as the given, it can be stated in three parts.

Part 1 – Source: There is a top life field described here as Source. It is a singularity functioning as the reality field. Source is intelligent, self-aware and curious about itself. To satisfy that curiosity, Source has expressed aspects of itself to explore the reality field, gain understanding about it and return that understanding. Important Elements are:

1 Reality consists of life fields and expressions of life fields (aspects and ideoplastic structures). As such, fields are the building blocks of reality. See the Etheric Fields Discourse.

2 Source can be considered the top life field, and as a singularity, the reality field. See the Source and Life Fields Discourses.

3 The range of influence of life fields is a function of attention on an imagined result with the intention that it be so (Perception and Expression Discourse).

4 Source expresses aspects of itself to gain understanding of its nature. See the Perceptual Agreement Discourse, especially the Anticipation Corollary [128].

5 Reality is ordered by way of Organizing Principles emanating from Source and which are based on Source's worldview. See the Reality and Organizing Principles Discourses [56].

Part 2 – Venues for Learning: At least one way in which understanding about the nature of Source is gained is by experiencing various local realities (venues for learning) as they are maintained in the imagination of

[12] According to an old Zen Buddhist saying, *"Before enlightenment chop wood–carry water, after enlightenment chop wood–carry water."* One must do the work to learn. One must do the work to test. One must do the work to understand. Study the material and do not expect to understand all of it with one reading. Contemplate your worldview.

personalities (local sources). These personalities are, themselves, aspects of Source probably many times removed. They act as organizing personalities for collectives of personalities they express into the venue they imagine. Important Elements are:

1 Venues for learning such as the physical are imagined with specific traits intended to enable experiences which will lead to desired understanding.

2 Expression of these venues may be a cooperative effort by many local creator entities.

3 A collective may be thought of as all of the children aspects of a local source personality.

4 The creative expression produces a thread of entanglement between the source and its expression representing intention and inherited purpose. Thus:

a As children aspects gain in intended understanding, they are attracted toward their source. See the Perceptual Agreement Discourse.

b Children aspects benefit by cooperating with other members of the collective.

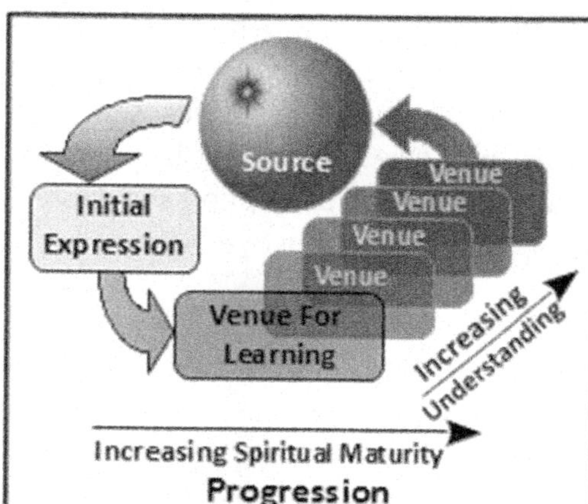

Part 3 – Person: An aspect of a local source is expressed into this physical venue for learning to be entangled with a human avatar at the moment of the human's birth. The resulting person has self-determination bounded (informed) by an inherited urge to gain the understanding intended by its source. When the human is no longer able to sustain the personality's conscious self in the physical, depending on its perception as moderated by worldview, personality either returns its understanding to its source personality or enters into a new lifetime in the physical or an alternative venue via a process described as transition. Important Elements are:

1 As avatar, the human's instincts are part of the person's worldview.

2 Maturity is a function of how well the human instincts are managed.

3 Perceptual agreement is the measure of personality's progression back to Source; the extent to which perception of reality is in agreement with worldview.

4 The anatomy of a person can be described as three major functional areas, all of which are nonphysical except for the human body:

   a The core intelligence organizing the life field of the person is the mostly unconscious personality. It is the immortal aspect and monitor of inherited purpose. This is the *I am this* of the person.

   b The conscious self of the person is the *I think I am this* aspect. We are taught from birth that we are our body, and have learned to think of what we sense with the body's senses as the entire reality. Conscious self is the source of intention which is the primary modifier of the perception and expression process.

   c The mostly unconscious perception and expression functional areas are informed by worldview to express the person's learned sense of reality to the conscious self. Intention is capable of modifying worldview. Lucidity is a measure of how aware conscious self is of the perception and expression process. See the Personality-Centric Perspective Discourse.

Read On

The Implicit Cosmology applies to every aspect of who we are, the relationship of our mind with the body we inhabit during a lifetime, the relationship of conscious self as *I think I am this* with unconscious mind as *I am this* and the nature of the reality our immortal aspect inhabits. That is a lot to condense into three parts, but they do outline the essence of what you will learn in the rest of this book.

An intellectually lazy person will say that *"We can never know."* As you read on, you will see that believing in this self-defeating phrase simply gives away your personal power and defeats any hope of fulfilling your purpose.

Implicit Cosmology

Discourse 3:
Organizing Principles

Orientation

Before I focus on the mechanics of reality, it is a good idea to describe how reality is modeled in general terms. Source is seen as a singularity in the form of a life field. That life field is our reality field. What that means is that everything is *in* Source, and the operation of Source is the operation of reality. You and I are products of that operation as is our physical universe.

Every instance of life is modeled as a life field. As such, life fields are fractals based on the Source life field. Life fields have functional areas which are more or less active in each instance of life. The Life Fields Discourse explains these functional areas. The Perception and Expression Discourse explains the interactions of life field functional areas.

As building blocks of the physical, atoms have component parts that are more or less used, depending on the atom's function (oxygen, iron, uranium). In the same way, life fields are building blocks of the greater reality and their component parts are more or less used depending on the life field's function.

As with the principles governing the behavior of atoms, an assumption that follows from the Implicit Cosmology is that the operation of reality is governed by principles which are the same throughout reality. In a matter of speaking, this is the *god assumption* shared by most if not all religions. The difference is that Source is not modeled as a divine being. It is seen as being impersonal, except that life inherits purpose and functional characteristics form Source.

The functional characteristics are inherited from Source by way of the process of aspectation. Via that process, there are child versions of Source which are in every way the same as Source. (Life is a fractal and Source is the top fractal.) The only difference appears to be that Source initiated purpose and its children inherited that purpose with the added urge to return understanding about the nature of reality to Source.

The fundamental form in reality is the field. Source is a life field which contains functional areas which are also expressed as fields. You will see this if you review the Functional Areas of a Life Field Diagram [xxiii].

Given the assumption that Source seeks to understand its nature, as Source gains understanding of itself, it is reasonable to expect these Organizing Principles to evolve. However, for our purpose, they may as well be immutable. Considered in a different way, if reality is ordered, then there must be influences that accomplish and maintain that order.

This discourse describes the principles that appear to control the organization of reality. I have made no attempt to incorporate the usual principles of Natural Law. One reason for this is that this cosmology is designed from a personality-centric perspective, and the principles of Natural Law are designed more from a person or body-centric perspective. The systems should be complementary.

Principles by Category

Given that there is a **Source** [55], then the following principles logically follow as organizing influences:

Source

As first cause, Source is a singularity modeled as a life field, the personality of which is both the experiencer and formative agent of the reality field. The expressions of that initial personality are aspect life fields which inherit purpose and understanding concerning the nature of reality. To provide a boundary for the model, the Implicit Cosmology begins with this source as an agent of creation that is considered the initial cause from which all else has evolved. Source is not a father god. Alternative names for Source include Infinite Intelligence, God (not an anthropomorphic god) and First Cause.

Source is assumed to be the top life field, and as such to have at least the fundamental characteristics shown in the Functional Areas of a Life Field Diagram [xxiii]. How Source came to be is beyond the scope of this cosmology; however, one might speculate that Source is itself part of a collective or that it began as an undifferentiated etheric field.

In effect, Source is reality, and therefore, the principles governing the operation of reality are assumed to also govern the behavior of Source. For this reason, it is believed that Source is continuing to evolve by way of its evolving aspects.

The basic properties and principles characterizing reality are assumed to be the result of Source's expression. That is, the rules governing the nature and behavior of the primal etheric field (reality field or Source's life field) are embedded in the field itself. Some principles can be seen as being directly attributable to the properties or other principles; however, the origin of others is less obvious.

The model for Source is the relationship between personality and its avatar. From this, Source is assumed to have the characteristics of curiosity and intention. In the same way that a person creates personal reality, Source is hypothesized to have created aspects of itself to explore its space; its local reality. The new space resulting from this initial aspectation would be a subset of what Source believes reality to be. This would be based on Source's worldview. See the Principle of Curiosity.

As shown in the Progression Diagram [51], the initial expression is thought to have been a relatively naive and uncomplicated expression representing Source's nature at that time. Personalities resulting from the aspectation would be relatively immature in their understanding of reality; however, subsequent aspectations would be based on an increasingly *correct* understanding of the nature of reality (the actual nature of Source).

Category 1: Organizing Principles of Reality

Reality is the life field of Source. In the Implicit Cosmology, the hierarchy of nested fields and their expressions compose all of reality.

The Implicit Cosmology represents an important departure from the physical references to which you may be accustomed for metaphysical cosmologies. Because it is based on a personality-centric perspective rather than a body-centric one, terms such as *energy* and *objective* are better understood as *influence* and *expression*. With that perspective, at the most fundamental level, reality consists of life and the expressions of life. The perspective sets the foundation on which the dynamics of the interaction of life with the environment can be better understood.

Early efforts to model reality included parallel hierarchies, one as the experiential aspect of Source and the other as the formative aspect. People were in the experiential aspect and it was their imagining that became the matrix around which the formative aspect (nature spirit or Diva) formed components of reality. This approach was partially an effort to explain reports of nature spirits and guardian angels. However, more recent understanding points to the idea that the Life Field Complex [xxv] adequately explains both the experiencer (people) and subsequent expressions coming from current understanding (worldview).

As for the nature spirits, this cosmology does not address the nature of the collective intelligent core represented by morphic fields [103]. For instance, do we sense the individual or the species collective when we turn our attention to plants and animals? There is much research needed to address that question.

A third characteristic of reality is the set of Organizing Principles that govern the behavior of life fields and their expression. From a body-centric point of view, the trilogy of body, mind and spirit might be translated from the Implicit Cosmology perspective as avatar, personality and organizing principles. See The Name of Reality in the Reality Discourse.

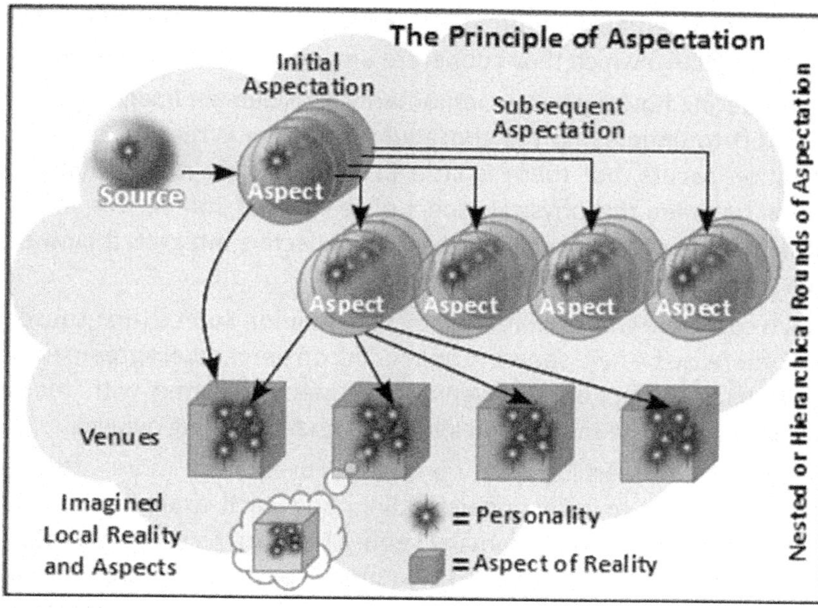

Collective

Personalities related by a shared source tend to cooperate to favor the progression of mutually entangled individuals. As noted in the discussion above about reality, life fields are aspects of Source, and in turn, are source for other life fields in a hierarchy of nested fields. The assumption of this model is that Source is curious about itself, and so each aspect (*little me*) inherits an urge to gain understanding. Many aspects are thought to be expressed by Source but not via the same instance of expression. Each of the aspect's inherited urges is expected to be slightly different from the others. This is shown in The Principles of Aspectation Diagram above as the *Initial Aspectation*.

Fractal theory is used here, so as fractals based on Source, every life field has the same qualities (functional areas as shown in the Functional Areas of a Life Field Diagram [xxiii]). However, they express those qualities differently, depending on their worldview and inherited urges.

Each *daughter* aspect of Source is expected to explore reality via experience, and that experience is via the expression (imagining) of a venue for experience and a *little me* aspect of itself to explore that venue. This is shown in The Principles of Aspectation drawing above as *Subsequent Aspectation.*

Because of entanglement, daughter aspects are thought to have a common objective which they cooperate amongst themselves to achieve.

Considering how a person momentarily considers an imagined situation in an effort to understand the consequences, there is reason to argue that aspectation occurs, but there is little to indicate how many rounds there might be between this physical aspect of reality and Source. The important point is that daughter aspects represent a collective integrated by the local source personality.

Each collective would be related by the common Source (first cause), but would otherwise be held separate depending on perceptual agreement. Also, initially sharp distinctions between collectives associated with the same venue would likely be moderated as shared understanding evolves.[13]

Venues for learning are probably used by more than one aspect/personality to gain understanding, and each aspect/personality is probably simultaneously using many venues to learn. As such, it is likely that other people in the world might be in other collectives. If so, it would be reasonable to think a person might have more rapport with people who are in their collective. This is very speculative but does help explain why some people are like our *soul brothers and sisters*. Also see the Principle of Cooperative Communities.[14]

---

13 **Foundation**: This is speculation based on the frequent assertion via transcommunication that the communicator is part of a group. In some instances, such as the Imperator Group (42) which has apparently made contact with many different mediums, the group presents itself as individual personalities, some of which claim to have had a physical lifetime. In others, such as Jane Roberts' Seth (26), the communicator discourages use of a name and identifies itself as a collective.

An important, practical result of this property is that the etheric personality of the avatar relationship is primarily associated with and concerned with an experience of the person's lifetime. It is part of a collective and is able to access experiences and understanding of other personalities in the collective, depending on perceptual agreement. Such access to potentially wide-ranging experiences could be a potent tool for personal progression.

14 A common message via mediumship is that a discarnate loved one will often claim to be *"always with you."* Hans Bender, speaking through the FELIX Experiment Group's medium, explained that loved ones are closer to us after they make their transition. (43) He also said that we project our presence into the physical from our natural place in the etheric.

The important message in Han's words is that, as is described in the Trans-Survival Hypothesis, our personality exists in the etheric. When our loved ones transition out of this lifetime, their attention is freed from their avatar and they are better able to focus on their etheric personality. So in that way, their perception is closer to us as a personality-to-personality relationship, if only we have taken the time to make our conscious connection with our personality more lucid.

Entanglement

Personality remains associated with its expressions via attention. In generic terms, entanglement is seen as a property inherent in the process of expression. It is hypothesized that the resulting aspect life field or ideoplastic structure remain associated with the source personality to the extent the personality intends the expression's continued existence.

This is seen as an ordering property which relates subfields with an organizing field. The *thread of attention* represented by the property of entanglement is seen to enable a top field to organize other fields to form and remain as intended.

For instance, it is hypothesized that a human body has an organizing field which maintains the human body image and the necessary information to gather subfields representing the component parts of a body such as skin, bone, cells and enzymes. (9) It is the property of entanglement that assures the necessary association of fields.

In principle, there would be a network of links of entanglement associating Source with all of its sub-aspects, and which carries its intention. In the same way, entanglement also functions as the link of influence between the etheric personality and its physical body, the avatar. Also see the Principle of Evolving Influence.

Fields

Fields are the building block of reality. As explained in the Etheric Fields Discourse, reality is modeled as a field which is Source's life field. It is in effect, the top field in a hierarchy of nested fields.

Fields are the expression of personality and occur in such forms as reality field (Source), venue fields (such as the physical), life fields, thoughtforms, intention fields, organizing fields, morphic fields (physical organism) and ideoplastic structures. See the Etheric Fields Discourse.

In parapsychology, etheric fields are known as psi fields. The psi field is usually modeled in parapsychology as a physical field of as yet undefined properties that is part of but overlays physical reality. In that definition, there is seldom reference to a greater reality. The name is used here for continuity. Technically, all fields are ideoplastic structures but the term is usually only applied to expressions of a life field. As such, reality consists of life fields and their expressions as ideoplastic structures.

Conceptually, a field represents a set of elements with common characteristics bound together by a common influence. Etheric fields are regions of reality that can be modeled as a node or nexus (attractor) entangled with a set of related elements. It is useful to model reality as a single etheric field expressed by a first cause or Source. In this view, the reality field is the top etheric field, from which all other fields have been derived via aspectation. In this way, the reality field has been differentiated as a hierarchy of nested etheric fields.

Evolving Influence

Span of entanglement increases as aspects gain in understanding. Considering the property of evolution, below, it should be noted that the human body was probably incapable of being an avatar of any consequence until recent times. As a speculation, Homo sapiens were probably the first organism in human evolution able to function as an avatar for an etheric personality. The first entanglement was probably rudimentary as compared with today. It is reasonable to expect that future generations of humans will be able to support even more lucid association between conscious self and etheric personality.

Clarity of the thread of entanglement is described as lucidity. Based on the comments below about evolution, it can be seen that people who work to develop awareness of their unconscious mind are increasing lucidity, and it is hypothesized that this will tend to evolve the human's morphic field so that future generations will more easily be aware of their etheric nature by way of greater psi functioning.[15]

[15] **Animals as Avatars:** It is unreasonable to assume humans are the only avatar species on earth. An example of an apparently self-aware, rational animal is the language-using parrot N'kisi. (44) The bird also appears to exhibits psychic ability suggesting greater lucidity than exhibited by most human avatar relationships.

Hierarchy

A hierarchical relationship exists between Source, aspects of source and subsequent expressions of those aspects. Please refer to the comments associated with the Collective principle, above.

A physical example of a hierarchy is the world, nation, state, city, suburb, home, family, individual. In the Implicit Cosmology, the hierarchy of interest is the relationship amongst Source, its aspects, its aspect's aspects, ... and so on for potentially many rounds of aspectation. There is a many-to-one relationship in the etheric hierarchy associated with Source and its associated aspects.

Some characteristics of Source are inherited by its aspects, just as some of the characteristics of those aspects are inherited by their aspects, ... and so on. In the concept of morphic fields, fields are also thought to combine as hierarchical collectives, in that an organizing field becomes entangled with sub-fields to produce organisms such as the human body.

Organizing Principles

Reality operates according to a body of Organizing Principles which are inherent from Source's creative expression. As previously stated, Organizing Principles can be compared to Natural Law; however, the perspective from which Organizing Principles are defined is more etheric personality-centric while Natural Law is clearly physical body-centric.

Source is modeled here in the sense of *from whence it came* rather than as a father god. Since reality is Source's life field, everything in reality is governed by the same rules of behavior governing Source and its expressions. In the same sense, it is accepted in mainstream science that the physical universe is governed by knowable principles without the need to question why or from whence the principles came.

Given the existence of Source and Organizing Principles, a logical argument could be composed to describe how the whole of reality might *self-organize*. How a very complex local reality can be developed based on a simple equation and a few simple rules is illustrated in the Mandelbrot Set. See the Cosmology of Imaginary Space Discourse.

Prime Imperative

Aspect personalities inherit purpose from their source. A foundation assumption for the Implicit Cosmology is that Source is curious about its nature. As described in the Perception and Expression Discourse, it is also assumed that Source responded to its curiosity by imagining itself in various situations so as to explore their implications. This assumption is based on the mental process involved in the Attention Complex which enables a person to consider and respond to environmental signals. See the Principles of Aspectation and Curiosity.

A personality entangled with a human for a lifetime is said to have inherited Source's urge to understand itself, but with a focus on uniquely different aspect of the Prime Imperative.

The Prime Imperative is related to the Principle of Perceptual Agreement, in that an aspect of a personality is created with an inherited urge to gain specific understanding. The inherited urge is expected to be based on an unanswered question.

An unanswered question is modeled here as a difference in perception between the personality with the question and the aspect of the personality expected to find an answer. As the aspect gains in understanding related to the question, it becomes more in agreement with its source. The aspect is not expected to be able to return until expected understanding has been acquired. In practice, an aspect might experience many lifetimes in one or more venues as it gains in understanding.

As a practical matter for Mindfulness, the Prime Imperative of a person is to align personal reality with the actual nature of local reality.

Time

Processes are time dependent while time is relative to perceptual relationship with Source. Time is addressed here in terms of perception. The perceived rate of time appears to be relative so that a personality experiences sequential time in its venue, but otherwise senses its influence as all time is now. In the physical region, for instance, time exists for persons perceptually within the physical, but physical time is thought to be irrelevant for personality *outside* of the physical.[16]

---

[16] Popular wisdom holds that all time is now (nontemporal); however, as I discuss in the Etheric Fields Discourse, there is no apparent way to get around the need for intervals separating concatenated events (streaming).

It is entirely possible that I am unable to visualize alternative solutions for sequential processes. I will maintain the assumption that time exists as a relative phenomenon in this book, and encourage you to find an alternative explanation that works in the context of the Implicit Cosmology.

Category 2: Organizing Principles of Formation

Formation is concerned with the creation process. The following group of Organizing Principles addresses the nature of that process.

Aspectation

The influence of intention on an imagined result expresses aspects of reality which are a subset of personality's personal reality. Source's intention to understand itself is the question referred to here as the Prime Imperative Principle. It is both the source of reality and the purpose for everything in reality. In this view, all expression from personality, including the personality entangled with our conscious self, exists to answer the question.

When a person imagines driving a new car, the car, its environment and the little personality in the car are all aspects of the person's personal reality (worldview). In practical terms, the imagined venue is functionally a subset of the person's understanding of the nature of its environment. The *little me* personality in the car is an aspect of the person as the person imagines him or herself to be.[17]

As modeled here, the expression of an aspect that inherits a degree of self-determination inherits the potential to express aspects to further explore the question. The source personality and its child aspects constitute a *round of aspectation*. The child aspect and its children represent an additional or n+1 round of aspectation. See the Principle of Aspectation Diagram [57].

Repeated rounds of aspectation necessarily result in a hierarchy of aspects with an increasingly smaller subset of inherited characteristics relative to the source. Because of this, children aspects may not be the same as their parent. A source personality must have a reason or purpose (a question) for extending an aspect of itself while the resulting aspect must have the purpose of satisfying that reason.

---

[17] Here, the process of creating an aspect of something is referred to as differentiation. At one time, differentiation was treated as a principle, but the concept is better treated under the concept of aspectation. The difference is that an aspect of something is a subset of that something. Differentiation is a process that produces aspects.

Aspects are thought to exist so long as the source personality holds them in its attention, and until they have satisfied their purpose.[18]

Attraction

The degree of perceptual agreement determines the extent of mutual attraction associating aspects of reality. Personality inherits a subset of its source's characteristics. Personality's perception evolves with increased understanding. A personality and its aspects is a collective, the individuals of which are bound, depending on the degree of perceptual agreement. Within the hierarchy of personalities, aspects of different collectives are bound, depending on the degree of agreement.

The term, *like attracts like* has been used to describe this principle. It is correct in the sense that the degree of attraction is determined by each person's worldview, which determines the person's perception.

Attraction is not the same as entanglement. Attraction represents a mutually attractive influence based on shared understanding. Entanglement is the influence associating personality with its expressions.

Cooperative Communities

An effort to express understanding is necessary for progression. Collectives are inherently cooperative communities. A person is attracted to communities of like-minded people cooperating to facilitate progression.

Note the distinction between collectives and communities. There is evidence that groups of personalities sharing the same source cooperate as a collective to fulfill the intention of their source. This cooperation is part of the response to the Prime Imperative.

---

[18] A common problem in metaphysics is the tendency to expect the rest of reality to be like the physical. For instance, attempts to apply quantum theory to etheric concepts may only make sense if reality was like the physical. While such concepts as nonlocality, entanglement and collapsing waveforms do sound a lot like experiences associated with the etheric, the position argued in Trans-Survival is that the physical is derived from the etheric via intended order.

The quantum aspect of the physical may represent the etheric-physical interface, but much more must be learned before it is safe to say that quantum principles effectively describe the etheric. Instead, it is best to say that they represent a model which illustrates how the etheric might behave. In other words, use the comparisons as learning tools, but do not expect them to accurately represent reality.

In the physical, we are attracted to communities of like-minded people for mutual cooperation toward gaining understanding. Cooperative communities likely support people from different collectives with rather different inherited objectives. The intention to gain understanding is a common factor. Cooperation is facilitated by rapport which is established between personalities as they interact in the community. The instinctual urge to cooperate is expected to be very strong in a collective; however, attraction to and cooperation in a community is secondary to normal challenges of living.

As an old adage goes, *"Our lot is to learn, and having learned, our lot is to teach."* Each member of a cooperative community fills both the role of seeker and that of teacher, simply by interacting with other members of the community.

The Golden Rule states that we should *"Do unto others as we would have them do unto us."* Virtually every religious organization has the Golden Rule somewhere in its literature. Considered from the perspective of the Implicit Cosmology, the meaning is that people should cooperate with one another to achieve their true purpose. Teachers come from those who have been taught. Perhaps a better way of stating the Golden Rule is: *Teach me as I teach you.* Also see the Collective Principle.

Continuity

There is one reality and it exists everywhere. There is a continuum of existence from here to Source. There are no gaps in this continuum; however, there is change from point to point along the continuum and this change tends to be stepwise. With Source as the origin, every point along the continuum is relative to the preceding point (inheritance, Nature's habits) and informs the following point (Expression based on worldview). See Hierarchy of Nested Fields [100] in the Etheric Fields Discourse.

Emergent Order

Elements of a field tend to self-organize. Borrowing from the study of chaotic systems, elements tend to be attracted to less uniform regions of a system. (45) The result is regions of order which potentially exhibit characteristics not found in the individual elements.

One result of this principle is the idea that reality is bunchy. That is, elements of reality tend to congregate. This is a little beyond my training, but

I believe the stepwise behavior so often found in nature is a direct result of this principle.

Another important result of this concept is the idea of cycles. We experience stepwise behavior in living systems as birth or regeneration, living or activity, maturity, rest and transition to begin a new cycle. Of course, a lifetime is an expression of this.

Evolution

Everything is subject to change based on history and current challenges. Nothing remains at rest indefinitely. Everything in reality changes through processes which may be orderly, but evolution is more likely to be punctuated throughout the process with catastrophic change in response to environmental irregularities.

The Implicit Cosmology incorporates the concept of morphic fields proposed in the Hypothesis of Formative Causation. (9) These fields are also more generically referred to as etheric fields in the Implicit Cosmology.

The formative information or *blueprint* contained in morphic fields is described by Sheldrake as *Nature's habit.* The fields cause formation of their associated organisms based on how the organism has *always* been formed. Without extenuating influences, a human morphic field should always produce identical bodies. The additional influences of parental genes and environmental forces results in the variety we see today.

Formation is considered a two-way process in that Nature's habit are able to record how the organism adapts so that learned behavior modifies habitual behavior. The mechanism enabling new behavior to modify Nature's habit is not clear, but there is likely a difference in the lasting influence of accidental changes such as the loss of a limb and intended changes such as learning to avoid an unpleasant experience.

The process of increasing understanding of the nature of reality depends on the Principle of Evolution. Expression also depends on it as worldview represents the organism's habit in that lifetime. Parts of the worldview clearly follow conscious self into transition as it *recovers* from a lifetime. This residual body-centric perspective is thought to affect how well conscious self adapts to the etheric.

As is noted above in the Principle of Evolving Influence, the link between personality and body in the avatar relationship has likely become possible via evolution of the human. A primitive link was probably established and then evolved as humankind became better able to comprehend abstract concepts.

Punctuated Evolution

Evolution is generally considered in terms of change by way of small increments; however, there is reason to speculate that trans-etheric influences might produce larger increments of change than expected.

Instrumental TransCommunication is a well-documented from of trans-etheric influence in which an intelligent influence (expression) causes a change in the physical. Morphic fields are etheric fields which are entangled with the physical organisms they govern. It is reasonable to argue that the influence wielded by personalities and morphic fields on physical processes is able to produce changes that would be seen as being beyond what can be reasonably expected from chance evolution dependent on mutation or survival of the fittest.[19]

Life Field

Life consists of Personality, Attention Complex and Conscious Self organized as an etheric field by Core Intelligence associated with Personality. Reality is modeled as a hierarchy of nested life fields with Source as the top field. In reality there are life fields and expressions of life. Source is the top life field. As a general statement, the Prime Imperative of a life field is to align personal reality with the perception of local reality.

The functional areas of a life field as modeled in this cosmology are shown in the Functional Areas of a Life Field Diagram [xxiii]. A detailed explanation is provided in the Life Fields Discourse. Also see the Personality-Centric Perspective and Perception and Expression Discourses.

Perceptual Agreement

Personality must be in perceptual agreement with the aspect of reality with which it will associate. This principle provides a mechanism by which order can be imposed on a system without many complex rules limiting behavior. Rather than some ethereal being telling us what we can and cannot do, or

[19] The central concept of Darwinian Evolution is that changes in nature are random and the beneficial changes perpetuate the specie. As has been stated a number of times in this book, random processes are conducive to trans-etheric influence (intended order).

This is not an issue of belief. The influence of intentionality on random processes is well-documented. Also, there is nothing in the Implicit Cosmology that makes a distinction between a naturally random process of nature and a random process produced by a random event generator.

where we can go in *heaven,* this principle assures that we will gravitate toward aspects of reality which we are best able to understand.

The practical consequence is that freedom of access to the greater reality is increased by gaining understanding which is in accordance with the operation of reality. Comprehending the intention of the principles listed here is a good beginning.

The self-limiting behavior based on understanding appears to be made practical by the *maybe* state of the Perceptual Loop shown in the Basic Functional Areas of Perception and Expression Diagram [xxiv]. New understanding occurs when new input is reasonably close to prior understanding. When it is, the new input may be allowed into worldview, albeit after it is visualized in the Attention Complex based on prior understanding. In this model, in response to the inherited urge to gain understanding, personality gravitates toward that which is familiar but with a preference for new experiences.

This principle results in the existence of the Veil of Forgetfulness which is the practical consequence of the etheric-physical interface. When a personality creates an aspect of itself to gain experience, the new aspect inherits a subset of its source's understanding. It is created with the imperative to *return* to its source and does so by gaining understanding.

See the Cosmology of Imaginary Space Discourse for discussion of how this might look as an alternative to usual cosmologies.[20]

Process

Reality is modified by way of ordered sequences of events. As a fundamental aspect of reality, process has a beginning, which may be the spark of inspiration to do something and one or more intermediate steps such as comparing an imagined outcome with worldview. The process ends with the expression of intended order. See the Perception and Expression Discourse.

Rapport

Personalities are interconnected by links of cooperation (influence) forming a matrix of relationships (cooperative community). One personality's awareness of another personality manifests as a link of influence between

---

[20] There is a requirement in Rupert Sheldrake's Hypothesis of Formative Causation (9) for *Nature's habit* to be evolved by creative solutions to environmental problems. The *maybe* solution in the Perceptual Loop provides this mechanism.

the two life fields that is related to visualization, attention and intention. The nature of this link of rapport depends on the clarity (intensity) of awareness and the reason for the awareness. These links are dynamic and are thought to facilitate cooperation.

Perceptual agreement is a moderating factor for rapport. Presumably, the more two people agree on the nature of reality, the stronger the link of rapport. This is something that can wax and wane from moment to moment, as one or the other party is more or less at ease with the moment. This idea is well-characterized by the book, *I'm OK–You're OK.* (46)

See the Principles of Entanglement, Cooperative Communities and Reciprocity.

Reciprocity

As personality influences reality, reality reflects that influence back to personality. This reciprocal influence is similar but not necessarily equal to the initial influence. Personality develops a personal reality (worldview) based on perception. When personality expresses a change of some form *into* the environment, the Attention Complex introduces the expectation to experience that change.

A practical example of this may be seen in healing intention, in which the practitioner may sense, even benefit from the influence of expressed intention. Another important example of this may be found in the *negative* influence that is experienced by a person who expresses *negative* thoughts toward another. In this case, *negative* means disagreement between a worldview which enables a person to imagine the negative influence in the first place and the actual nature of reality. See the Principle of Balance.

Category 3: Organizing Principles of Personality

Personality is the immortal aspect and the true *I am this* of the life field. It expresses dual characteristics of experiencer and formative agent.

As experiencer, personality expresses inherited purpose to and incorporates acquired understanding from the Attention Complex [110]. It is often referred to as a person's higher self.

The formative agent is personality's intelligent core which represents the organization of the life field. This is comparable with the autonomic system for a biological organism. From the perspective of conscious self, Personality and Intelligent Core are typically indistinguishable and are mostly unconscious regions of mind.

These are the principles thought to moderate the behavior of personality.

Attention

Attention is a fundamental aspect of life necessary for perception and expression. As is illustrated in the Functional Areas of Perception and Expression Diagram [xxiv], the Attention Complex enables a number of important functions of life, of course including perception and expression, but also the mechanism necessary to evolve worldview. See the Principle of Curiosity below.

In First Sight Theory (13), attention is an unconscious state. As shown in the Functional Areas of Perception and Expression Diagram [xxiv], the Attention Limiter acts as a filter which can block environmental information if it is tested as having no interest to the organism.

Balance

The natural order is the middle way in which actions are in agreement with Organizing Principles. Balance is a way of looking at living in accordance with nature. Personality must expend effort to maintain an extreme state. It is hypothesized to be as difficult to sustain extreme positive as it is to sustain extreme negative. It is likely that true evil [154] in the religious sense is not really possible. In theory, the state would too quickly dissipate as the system is *attracted* to a balanced or more neutral state. See the Mindfulness Discourse.

Curiosity

Curiosity is the source of attention. An important assumption of the Implicit Cosmology is that a fundamental characteristic or property of the life field is curiosity. The motivation for attention is seen as curiosity about the nature of the environment. It is proposed that specific curiosity is inherited from Source. See the Principle of Attention above.

The Anticipation Corollary of First Sight Theory (13), reasonably translates as curiosity. I am modeling curiosity as a basic characteristic of life so that curiosity leads to attention and purpose.

Discernment

Believe what you wish but understand the implications of what you believe. While worldview informs perception and expression, the conscious intention to align perception and expression with the actual nature of reality is discernment. Suspended judgment allows personality to consider experiences without incorporating them into worldview.

Discernment is possibly the most important quality one can intentionally cultivate. Those who have consciously stepped on a path toward understanding make little progress until they have learned to experience life without the need to explain or rationalize experience. Understanding is not a list of *this is true* and *that is not.* It is the comprehension that this is related to that in some way.

Experience

Personality inherits the urge to seek experiences which might lead to understanding the nature of reality. It is likely that a person's entire reason for existence is to gain understanding of a certain aspects of nature which might only be accomplished by direct experience. In a very real sense, experience is the purpose of life, but acquisition of experience should be unconsciously or consciously informed by the Prime Imperative.

Expression

Reality is expressed via personality's attention on an imagined outcome with the intention to make it so. Intended order is the influence expression has on the environment. It may be in the form of conscious awareness of an output of the perceptual process, speech, order impressed on a chaotic process or an ideoplastic form. Expression is initiated by conscious intention but the formation of expression is a mostly unconscious process of the Attention Complex. See the Perception and Expression Discourse.

Focus

The ability to manage expression is, in part, a function of how well conscious self is able to focus the expression of intention. Intention is an influence which has the quality of focus and purpose. If the person is uncertain, distracted by environmental intrusions or harbors an unconscious fear or doubt, intention may be conditional, resulting in a poorly focused influence on the Attention Complex.

Intention

Intention is the motive influence of expression. Expression is usually described as attention on an imagined outcome with the intention to make it so. The coherence of the expression, and therefore its effectiveness, depends on clarity of visualization, but also on clearly expressed intention.

A useful way to consider intention is to think of expression as intended order. In trans-etheric influences, intended order is a way of describing the influence of expression on chaotic systems as a transformation according to intended order.

The Visualization Function in the Attention Complex is informed by the nature of intention signaled by conscious self, but the actual form of the expression from the Attention Complex is moderated by Worldview. See the Perception and Expression Discourse and the Perception and Expression Diagram [xxiv].

Local Reality

Reality is bound for a person by the limits of experience. Following the idea that we can only comprehend what we are aware of, local reality is that part of the reality field of which a person is aware. For a person as an entangled avatar, local reality is always a subset of the physical universe so that home, community, town and maybe country region are relatively clearly visualized by a person while a different city and the surface of Mars might be just concepts. Local reality is seen as a true region of the greater reality, so that understanding local reality is the same as understanding a part of the greater reality.

It is important to distinguish between local reality and personal reality. Local reality is a subset of reality, but personal reality is colored by worldview in the perceptual process so that it only represents what the person thinks of local reality. See the Principle of Personal Reality.

Perception

Experience is a function of intent, attention, external influence and worldview. Personality is shown in diagrams such as the one for the Perception and Expression Diagram [xxiv], as the Intelligent Core which manifests as Attention. Attention is a required state for perception, but the Perceptual Loop process determines what is allowed into worldview, what is consciously experienced and what can be expressed. In this model, personality can only become aware of what is expressed from the perceptual process. See the Perception and Expression Discourse.

Personal Reality

Perception of reality defines personal reality. Currently held beliefs influence perception of the nature of reality. A place is neutral, but two people might experience the same place in very different ways. That difference in perception is the difference in the two people's worldview. The Prime Imperative of the life field is for conscious self to align personal reality with the actual nature of local reality.

Point of View

Point of View is the underlying assumptions held by personality when determining how to relate to its environment. Conscious self's point of view is based on its worldview and understanding. While every experience goes to modify worldview, worldview colors perception of all subsequent experiences. See the Point of View Commentary.

Progression

Progression is a measure of personality's understanding. A fundamental assumption of the Implicit Cosmology is that personalities produce aspects of themselves to gain understanding about their environment. A measure of understanding is progression from Source's initial expression of curiosity to realization of understanding. From the perspective of Source, creation is expression with the intention to understand personal reality. Understanding is hierarchical; the nested fields are collapsed toward Source with each advance in understanding.

Based on the Prime Imperative, Source is still learning and doing so through its aspects. Given that understanding is an acquired characteristic of personality, the worldview of an aspect must include a subset of its source's understanding and the results of an urge to gain further understanding (targeted by its source). The theory holds that, as the aspect personality gains understanding, it becomes more perceptually in agreement with its source.

Self-Determination

Personality's behavior is limited only by the Organizing Principles. Self-determination can be understood as personal responsibly.

Suspended Judgment

Increased understanding may come from unexpected or unwanted outcomes. As we mature into adults, worldview—what we unconsciously think is true about the world—is mostly populated with cultural beliefs. These beliefs are too often local prejudices and baseless lore that, if acted on, may lead us into a less than beneficial way of living. From this perspective, the purpose of life can be described as a quest to align worldview with the true nature of reality; that is, to replace beliefs with *correct* understanding.

Consciously not making a decision about what we think of something is a good way to take control of our unconscious mental processes. This is a major part of mindfulness.

Suspended judgement involves avoiding good-bad, agree-disagree, believe-disbelieve decisions about experiences so as to allow time—sometimes months—for more information to develop. It is important to be responsible for our actions, including seeking to accomplish specific tasks. The idea is that realizing something other than an intended result does not mean failure. Instead, the unexpected result offers an opportunity to reevaluate the assumptions on which the actions were based.

Suspended judgment can be considered a more contemporary expression of the Law of Detachment. The idea of the Law of Detachment is that being set on a particular result of an action might blind us from seeing the potential benefits of alternative results. But there is a problem of literal interpretation, as people instinctively reject the notion that they should be detached when virtually everything in society tells them to become engaged.

The principle of suspended judgment supports a more general perspective. Our unconscious mental processes are hardwired to make decisions. That means there is an *accept, reject* or narrowly defined *conditional accept* decision for everything we do, sense or visualize. Consciously deciding to replace a *reject* outcome with a wait and see response teaches the unconscious mind to be more lenient in the *conditional accept* decision. This has the potential to make the content of worldview more accessible to evolution toward greater understanding.

Transition

Experience is gained in cycles which are delineated by periods of transition. For a personality entangled with an avatar in the physical, when the human is no longer able to sustain the conscious self, there is a process of release and movement out of that relationship. During this process personality's perspective changes from the physical venue for learning to some other venue. From the body-centric perspective, this is described as dying; however, it is believed to be more correct to think of this as the movement of the conscious self's focus of attention to a more natural environment.

Transition is a more accurate term for what is currently called death. Also see The Transition Experience [166] in the Progression, Teaching and the Community Discourse.

Understanding

Understanding is perception of reality as it is and not as it is believed to be, with emphasis on underlying principles. Experience becomes understanding as personality aligns personal reality with local reality. A fundamental concept in the Implicit Cosmology is that personality exists to experience, gain understanding from that experience and *return* that understanding to the Collective. Understanding is seen as being a quality of personality, and is therefore the quality determining perceptual agreement. Understanding is a measure of progression. Also see The Nature of Understanding [161] in the Progression, Teaching and the Community Discourse.

Visualization

Expression is based on the intention to express an aspect of reality based on informed visualization. The outcome of expression is dependent on a visualized objective based on current understanding which is held in worldview. It is necessary to have a clear idea of the desired outcome. More clarity in the visualization means a more specific outcome. The emphasis here is on informed. The more this understanding agrees with reality, the more realistic can be the imagined outcome.

Worldview

Worldview is a learned response moderated by understanding. As one of the functional areas of a life field, worldview appears as a thoughtform containing the memory, knowledge and point of view of the personality. It is much like a database. When a person is born the worldview database is essentially empty but it has a *front-end algorithm* which represents personality's maturity based on prior progression and an inclination toward experiences that offer the opportunity for specific areas of understanding.

The worldview database is populated by culture, family, media, religion and schools as these influences are biased by the *front-end algorithm*. It is the worldview that acts as a filter to determine how the person reacts to situations, prejudices, beliefs, acceptance or rejection of information. Worldview follows a person through transition and influences postmortem experiences.

Worldview resists change, but evolution of worldview occurs in small increments with ambiguous outcomes of the Perceptual Loop. Creative solutions to environmental challenges, if they do not represent radical changes, are expected to evolve worldview.

A person shares the Attention Complex with its avatar, which means the human's instincts are an important part of worldview. Human instincts are especially influential in the early part of the person's lifetime, but are expected to be less influential as the person gains in maturity.

More about worldview in the Point of View Commentary.

Implicit Cosmology

Discourse 4:
Source

Introduction

Source is the term used in the Implicit Cosmology as first cause for the creation of reality. Source is not considered a God in the biblical sense. A foundation assumption of the cosmology is that reality is knowable. As such, the cosmology is used as a predictive model to propose characteristics of Source that are consistent with an evolved reality.

Source is defined in the Organizing Principles Discourse as:

> *As first cause, Source is a singularity modeled as a life field, the personality of which is both the experiencer and formative agent of the reality field. The expressions of that initial personality are aspect life fields which inherit purpose and understanding concerning the nature of reality.*

The important concepts in this definition that must be developed in the cosmology are:

Source is the Reality Field

Reality is modeled as a conceptual domain, and a field is characterized as a concept's zone of influence. Fields are modeled as the building blocks of reality beginning with Source as the top field. As such, Source's life field and the reality field are the same.

The field concept is addressed in many parts of this book, but you might begin with the Etheric Fields Discourse.

Personality and Intelligent Core Concepts

The personality is the experiential part of who we are, while the Intelligent Core is the formative part of who we are. In our human body field (morphic field), the body image is the intelligent core while the personality is the expression of purpose.

These concepts are explained in the Life Fields Discourse.

Expression and Perception

A life field interacts with its environment by sensing its influences as perception and responding to those influences as expression. The two modes are governed by a functional area described as the Attention Complex. The functioning of that complex is informed by personality's intention, but it is moderated by worldview. These functional areas are explained in the Life Fields Discourse. They are applied in this discourse to explain the actions of Source.

Aspect, Aspectation and Understanding

In a practical sense, reality is a thought exercise. We sense the world around us as a mental translation of various psi and physical signals that is expressed to our conscious self via our Attention Complex. In a sense we create our world. This is explained in the Perception and Expression Discourse.

The mechanism with which personality explores reality is referred to in this cosmology as aspectation. Aspectation is the differentiation of something to produce an aspect of that something. Probably many times removed, we are an aspect of Source. This is explained in the Perception and Expression Discourse.

Also in this cosmology, the purpose of aspectation is to gain understanding. In that view, we are the imagining of a personality. We have inherited purpose from that personality, and we will satisfy that purpose by gaining understanding about that purpose which will become part of the personality that imagined us. This is discussed in the Perceptual Agreement Discourse.

All of these concepts help to define Source. I have integrated them into a single view for this discourse, but as I have explained before, understanding these concepts depends on understanding concepts that will be explained in other discourses.[21]

Not Proof but Reason to Consider

My motto for EthericStudies.org is *Believe what you wish but understand the implications of what you believe.* When I considered the concept of Source

[21] For me, it has been a grand learning experience. Almost every day has produced an *aha moment* as I realized that a concept really means more than I had thought. Please do not be frustrated if you must experience the same sort of discovery. It is worth the journey!

with this point of view, it became clear that an objective explanation requires that Source be bounded by some outer limit. The infinity concept does not really fit into this cosmology, so it is necessary to simply say that Source is a singularity that may or may not be amongst other singularities. For our purpose, reality begins with Source.

While no single concept explains Source, the ones offered here are complementary so that together they provide a reasonably clear model describing the nature of Source, which is, in turn, the nature of reality.

As Above, So Below

This cosmology includes the assumption that reality is homogenous, in that knowing the nature of one part indicates the nature of the rest. Hermes is said to have taught this fundamental concept around 5,000 years ago via the *Emerald Tablet*. (47)

Here are the first two lines of the *Emerald Tablet*:

1 *It is true and no lie, certain and to be depended upon, that which is above is as that which is below; and that which is below is as that which is above, for the performance of the one truly great work.*

2 *And as all things are from only one thing, by will of the one God, so all things have their origin in this one power, by adaptation to their individual purposes.*

The Great Work is the process of transmuting a faith-based worldview into one that is in accordance with the actual nature of reality. Of course, the *one thing* is God, but important to note is that, while Hermes has told us that all things are of God, he also said that each individual has been differentiated as a unique aspect of God. This echoes the concept of life as a fractal in which God is the top fractal.

As far as I can tell, Hermes relied on a cosmology that is very similar to the Implicit Cosmology. Of course, this is not proof in any way, but it is important in that modern Ancient Wisdom Schools are typically at least partially based on what is considered the Hermetic wisdom. There is some indication that such revealed information can be reliable: for instance, the evidence of ITC and analysis of some of the more prolific trance channels such as Jane Roberts' Seth. (26)

Fractals

Fractals are a useful way to visualize reality as it is modeled in this cosmology. A fractal is a pattern that is repeated at every scale. The Sierpinski Triangle plot illustrates how an equilateral triangle is a fractal of an equilateral triangle array. (Draw an equilateral triangle, put a point at the halfway point on each line and connect those; repeat indefinitely.)

In the Fractals Diagram below, I refer to the largest triangle in the Sierpinski Triangle as the *top fractal.* The same can be said of what is sometimes referred to as the Apple Man of the Mandelbrot Set. The third frame shows one of the Apple Man fractals which becomes visible as the scale is changed to magnify the set. The black area in the upper-right corner of the last frame is part of the top fractal. Also see the navigation example [173] In the Cosmology of Imaginary Space Discourse.

Relating the Sierpinski Triangle to the hierarchy of nested fields concept, the top fractal can be compared to Source and the reality field for that set. The first round of aspectation produced three triangles comparable to aspects of Source expressed to explore the reality field. The next round produced three triangles inside each of the first round triangles, and so on. Each set of three can be said to be nested within the previous triangle. In this way, there is a many-to-one relationship for each round.

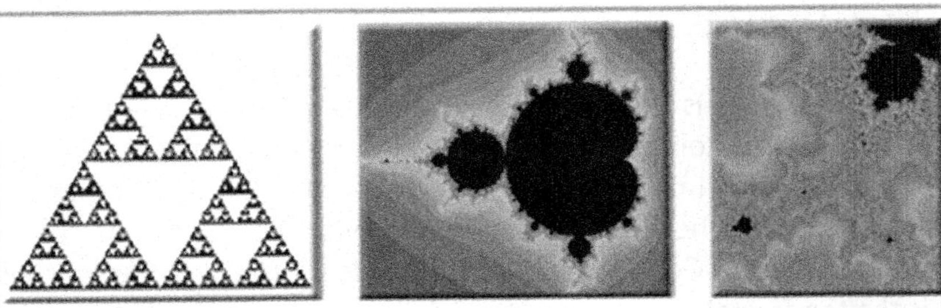

Sierpinski Triangle — Mandelbrot Set (Top Fractal) — Fractal (lower left) as it relates to top fractal (upper right)

Fractals: a pattern that is repeated at every scale

Each of the three triangles that result from the next round of aspectation will be within a source triangle. Compare that to yourself as part of a collective produced by a personality from a prior round of aspectation. It is probable that there are many different collectives sharing the same aspect of reality—in our case, for the purpose of gaining understanding through experience.

Life Fields

The model for a life field is based on current understanding about the way we think and on observed paranormal phenomena. The foundation assumption is that characteristics of the greater reality can be modeled based on what is known about the local aspect of reality (as above, so below). Information about local reality comes from mainstream science, parapsychology and transcommunication. The Trans-Survival Hypothesis is used as the beginning point for black box analysis [14] (Section I Intro.).

As explained in the Etheric Fields Discourse, a field is a set of elements with related characteristics which are bound into a system by a common influence (the attractor). In metaphysics, a field is bound by the extent of influence the attractor is able to exert.

The attractor for a field is probably best described as a concept. The influence a concept is able to express on the elements of its field can be described as purpose. In a life field, personality is the attractor. In the reality field, Source is the attractor. In a collective field, the local source for personalities in that collective is the attractor.

Morphic Fields

Rupert Sheldrake (9) deserves much of the credit for how the life field is modeled in this cosmology. As is discussed in the Etheric Fields Discourse, he has proposed the idea that living organisms are organized under the influence of morphic fields.

Paraphrasing with Implicit Cosmology terminology, for an organism, morphic fields are associated as a hierarchy of nested fields in which each aspect of the organism is managed by a field. For instance, muscle cell morphogenesis is managed by morphic fields based on how they have always been formed; skin cells are managed by skin cell morphic fields; and bone cells are managed by bone cell fields, to name a few. All of these are managed by a human being morphic field.

In that hypothesis, morphic fields have a sort of intelligence, memory which Sheldrake calls *Nature's habit,* a mechanism to express that information on the living process, and a means to modify Nature's habit as creative solutions for environmental challenges found by the organism. If you review the Life Fields Discourse, you will see equivalents for those characteristics in the functional areas of a life field.

An organism is physical while the organism's organizing influence is etheric. This is just like the relationship between our human brain which is part of a physical organism and our mind which is etheric. To manage the development of an organism, Sheldrake's model requires morphic fields to have the ability to express trans-etheric influences. My model for expression is largely based on that assumption.

Sheldrake's model also requires a trans-etheric means of sensing changes in the organism. That is part of the inspiration for the perceptual process described in the Perception and Expression Discourse.

An important process in both expression and perception is the means to compare proposed expression or perception with worldview. I have described that as the Perceptual Loop [112]. First Sight Theory [126], proposed by James Carpenter (13), provided the inspiration for the ruleset needed to make the Perceptual Loop work. First Sight Theory is based on meta-analysis of the very large number of psi functioning studies, and the non-psi parts of First Sight Theory are beginning to be recognized by mainstream researchers. (38) (11) (12) (48)

Unconscious Perception

Who we are is not who we think we are. (49) As the Functional Areas of a Life Field Diagram [xxiii] shows, the Attention Complex is an unconscious part of our mind. We only become consciously aware of the results of the processes represented by the Perceptual Loop. What we become consciously aware of is based on our worldview. (11), (39)

The importance of unconscious perception is that we use the same mechanisms that organize cell formation to develop an objective image of our physical world. This is an emerging realization of mainstream science; about which you can expect to hear a lot more in the future.

As I will explain later, our unconscious perception is complemented by our unconscious expression. This is expressed in the second assumption of Carpenter's First Sight Theory.

Source as a Life Field

Source is the top life field, and as such, it is also the reality field. In effect, everything in reality is *in* Source's life field by way of the link of entanglement between first cause and subsequent aspects. This influence of entanglement is a function of intention, attention and imagined outcome.

A morphic field is also modeled to place the organism it influences *in* the field. This is because a morphic field imposes an influence on the biological process. *In the field* means within the field of influence expressed by the morphic field or life field.

A Contemplation

A mental exercise explains the next step in the model. Read this paragraph and then close your eyes and relax for a few minutes. Imagine that you have just purchased a new car. Think of how it might look. See yourself driving the car. Perhaps you are driving it through your neighborhood so that your friends can see you. What does that feel like? Spend a little time driving your imaginary car and then return to this text.

This contemplation illustrates the process of expression. The car, *little you* and where you imagined driving the car are all expressions of the Attention Complex. All that you visualized was based on your worldview. That is, you are limited in what you can visualize by what you believe is true.

The imagined scene was an ideoplastic construct, meaning that you were forming objects of reality with your mind. The scene will exist as long as you keep attention on it. It will continue to exist in your memory indefinitely.

An important point is that you consciously intended to have the car-driving experience. That intention initiated a creative process in your unconscious mind which developed the desired experience and expressed it to your conscious self as an imagined experience.

Again, the imagined experience was limited by your worldview (personal reality). Except for minor modifications logically based on prior experience, everything in the fantasy agreed with your prior experience.

Emerging understanding in psychology about how people experience reality indicates that, when you experience something you think of as objective reality—say actually driving a physical car—that awareness comes to your conscious mind via the same unconscious-to-conscious process as I described here for your car-driving fantasy.

If you had not been prepared by prior experience, you would likely not have been able to tell the difference. An important rule of such imagining is to allow your imagined *little me* self-determination so that you can gain from it *as if it was real* experience.

It is difficult to sense what an imagined being is feeling, but with practice, this can be a powerful tool for gaining understanding. Just behave *as if* you sense what it is thinking and feeling. The only real hindrance is the autonomous perception/expression process you allow for your imagined self. You know you are doing it right when your *little me* surprises you in some way.

Note that the imagined person is an aspect of you. In terms of this cosmology, you are the source personality for *little me*. It is likely you have conducted many such exercises, spontaneously imagining what something might be like. You can probably recall such imagining, even from many years ago. All of those *little me* aspects continue to exist in your life field. You are the top life field for a collective of your *little me* Aspects. They exist within the influence of your life field as part of your personal reality.

Another characteristic to note is that your imagined experience occurred in your mind. It did not produce an objective experience as we think of an experience in the physical world. I imagine my life field as a kind of etheric balloon in which the product of all of my imagining produces a kind of contemplative reality complete with a venue for learning and aspects of me to explore the venue. Some important realizations from these imagined aspects have emerged into my awareness.

Personal Reality

Just as we can develop a mental image of *little me's* world, so do we create our personal reality based on what we believe to be true (worldview). Since we only become consciously aware of what our unconscious perceptual processes present to our conscious self, we literally experience objective reality as it is expressed by our beliefs. This is becoming established science. (11), (39) We manage our sense of what is real by learning to change worldview. One way to accomplish this is via mindful living. In effect, that is what we do while learning to understand our world.

If life fields are fractals in Source's reality, it is arguable that Source also has a personal reality and is seeking to align it with reality's actual nature. As such, Source and all the rest of us are still learning. That is what we do.

Modeling Source

As mentioned above, an important assumption in this cosmology is that life is a universal characteristic of reality. Also, life is modeled as a field which has an *I am this* sense of identity I refer to as Personality.

A life field also has an Attention Complex in which perception and expression are managed. It is in that functional area that the life field's memory is stored as worldview. This is the equivalent of Sheldrake's Nature's habit; however, it also contains understanding and acts as a moderator for perception and expression.

Populating Reality

To establish a beginning point in this cosmology, Source is modeled as having *somehow* begun as a singularity with the characteristics of a life field. This may have evolved from some pre-reality potential or it may be an aspect of yet a more conceptually remote intelligence. Without more input, it must only be assumed that Source exits.

A second assumption is that Source is curious about itself in much the same way we might be curious about how it would be to drive a fancy car. The act of learning about itself is modeled as the act of creating reality.

In effect, Source is contemplating itself. In doing so, it has imagined itself in what it believes is its world. Such contemplation would be multi-threaded as Source considers various aspects of what it thinks is real.

In the terms of this cosmology, to facilitate its self-contemplation, Source has created a venue for learning and created aspects of itself to have experiences in the venue and return understanding. Those aspects are life fields that have inherited from their source a degree of understanding and the urge to gain specific additional understanding.

The aspect's personal reality would be a subset of Source's personal reality; however, because of the aspect's experiences and eventual understanding, its personal reality would become different from Source's. In principle, its understanding about at least its small part of Source's reality would be more true to its actual nature.

The process of imagining venues based on worldview and exploring those venues with an aspect of self is a basic function of a life field. As was illustrated in the above contemplation, in practice, the question automatically created the venue; the intention to understand automatically produced the aspect personality.

Self-Organizing Reality

An interesting aspect of the Implicit Cosmology is the idea that reality is self-organizing. Source discovers itself. It does not create itself.

The Intelligent Core [110] of each life field establishes and maintains the structural form and function of the field. The rules for these are a natural result of purpose. By this I mean that requiring understanding means there must be a mechanism by which understanding may be gained. Understanding needs to be tested so there must also be a mechanism to express that understanding. And for that, there must be a process to visualize that understanding and compare it with prior understanding. These mechanisms would naturally evolve to become Nature's (current) habit.

The Organizing Principles are arguably the direct result of Source's curiosity. The Formation Principles naturally evolve from the Reality Principles and the Personality Rules automatically evolve from the Formation and Reality Principles.

In a cooperative community, the exchange of ideas amongst its members requires them to organize their thoughts before speaking or acting. The organization is based on worldview; however, expression produces feedback, in effect testing worldview. It is in this way that worldview can be better aligned with reality.

Based on this worldview testing mechanism, the rules governing perception and expression translate into a cooperative community rule of *"Teach me as I teach you."*

A self-organizing reality does not mean that there are no intelligent, purposeful influences in that organization. It just means that the purposeful influences are directed toward optimizing understanding through experience.

Implicit Cosmology

Discourse 5:
The Nature of Reality

Introduction

Popular wisdom has it that there is one reality. By definition, it is absolute; however, a second view is that reality is relative. The reality concept is a good example of the difference between body-centric (absolute reality) thinking and personality-centric (relative reality) thinking.

The Implicit Cosmology is designed with the assumption that perception of reality must be relative. That is because expression is always informed by perception. In every way that matters, reality is a function of our perception of the nature of the world in which we live.

Psychologists will tell you that our perception of time, distance, even meaning is a function of many factors such as personal stress and enthusiasm. All of these factors go to modify the content of worldview and the *Accept, Reject, Maybe* decision of the Perceptual Loop [112]. See the Perception and Expression Discourse.

The Name of Reality

It may be helpful to list the various types of reality discussed in this discourse. As you read this, the key point to consider is that reality is relative to worldview and worldview is relative to experience. The fruit of this dependency is understanding.

Baseline Reality

Source's perception of reality (the Reality Field) converges on a potential reality represented by organizing Principles. Perceived aspects of reality will converge on the baseline, but are not expected to coincide. Source's life field is the Reality Field and is Source's personal reality. Since the cosmology describes the education of Source about itself, the reality filed is expected to evolve as understanding increases. This should include an evolution of Organizing Principles.

As is illustrated in the Limits of Understanding Diagram [161], understanding is an evolving influence in the Reality Field. Presumably, there is a naturally occurring baseline or end stage in which Source's increasing understanding of itself converges in an asymptotic manner toward a form of stability in which the influence of all of the Organizing Principles achieve a stable state of equilibrium.

Perception of the Reality Field, how close it is to baseline reality, will not be uniform. Source will have a less informed perception of itself, while some aspects, probably many rounds removed, can be expected to have acquired a more sophisticated understanding of reality.

There does not seem to be sufficient information to speculate about what might happen when baseline stability is achieved. My sense is that the lifecycle of a person is something of a standard model for the life field fractal. If that is true, the stability means old age. At some point in Source's evolution, I would expect to see something that generates a next generation Source.[22]

Local Reality

Reality is bounded for a person by the limits of experience. We can only comprehend that which we are aware of. As such, local reality is that part of the reality field of which a person is aware. For person as an entangled avatar, local reality is always a subset of the physical universe. Home, community, town and maybe country and region are relatively clearly visualized while a different city and the surface of Mars might be just concepts. Local reality is seen as a true region of the greater reality, so that understanding local reality is the same as understanding a part of the greater reality.

Personal Reality

Worldview defines personal reality. Currently held beliefs influence perception of the nature of reality. For instance, a place is neutral, but two people might experience the same place in very different ways. That difference in perception is the difference in the two people's worldview. The Prime Imperative [61] of the life field is for conscious self to align personal reality with the actual nature of local reality.

I am saying that conceptual reality is real and the physical universe is an ideoplastic expression. If that is true, the degree to which a person is aware of the greater, conceptual reality represents a meaningful segment of personal reality. Considering the definition of local reality, this becomes a

[22] For those of you who are interested in Zero-Point Energy, the difference between local and baseline reality would represent potential future as it is being shaped by increasing understanding of the collective. It might be interesting to speculate about the nature of that potential.

fuzzy area of this cosmology since the person experiences, and is therefore, aware of some aspects of conceptual space.

Differential in Reality

An interesting speculation concerns Zero Point Energy (50) which is what remains after all other energy has been removed from a system. A possibly related term is the cosmological constant which is the value of the energy density of the vacuum of space. Both of these concepts suggest that the physical universe has a baseline of energy which *floats* above an absolute zero energy potential.

Is it possible that the difference between Source's personal reality and the physical universe is seen in this difference in potential? Perhaps each round of aspectation, with its resulting differential from the baseline aspect, added a little to this difference. If one were to go to some round of aspectation between *here* and Source, would the cosmological constant be less?

Entanglement

A characteristic of association amongst a hierarchy of personalities is entanglement, which is a bond established between personality and its expression. Entanglement is an important influence between personality and its aspect (aspectation). Conscious self is seen as entangled with its source personality, which in turn is entangled with its source and so on.

Parallel to the hierarchy of personalities are nested aspects of reality in which the reality field, local reality and personal reality of this scenario can be considered *nested* sub-realities or venues for learning. Some aspects of sub-realities would be shared venues for learning, providing shared experiences for some personalities or collectives. As such, the physical universe is seen as the physical aspect of reality in which entangled conscious selves (persons) share the general nature of the physical via their avatar but have personal points of view (personal reality) codified in their worldview.

If you think of the concept of entanglement inherent in expression as threads of influence, it is reasonable to imagine reality as a matrix of purposeful influences integrating all of the relative aspects of reality into a single filed of mutual intention. As a person, we would be a source of influence as our thoughts are expressed into the etheric, and in turn, we would be subtly influenced by that expression, perhaps as a reassuring background impression that we are part of a much greater whole.

Discourse 6: Etheric Fields

Introduction

A field can be described as a set of elements with related characteristics which are bound into a system by a common influence. Reality is modeled as conceptual, and the zone of influence of a concept is characterized as a field. Fields can be *within* fields and an element can be in more than one field.

Etheric is a generic term for the conceptual space of reality, as compared to physical space, which is an aspect of etheric space. It is useful to model reality as a single etheric field, which is referred to here as the reality field. The reality field is the top etheric field, from which all other fields have been derived via aspectation. In this way, the reality field has been differentiated as a hierarchy of nested fields.

The fundamental characteristic of the etheric was once referred to in this cosmology as etheric energy; however, after further consideration, *etheric field* is a more appropriate term. As will be shown here, an etheric field is very different from physical energy and has very different characteristics. See Energy Verses Field in this discourse.

Characteristics

Etheric fields can be characterized as:

Nonlocal

From the perspective of a physical observer, etheric fields are ubiquitous or everywhere at once. It may be correct to say that *everywhere is here*, alternatively, *here is everywhere*. See Nonlocal and Nontemporal in this discourse. The concept of nonlocality will make more sense as you read the rest of the discourses in this section.

Etheric

Since the etheric is conceptual, terms like *wave* and *frequency* have no meaning when discussing etheric fields. Also, there is no apparent direction or intensity of propagation. There are conceptual equivalents to physical terms. For instance, the physical concept of *direction* might be thought of in etheric terms as the influence of attention on visualized outcome. See the Perception and Expression Discourse.

Association is Limited by Perceptual Agreement

Since etheric space is nonlocal, all concepts coexist with personality, and movement from one to another is accomplished by changing focus of attention. This, in turn, is limited by the person's perception which is moderated by the person's worldview. Therefore, the ability to associate with any aspect of reality is seen to be limited by a person's sense of what is real and true.

Etheric fields are always an expression of personality, and are therefore limited by that which limits personality.

Influenced by Intentionality

Etheric fields are responsive to personality's intention to express intended order.

A Precursor to Objective Reality

Physical energy and matter is seen as a result of the influence of intention on an etheric field. That is, physical energy and matter are expressions of personalities.

As modeled in the Implicit Cosmology, Source, as the top life field, has expressed the rest of reality in the form of aspect life fields and ideoplastic structures.

Origin of an Etheric Field

Expression is basically the production of an effect via the intention to make an imagined outcome *real*. Here, *real* is as it is experienced by the creating personality. In effect, it is an intended change in the personality's personal reality.

The concept of etheric fields is closely associated with the idea of intended order which is a characteristic of expression. That is, when we say something, we have first ordered our thoughts, selected the appropriate words, and spoken with a tone and loudness that fit the situation. Doing so, we impressed order on the circumstance based on what we imagined and intended.

Etheric fields are also closely associated with the concept of aspectation. In that, a personality creates an aspect of itself to experience a specific venue for learning. Reality is modeled as being a hierarchy of nested fields, in that there are many rounds of aspectation as mostly autonomous aspects seek to further refine the venue for learning. Aspects remain associated with their source by way of a thread of influence described as entanglement.

Persistence of an Etheric Field

Expression may be as complex as creation of our physical universe or a life field. Behavior of an expression depends on the source's intention.

Expression is fundamentally a thoughtform. That is, it has a subject and one or more related references as purpose. It is a field in that the related references are dependent on the subject to have meaning. A thoughtform exists as long as the source has attention on it.

The lifecycle of a life field is limited by the achievement of understanding. When the reason for aspectation has been satisfied by the aspect's acquisition of appropriate understanding, the aspect's worldview and personality is merged with its source. Here, the understanding is that our sense of self does not cease but continues as part of our collective.

Mechanics of an Etheric Field

The way concepts associate as a field is rather different than what you may be familiar with in the more objective terms of physical space. To understand this difference, begin by considering how we store memories.

Remember a place you recently visited for vacation. The memory of that place exists in your mind as part of your worldview. Think of it as a set of elements such as name, location and quality of experience. All of the elements of your memory are colored with feelings, impressions and associations that are unique to you.

The memory of the location exists as a sort of gestalt [438] thought bundle which you can explore by accessing your memory of the vacation. The thought bundles remain as latent recollections, potentially forever. As with everything in your worldview, in some small way, they influence everything you do.

The elements of your vacation thought bundle are entangled with one another and with you. While all of the elements in the memory exist as you remember them, they could exist in another person's memory as a different set because that person would remember them differently.

The mental representation of the place in this thought exercise is the top memory and the attractor for the rest of the elements in the memory. This way of thinking is not important to the memory, but it is helpful in modeling the experience. In effect, all of the elements are attracted to the top; let us call it the *name memory*.

Entanglement is an interesting concept because it represents a link of influence that affects you and the element. The elements of your memory are entangled with you as the creator of their unique character. Another person remembering them differently will also be entangled with them, but the elements will represent a difference in character.[23]

The links of influence represented by the entanglement are not just one way. They are influenced by the way we think of the elements of the field. A contact field represents just such an accumulation of influences which colors the actual nature of the elements. That is, if enough people consider a place holy, then it begins to take on the characteristic psi signal of what people think is holy. Someone who is not aware of the holy nature of the place can be expected to at least unconsciously sense that signal.

Nonlocal and Nontemporal

Two characteristics of the etheric that need to be clarified are time and distance.

Nonlocal

Research into the psi field and psi functioning is showing that there appears to be no perceptual separation or distance between a healer-practitioner and sitter or remote viewer and target location. This *everywhere is here* characteristic of the psi field is referred to as nonlocality. (Actually, nonlocality is usually used in the context of *here is everywhere*.)

A similar effect is reported in quantum mechanics. And so the current model for psi is that it is nonlocal, meaning that one can think of someone in another city, and in principle that someone should be aware of this attention as it occurs.

Characteristics of Instrumental TransCommunication (ITC) [279] also support this nonlocal characteristic, making it reasonable to include the concept in the Implicit Cosmology. And, even though physical sensing is local

---

23 Here is a characteristic of our connectedness. Many people will have the same elements in their memory of a shared experience called *vacation*, but they will color them differently. In effect, each creates an aspect of that place as a different thoughtform. We all share the world, but each of us has a different sense of the world.

as our mind perceives it, the actual processing of physical information is also in the etheric, and therefore is nonlocal.

Nontemporal

A concept often included in the Super-Psi Hypothesis is the *nontemporal problem*. Like nonlocality, quantum principles seem to indicate that time is not relevant. This nontemporal characteristic is also seen in the many reported instances of precognition.

Researchers have shown that a person will unconsciously respond to a stimulus before it is delivered. The terms they use for sensing near-term events is *presentiment*. Also, the Global Consciousness Project (51), which is an array of Random Event Generators (REG), has detected changes in randomness moments before major events such as the 911 attack. These seem to indicate a global sense of dread before disasters occur. This is an indication of nonlocality and short-interval nontemporal characteristics of reality.

There are reports of astoundingly accurate precognition about events that came to pass far in the future. Let us call those reports *extraordinarily precognitive* and ignore them for the moment.

Research is showing that events produce psi signals and that people normally unconsciously senses psi signals. Current research also seems to indicate that events are accompanied by a probability signature; let us call those *potential fields*. The speculation is that any single event is likely preceded by many potential futures, one of which eventually becomes the event, and that these many potential futures are unconsciously sensed by people as psi signals. See the Perception and Expression Discourse and Mediumship Study.

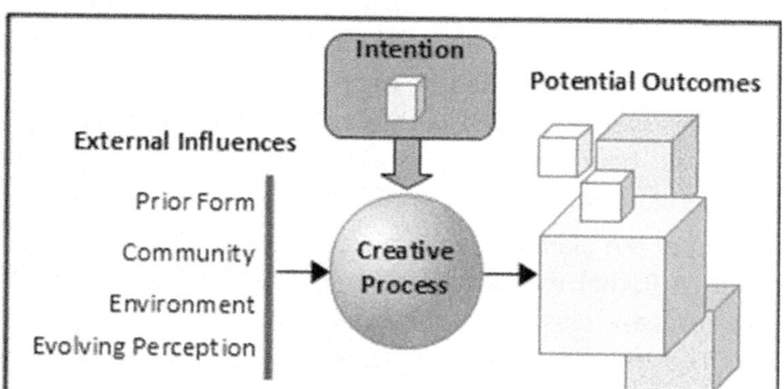

Potential Future (Potential Fields)

Potential fields would be sensed as more or less probable, depending on how imminent the related event is and the character of the influences shaping the event. An event may not occur, even though there were many potential fields. A person would sense a potential field just as any other field, even if the event itself is aborted. Sensing and reporting the potential for an event to occur would be seen as precognition. Also, a person would be drawn to the potential that most agreed with his or her worldview.

To emphasize this point, emerging understanding in consciousness research indicates that people normally, unconsciously sense psi signals from the environment. In the Implicit Cosmology, events are modeled as being sensed in the form of growing probabilities. A very substantial event is probably preceded much earlier in time with the emergence of related potential fields. Many of the extraordinarily precognitive reports did concern momentous moments in our history, such as the 911 attack and commercial airline crashes. It is worth speculating that these extraordinary demonstrations of precognition are actually normal sensing of potential fields. If so, then *the concept of a nontemporal nature of reality would not be supported*. All time is not now; process requires time.

Futurists, Long Rang Planners

A good, everyday life example of how emergent possibilities can be used to produce predictions of future events is the way long range planners produce strategic plans for business. You may have encountered people who claim psychic ability and market themselves to businesses as futurists. It is arguable that a futurist is just a long range planner who consciously emphasizes intuitive ability over corporate knowledge.

Part of my career in the communications industry was in network planning. I was indirectly part of a team assigned to produce long range forecasts involving consideration of currently known influences such as existing equipment, environmental hazards and capacity requirements. Those influences were modeled with a host of influences thought to be coming such as new right-of-ways, changes in technology and new government regulations.

The output of the planning process was a forecast based on the knowns and possibilities. Rather than a prediction from a psychic based on intuitive ability, the forecast was ostensibly based on cooperate and industry knowledge. The results were the same. Even though forecasts were objectively supported by objective facts, it would be easy to characterize such a plan as a precognitive prediction. It all depends on your perspective.

A good planner is able to synthesize a lot of information into a mental image and clearly express that to others as a general picture. From the perspective of the Implicit Cosmology, the difference between a good planner and a mediocre one is lucidity [151]. Just like a psychic predicting the future, a planner's ability to sense the psi fields associated with a target forecast is likely a deciding factor in its usefulness.

From experience, the common factor for psychic prediction and planner forecast is the ability to sort through all of the environmental signals to focus on those which are most likely meaningful for the prediction of forecast. Neither involves seeing the future as an isolated event. Future events are the product of processes shaped by environmental influences which can be sensed during the process.

Energy Verses Field

The concept of energy as it is applied in the physical does not appear to apply to nonphysical aspects of reality. Instead, what stands for differences in potential is the influence of intentionality on potential characteristics of ideoplastic structures. That is, form evolves out of the expression of a visualized order. It is the influence of intention to experience a visualized order that provides the template for formation. This is essentially the same concept as used in quantum theory in which the *"quantum wave function is collapsed by observation,"* as it was described by Radin, et al. (41)

Commonly used terms, such as *spirit* and *spiritual energy*, can be considered synonyms of *psi* as they are intended. Terms such as *psi energy* (sometimes referred to as *psychic energy*), *auric energy*, *ectoplasm*, *orgone energy* and *od* refer to forms which are also seen as synonyms for psi. *Psi* or

Physical	Etheric
Electrons/Magnetism	Attention
Potential	Intention
Propagation	Expression
Work	Visualization

Compare Physical and Etheric Influences

psi field is used here as it is used in parapsychology; however, when the term *psi energy* is encountered, it should be read as *psi influence*.

Hierarchy of Nested Fields

As stated for the Principle of Aspectation [63] concerning the expression of an aspect of itself by a personality, the aspect inherits a degree of self-determination and is able to express aspects of itself to further explore the question. The source personality and its child aspects constitute a *round of aspectation*. The child aspect and its children represents a subsequent or n+1 round of aspectation. Repeated rounds of aspectation necessarily result in a hierarchy of aspects with an increasingly smaller subset of inherited characteristics relative to the initial source.

Using the fractal as a model, fields are attracted to primary fields in a hierarchical arrangement of nested fields. As an example, and using the Hypothesis of Formative Causation as a model (9), a human body is composed of a top field which has the memory of how humans are formed. That field acts as an attractor to organize a hierarchy of fields representing the formation of major parts of the body (bones, heart, skin). Those sub-fields act as attractors and organizers for supporting formation of cells, enzymes and DNA.

Fields in the Form of ...

The field concept is applied throughout the Implicit Cosmology discourses, but usually as a reference to specific subcategory of the fields concept. Here then, is a brief list of subcategories:

Etheric Field

An etheric field is a region of reality that can be modeled as a node or nexus (attractor) associated with a set of related elements, a shared influence.

Reality Field

As stated above, reality is the *body* of Source. As such, Source as the top field is considered the reality field in this cosmology. While there may be something *outside* of the reality field, the scope of this cosmology is limited to *within* the field.

Venue Field

In this model, any field that becomes a *container* for experience is considered a venue field. It inherits characteristics of its creator's life field based on habit and differentiated by intended outcome. A venue field may be as simple as that created by a person imaginatively exploring how it might be to drive a new car. See Venue for Learning [137] in the Perception and Expression Discourse.

A venue may also be developed by a collective. The physical universe is an aspect of the reality field. It has been differentiated as a subset of a personal reality held by one personality or a collective of personalities (life fields). While we have been taught to think of these things in the physical as objective, from the etheric perspective they are ideoplastic structures. See the Perception Diagram [xxiv] and the Perception and Expression Discourse.

Life Field

Life fields are expressions of Source and inherit their formative image (Nature's habits) and instincts (urge to gain understanding) from Source. Life fields are entangled with Source. They are etheric fields which have a personality as the experiencer, an intelligent core as a formative agent, and a worldview representing their personal reality.

In the Hypothesis of Formative Causation (9), Nature's habit which guides a morphic field would be the worldview. The core intelligence of the morphic field would be that which manages perception and expression. The fields are nested in that a human organism is formed as a hierarchy of life fields. See the Functional Areas of a Life Field Diagram [xxiii].

Thoughtform

As a rudimentary etheric field, thoughtforms contain information about a specific subject. They may be nested so that a transportation thoughtform, for instance, may contain information about all forms of transportation, while a red wagon thoughtform would contain information specific to red wagons and would be a subset of the transportation thoughtform.

For perspective, elements of a thoughtform are not piled into a bundle as we have been taught to visualize physical fields. As sets of related concepts, they are entangled with a core concept (wagon, red wagon), but might originate from any source in reality. The only rule for association appears to be perceptual agreement [67]: although all related concepts would be attracted to the core concept, personality would only be able to access those it is able to perceive (comprehend). The elements of a thoughtform are nonlocal.

In a strict sense, all etheric fields are thoughtforms, but in a practical sense, there appears to be a difference in coherence between a thoughtform as a life field and one that is an ideoplastic expression of a life field.

As an aspect of Source, a life field has a personality as experiencer and an intelligent core as a formative agent. Thoughtforms are modeled as an expression of a life field but not an aspect. As such, they depend on the continuing intention of at least one personality to exist.

An important distinction between life fields and thoughtforms is how the elements of the field are entangled. Elements of a thought form may have many threads of entanglement such as the originating life field and many category relationship associations. The elements of a life field are directly associated with the Intelligent Core [110] while the life field itself will have more defined association with parent and child fields.

Contact Field

The Principle of Rapport is defined as *Personalities are interconnected by links of cooperation (influence) forming a matrix of relationships (cooperating community)*. Rapport can be considered a special case of entanglement which produces a form of influence that facilitates trans-etheric influences.

A contact field can also be thought of as a type of thoughtform which becomes entangled with a group of people. Such a thoughtform may support similar trans-etheric influences which otherwise require the entanglement of a personality-avatar relationship. In one form, this is seen in reportedly haunted places, but more generally, it is seen in churches and places often considered with reverence. As we learned in the ATransC, a distributed community is capable of building a contact field.[24]

Intention Field

Expression of intention (intended outcome) is generally thought of as a thoughtform which acts as a formative influence for physical action (motion, voice), conscious thought or an ideoplastic structure. Intention fields persist as long as the personality maintains attention on the intended outcome. See Intention Field [100] in the Perception and Expression Discourse for more on this.

---

[24] Transcommunication development circles can develop substantial contact fields by way of rapport amongst member sitters, focused intention and long-time association. It is thought that this accumulated field facilitates trans-etheric influences.

Potential Field

Events are accompanied by a probability signature referred to here as a potential field. In this concept, an event might exist in the etheric as a potential to occur. Each potential field represents the possible characteristics of an event, so that an event that has occurred may have been preceded by many potential fields representing different outcomes.

Events are a product of expression and the potential field is characterized according to quality of attention, clarity of visualized outcome, and extent of intention to make it so. The lifecycle model for an event begins with the inclination to do something. The actual decision to act may never come and the possible way to act may take many forms. Each possible way to act that emerges into the creative process is thought to produce a potential field. How far into the process the event progresses determines the influence of the potential field.

Psi Field

Psi is a placeholder term used in parapsychology for an as yet undefined characteristic of reality which is experienced as a medium for propagation of intentionality. It is identified with such extraordinary human abilities as remote viewing and distant influence of intentionality.

While physical energy is local and time dependent, as defined in parapsychology, the psi field is experienced as being everywhere at the same time (nonlocal) and also as being independent of time (nontemporal). The term *psi* is often used in the Implicit Cosmology as it is intended in parapsychology to facilitate communication. It is modeled as interconnected etheric fields, each instance being associated with a life field's perception.

Organizing Field

An organizing field is seen as the top formative field of an object of reality. If the object is a physical organism, then the morphic field model is used. If it is an ideoplastic structure, then it is seen as the field which maintains that structure as it is sustained by intention from a personality.

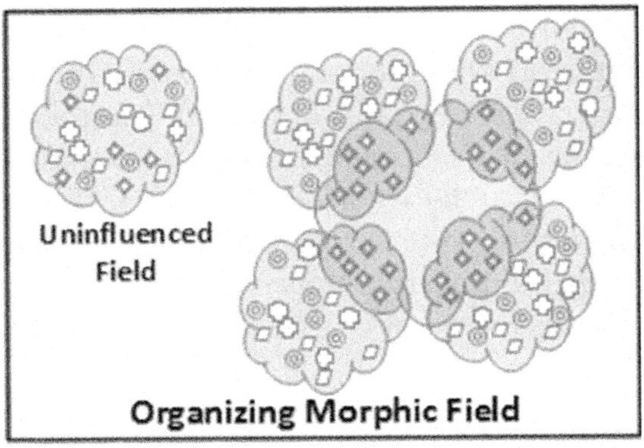

The morphic field concept depends on an interesting characteristic shown in the Organizing Morphic Field Diagram. A morphic field for hair, for instance, may produce many different versions of hair. In this example, the five smaller clouds in the diagram represent morphic fields for single instances of hair. The various shapes represent the types of hair the fields have the potential to produce. The larger, overlapping cloud represents an organizing field for, let us say, an arm. In the hierarchy of fields for an organism, the arm-organizing field attracts the kind of hair intended for the arm based on the sub-field's potential and Nature's habit for that organism.

The selection of form based on *higher* fields offers a way of linking of the concept of thoughtforms with organizing and morphic fields. They are conceptually the same because both are etheric in nature and both contain information associated with a specific aspect of reality. It is the hierarchical nature of fields and effect of intended order that determines which potential in a field is to be realized.

Morphic Fields

As defined in the Hypothesis of Formative Causation (9), and in terms of transcommunication, morphic fields are etheric fields that directly influence physical processes to produce and maintain physical organisms such as plants, insects and people. In more generic terms, morphic fields are organizing fields.

Terms for Morphic Fields

The etheric field concept is influenced by the work of Dr. Rupert Sheldrake who has proposed the Hypothesis of Formative Causation. (9) Since this is likely a new concept for many readers, it is important that the new terms are understood as Sheldrake intends. The following definitions are from Sheldrake's Glossary. (52)

Field

A region of physical influence. Fields interrelate and interconnect matter and energy within their realm of influence. Fields are not a form of matter; rather, matter is energy bound within fields. In current physics, several kinds of fundamental fields are recognized: the gravitational and electro-magnetic fields and the matter fields of quantum physics. The hypothesis of formative causation broadens the concept of physical fields to include morphic fields as well as the known fields of physics.

Holon

A whole that can also be part of a larger whole. Holons are organized in multi-leveled, nested hierarchies or holarchies. This term, due to Arthur Koestler, is equivalent in meaning to morphic unit.

Hypothesis of Formative Causation

The Hypothesis of Formative Causation argues that organisms or morphic units at all levels of complexity are organized by morphic fields, which are themselves influenced and stabilized by morphic resonance from all previous similar morphic units.

Morphic Field

A field within and around a morphic unit which organizes the unit's characteristic structure and pattern of activity. Morphic fields underlie the form and behavior of holons or morphic units at all levels of complexity. The term morphic field includes morphogenetic, behavioral, social, cultural and mental fields. Morphic fields are shaped and stabilized by morphic resonance from previous similar morphic units, which were under the influence of fields of the same kind. They consequently contain a kind of cumulative memory and tend to become increasingly habitual.

Morphic Resonance

The influence of previous structures of activity on subsequent similar structures of activity organized by morphic fields. Through morphic resonance, formative causal influences pass through or across both space and time, and these influences are assumed not to fall off with distance in space or time, but they come only from the past. The greater the degree of similarity, the greater the influence of morphic resonance. In general, morphic units closely resemble themselves in the past and are subject to self-resonance from their own past states.

Morphic Unit

A unit of form or organization, such as an atom, molecule, crystal, cell, plant, animal, pattern of instinctive behavior, social group, element of culture, ecosystem, planet, planetary system, or galaxy. Morphic units are organized in nested hierarchies of units within units: a crystal, for example, contains molecules, which contain atoms, which contain electrons and nuclei, which contain nuclear particles, which contain quarks.

Ideoplastic Structures

The term *ideoplastic structure* was intended to be used for objects formed during a physical séance. For instance, a séance might begin with *spirit* lights and raps and then things might move. In some rare demonstrations, ectoplasm might be seen.

These objective darkroom séance phenomena are thought to be caused by ideoplastic structures produced by the circle's etheric control personality in cooperation with the medium and sitters. These structures cease to exist when the control removes attention.

In concept, the ideoplastic structures of the séance room and physical objects of our everyday physical universe are the same, the difference being the degree of intention applied to their formation and persistence of attention on them by their source personality.

As the term is used in this cosmology, ideoplastic structures are normally thought structures, but they can become physical via the application of physical effort. For instance, an imagined birdhouse can be made very real in the mind of the person as a thoughtform.

If the person expresses the intention for the birdhouse to be real to the point that it is real in his or her mind, then it would be thought of as part of the person's worldview or personal reality. It would be an imagined birdhouse that would be called a delusion if the person physically acted as if it were real.

If the person applied physical effort to make the imagined birdhouse physically real, that is, if the person built it, then the ideoplastic structure would be thought of as a physical thing.[25]

Limiting Association via Perceptual Agreement

The Perceptual Agreement Discourse describes self-organizing influences that moderate how life fields are able to associate. In this concept, there is a natural process that organizes reality based on the degree of understanding represented by the worldview of a life field's personality. That is, life fields are discrete units defined by their perception of reality (worldview) and are able to gravitate toward a common nexus with other life fields representing a similar perception of reality. This relates to the *Like Attracts Like* concept posed in the Hermetic Wisdoms.

The Cosmology of Imaginary Space Discourse includes a discussion of how the simple equation used to plot the Mandelbrot Set can produce a very complex object in imaginary space. It also explains how the equation can be compared to perceptual agreement. *Movement* within that space is based on how quickly the result of a simple calculation approaches infinity, which is a reasonable analog for progression.

---

[25] The need for physical effort to embody an ideoplastic structure is a traditional body-centric view. It is practical to accept this as fact, but there are forms of trans-etheric influence that appear to be exceptions to this rule. Precipitation mediums appear to be capable of producing physical objects based on the influence of intended order on chaotic forms of material.

Precipitated art is a physical-to-physical process that appears to involve the same mechanism as apports. One difference is that, in precipitation, there appears to be a chaotic phase of the apport in which physical material is disintegrated into a chaotic state and then reintroduced into the physical in a new, intended form. The work of precipitation medium Hoyt Robinette (31) is a good example of this.

Discourse 7:
Life Fields

Introduction

The idea of life as the building block of reality is not new. What is new, I think, is that the nature of life is modeled here from a survival of consciousness point of view. Here, I am saying that life is reality, either as life itself or as expressions of life. That is one of the fundamental concepts of this cosmology.

Understanding the functional areas of a life field and how they interact will aid every aspect of your pursuit of self-improvement (progression). Current understanding of human nature tells us that we are only conditionally conscious; understanding is relative; perception is based more on habit than on sensed information; and much of what is sensed by our unconscious mind never reaches our conscious self. The reasons for these characteristics are evident in the functional areas described in this discourse.[26]

Self-organizing Fields

Life fields are modeled as regions of influence originating with the Intelligent Core aspect of personality. Life fields are composed of a set of functional

---

[26] A good subtitle for this Discourse would be *The Anatomy of a Life Field*. It is necessary to understand the component parts of the life field as it is modeled here to understand your nature. The Life field [xxiii] and Life Field Complex [xxv] Diagrams will make little sense if you do not. At the same time, understanding the Life Field Complex Diagram will give you important tools for managing your progression.

Again, I should say that the two functional diagrams are probably not something a person with an advanced psychology degree would show you. They are *as if* diagrams that satisfy Black Box Analysis [14]. Any proper functional diagram of your etheric nature should at least account for the functions I discuss here. Take your time in this section, contemplate the diagrams, and think of ways they integrate into your way of thinking.

I just had a conversation with another person who made it clear to me that he knew all about metaphysics. Don't be that person! It is a reasonable wager that your understanding is out of date.

areas which adhere to a set of Organizing Principles. Life fields occur as elements in a hierarchy of nested fields emanating from Source. All life fields have the same functional areas but might express the characteristics of those functional areas more or less than life field in other parts of the hierarchy.

As the beginning point for reality and top life field, Source has all of the characteristics of a life field. The hierarchy of life fields is modeled as aspects of Source, and therefore, inherit those functional areas and an urge to understand based on Source's supposed curiosity about itself. In this way, progression of a life field is described as increasing understanding about the nature of reality (nature of Source), resulting in increasing perceptual agreement with Source and the eventual merger of that understanding with Source.

As a general statement, the Prime Imperative of a life field is to align personal reality with local reality.

Functional Areas of a Life Field

The functional areas of a life field are shown in the Functional Areas of a Life Field Diagram [xxiii], below, and are explained as follows.

Personality

Personality is modeled as the intelligent core and source of purpose for a life field. It is who a person really is as the *I am this* and what survives after transition.

You may be more familiar with the idea of mind as conscious and unconscious states. The idea of mind is a body-centric construct. The Implicit Cosmology addresses the mind concept as personality representing the immortal aspect of its life field and the conscious self or the *I think I am this* of you peering out at the world from the perspective of your human's eyes.

You will notice in the life field diagrams that a functional region titled Attention Complex [110] is situated between personality and conscious self. That is a normally unconscious part of a life field that supports perception and expression. I say *normally unconscious* because during increased lucidity, conscious self is able to sense more of the Attention Complex functions.

Since the Attention Complex is the functional area of a life field that supports sensing environmental signals, it is also the link with the collective representing all of the personalities associated with a common (local) source.

While personality is modeled as being entangled with its source and related collective, it also interfaces with unrelated personalities.

Entanglement may be involved with these unrelated personalities, but a more useful model is rapport.

Note that the perception of environmental signals, including etheric personality-to-personality exchanges, are filtered by the Attention Limiter in the Attention Complex. This limiter is informed by worldview, and it is this filter function that determines the life field's relationship with the Principle of Perceptual Agreement: *Personality must be in perceptual agreement with the aspect of reality with which it will associate.*

As understanding increases for the personality-life field complex, lucidity increases for the conscious self-personality channel. The three states we should track are:

Little or no understanding

Personality entangled in an avatar relationship: the personality-conscious self relationship is dominated by the Attention Complex which includes little or no lucidity.

Some understanding

Personality in transition out of an avatar relationship: the personality-conscious self relationship continues to be dominated by worldview in the Attention Complex. Presumably some understanding has been achieved in the last lifetime and there is some increase in lucidity. (Worldview follows the conscious self into the etheric.)

Understanding converging on intended purpose

Lucidity verges on a clear and open channel between personality and conscious self.

Presumably, the purpose for which personality was expressed as an aspect of its source personality is to gain understanding about specific aspects of the source's personal reality. The expectation is that increasing understanding tends to merge reality fields as understanding approaches first cause Source's expectation of understanding about its nature.

As this collapsing of personal reality fields continues, the child personalities are expected to maintain an *I am this* sense, but to be otherwise immersed in the accumulating collective.

Intelligent Core

The intelligent core of a life field is the formative aspect of personality while personality is the experiential aspect. The intelligent core for personality is comparable to the avatar's Body Image functional area of the Life Field Complex Diagram [xxv].[27]

Attention Complex

Personality holds the initial imperative to gain experience, and this informs the rest of the life field in the form of attention. As such, attention is based on the decision to understand or take action. The Attention Complex includes Worldview, Perceptual Loop, Attention Limiter, and the Intention Field, which includes Visualization and Perception. The complex of functional areas is modeled as the focus tool for sensing the environment and personality's expression into the environment.

This is a potentially confusing part of this cosmology. The personality functional area is sentient and presumably capable of expression independent of conscious self, but all of its perception and expression is by way of the Attention Complex. This interface with the environment would normally be unconscious.

Conscious self uses the same Attention Complex. For a person, the human avatar also used the complex.

To be clear, both the personality and conscious self are isolated from the environment and other personalities by the Attention Complex. Since worldview is the dominant influence in perception and expression, it is this isolating filter that assures compliance with the Principle of Perceptual Agreement.

---

[27] Most, if not all, cosmologies I have studied describe formative and experiential aspects of Source. Each are usually described in terms similar to *parallel hierarchies of nested fields*. For instance, one hierarchy is of nature spirits or devas with responsibility for care of the various kingdoms of life (mineral, plant, animal). The other hierarchy is usually of people experiencing lifetimes to gain understanding.

I repeated that cosmology in the *Handbook of Metaphysics* (6) , but now I see that both aspects of life are part of the same functional area with one part emphasizing the mechanics of existence and the other focusing on the purpose of existence. This coexistence of the two influences is comparable to a corporation's Chief Financial Officer and Chief Operating Officer. Both share the same goal of a successful business but focus on different aspects of that effort.

Attention Limiter

Personality's awareness of the environment is limited to what is of interest to it. If an external input has no interest to personality or the avatar, it will probably be ignored. As such, an external threat perceived by the human would be of interest, as would something pertinent to personality's objectives. All else is likely to be ignored. See First Sight Theory [126] in the Perception and Expression Discourse.

Because there is no apparent attenuation of a psi signal between life fields, at least in principle, all information exchange is detected by each life field, but only that which is pertinent to a life field is actually processed as an environmental signal. The Attention Limiter is comparable to the network address function in data communications.

Even though the Attention Limiter is an unconscious function, it is informed by intention expressed by the conscious self. See the Mindfulness Discourse.

Worldview

Think of Worldview as the life field's database which contains personality's perception of reality. When in an avatar relationship, it is shared with the human avatar and also contains the avatar's perception of reality.

As this cosmology is written, personality is an aspect of Source, possibly many generations removed. When personality was differentiated from its local source personality, it inherited a subset of understanding and the urge to gain specific understanding.

When personality is entangled with an avatar, conscious self is associated with its host and its worldview is modeled as being shared with the avatar. This relationship is shown in the Life Field Complex Diagram [xxv].

As shown in the Functional Areas of Perception and Expression Diagram [xxiv], information that passes the Attention Limiter is subjected to the Perceptual Loop. If the information agrees with Worldview, it will become part of Worldview and be presented to the Perception Function to be expressed as a thought, action or ideoplastic structure.

The visualization process may produce a sufficiently familiar characterization of the information for it to produce a *Maybe* result in the Perceptual Loop. Such information, as it is characterized in the Visualization function, can be incorporated into Worldview.

It is important to note that it is by way of the *Maybe* results from the Perceptual Loop that the Worldview Function can evolve. There is more about worldview in the Point of View Commentary.

The avatar is an organism that has evolved in the physical, and following the Hypothesis of Formative Causation (9), has a morphic memory *database* that is not shared by the entangled personality. However, the avatar's morphic memory does influence the life field worldview, specifically with the avatar's instincts. This is shown in the Life Field Complex Diagram [xxv].

Perceptual Loop

An important product of the Attention Complex is conscious perception that is developed by the Perceptual Loop. Notice in the Functional Areas of Perception and Expression Diagram [xxiv] that visualization informed by worldview is part of this process. It is here that the way we experience the world is shaped. Our personal reality is formed as environmental information is translated based on worldview and presented to us as conscious perception.

The perceptual loop is probably a continuous cycle. It is likely that it produces a stream of consciousness form of perceptual noise when the Intention Channel is not impressed with intent. Such perceptual noise sounds a lot like dreaming.

The mind attempts to identify otherwise indistinguishable stimuli by suggesting perception of something familiar. This is why, for instance, people hear voices or music when there is only noise.

As discussed in the Perception and Expression Discourse, First Sight Theory (13), can be applied here as a rule set for the operation of the Perceptual Loop.

Intention Channel

Intention is the motive force for expression, and based on a clear understanding of true desire (objective), it can be a very powerful tool for living. As a comparison, an imagined outcome without the intention to make it so is just a fantasy.

Think of this as the influence which will drive the accumulated potential toward the desired result. It is also the mechanism by which the Attention Limiter can be modified. Intention is also the influence that begins the process of expression.

Visualization

The Visualization Function is used to help identify information that is submitted to the Attention Limiter. It also helps to formulate the life field's expression back into the environment or perception to the conscious self.

Here, *visualize* is used as a conceptual term that, if physically sensed, would be more a gestalt understanding rather than a mental picture or sound. *Render* might be a better term. As a frequent out-of-body experiencer, Robert Monroe described this form of information exchange as a *Rote* (53), which is a thought ball or packet of understanding. This includes the thought, associated memory, understanding, experience and history as a complete concept.

The Perceptual Loop is seen as an iterative process. As is shown in the Perception and Expression Diagram [xxiii], when there is an intention to express something, it must first pass through the Perceptual Loop. This process is imagined as a rapid cycle of repeatedly comparing and modifying a rendering in the Visualization Function until it agrees with worldview. Ideally, the output is a clearly imagined rendering of what is intended. This then becomes perceived by conscious self and is potentially expressed into the environment.

If the visualization is not accompanied by intention, then no expression will occur. The visualization will be little more than fantasy.

Visualization can only be as clear or correct as worldview allows. If the personal reality contained in worldview is very different from local reality, the rendering will be equally incorrect and if expressed, may not be possible or will be distorted.

Perception

Perception informs the conscious part of an otherwise unconscious process represented by the Attention Complex. As such, the output of the Perception Function is conscious awareness of the environmental information that came through the Attention Limiter and that agrees with worldview.

Expression

Expression is a creative process that manifests concept into form. What is typically thought of as energy in the physical is better modeled as potential expression in the etheric, which is managed with attention, visualization and intention.

A life field's expression is seen as the precursor to all physical energy, meaning that the imagined future exists as an etheric or conceptual thing that must be transformed into physical, objective form. During the physical expression of the intended result, physical action must be involved. The objective energy may be money, fuel or even labor. See Contact Field [100] for a discussion about an alternative to the avatar relationship requirement for physical expression.

Relevance of Life Field Model

Keep in mind this is a model based on an engineer's reckoning using black box analysis. The functional areas are useful for such purposes as mindfulness and the study of transcommunication but may not resemble what your neighborhood psychologist might think is right.

If you spend a little time contemplating the functional areas discussed here and in the Perception and Expression Discourse, you should begin to see how they might have profound implications for the way you study things paranormal.

Implicit Cosmology

Discourse 8:
Personality-Centric Perspective

Introduction

This discourse is about our nonphysical nature: our conscious mind, memories, and that unconscious part of our mind we sometimes meet in our dreams. It is for you to decide, but it has been my experience that our happiness and progression depend on our informed life choices. If the Implicit Cosmology is reasonably correct, it must be understood that what we do in this lifetime will affect the rest of our eternity.

This subject is important because understanding the more universal nature of who we are will have an important influence on how we understand the implications of our life experiences.

Put a different way, we have been taught from birth that the limits of our word are what we can see, smell, taste, feel and hear. In fact, our physical senses only mark the limits of our body's physical world. Researchers are now telling us that we psychically sense the world before our body's five senses check in.

Those scientists who are willing to at least consider the nonphysical nature of mind tell us that even information from our five senses must be translated into the same psi form before their information can be considered. This means that our real sensory functions are better related to our core personality than to our physical body.

If this is true, then we have spent a good part of our lifetime totally misunderstanding what *real* actually means. That is a pretty shaky foundation on which to make life-changing decisions. The solution, of course, is to change our perspective through education.

Perspective

The perspective from which a question is considered has a lot to do with how it is answered. Our normal perspective is that of being our body. We look at the world through our body's eyes from the perspective of within our body's head. This is natural. We have had that perspective since birth. However, if we are immortal personalities temporarily hosted by our physical body, the more correct perspective is to say *"This is my body but my true self is not physical. I am not my body."* So we might say that the usual perspective is *body-centric* but that the more correct perspective is *personality-centric.*

Our Body is a Complete Organism

As an organism that has evolved on this planet, our physical body is probably fully functional without our entanglement. It is proposed in the Hypothesis of Formative Causation (9) that the morphogenetic development of physical organisms, such as our body, is organized by a morphic field which supports the body image and the necessary rules for its development.

As described for physical organisms, morphic fields are life fields that are explained under a different naming system. They include functional areas for worldview (memory, Nature's habit), perception, levels of consciousness and a means of expression. As such, the human body life field is essentially the same as our personality life field. By that, I mean that our worldview is directly comparable with Nature's habit. The one major difference appears to be that our instinct is to gain understanding through experience and our body's instinct appears primarily to survive in the physical world.

The message is that, during our physical lifetime, we share worldview and many of our perceptual and expressive abilities with our physical body. However, as shown in the Life Field Complex Diagram [xxv], the morphic field for our avatar also has a functional area representing *Nature's habit*. The Body Image functional area is comparable to our personality's intelligent core.

Our body as Avatar

Based on current understanding, our body has a consciousness which is not completely suppressed during a lifetime. In fact, the relationship, is at least to some extent, a symbiotic one in which our daily choices are often greatly influenced by our body. A person gains in maturity by learning to cooperate with the body consciousness while remaining true to the ideals of balance and progression toward greater understanding.

It helps to understand how to distinguish between our body's instincts and our understanding. To do this, it is necessary for a person to recognize the difference between the inherited urge to understand and the body's instinctual urge to survive and perpetuate its kind.

It is also a challenge to distinguish between the beliefs that are taught by our local culture and the actual nature of reality as it is understood when those beliefs are superseded by understanding. This can best be accomplished when we are aware that we are not our body.

Avatar

In Hinduism, an avatar (from Sanskrit for *descent*) is characterized as a deliberate descent of a deity to Earth. The term can be translated into English as *incarnation*. A person is an etheric personality entangled with a human body in an avatar relationship.

In the avatar relationship the personality remains associated with the etheric aspect of reality but its perspective is the personality's conscious self as it thinks it is the physical body. In trying to understand this relationship, the most important thing to remember is that we do not know much about how humans might behave if they did not have an entangled personality. While it is necessary to make a few assumptions for informed speculation, the underlying rule should be that humans are life forms which deserve respect and good care as hosts that enable our existence in the physical.

The relationship between etheric personality and the human avatar is shown in the Life Field Complex Diagram [xxv]. Notice in the diagram that the Attention Complex (middle) is shared by the etheric personality (upper-left) and the complex representing the human body consciousness (upper-right). The actual human body is in the box marked Avatar (lower-right). This relationship is discussed below.

Cooperation Between Personality and Avatar

At the time of personality's entanglement with its human avatar, its worldview is populated with a subset of understanding and the urge to gain further understanding; all inherited from a source personality. As is shown in the Life Field Complex Diagram [xxv], the conscious self and its human avatar share worldview. This shared memory is the main point of this discussion, because beginning at birth, the dominant conscious self must learn to manage the human influence, even as it learns to adapt to its local reality.

An example of the human influence's persistence is demonstrated by the way we frequently hear in transcommunication that newly transitioned communicators still identify with their physical bodies. Some communicators even report something like a *getting well* period as they become accustomed to healthy mind and body. This suggests that people remain under the influence of the human body image well into transition.

Understanding the avatar relationship can help us live a productive life from the perspective of following inherited urges from personality while managing our body's instincts. The challenge is in learning how to distinguish which are our body's issues and which are real to our personality. The Mindfulness Discourse addresses this.

Each avatar relationship begins with a blend of personality and human body traits. One way to make an informed guess as to what those traits are is to look for those that we were born with that influence our behavior today. If the avatar hypothesis is correct, then the human part of our worldview should be memory, beliefs and instincts related to survival. Understanding and inclinations inherited from our etheric personality should be related to increasing understanding or progression of personality. For instance, responding to peer pressure would support herd or tribal safety but it would also tend to restrict learning. In a similar way, compulsive behavior would seem to suggest instincts while obsessive curiosity would seem to support learning.

Survival of Body Mind

It is difficult to say that the human exists just as an avatar for etheric personalities, and then to say that all life forms have a personality. The collective model probably applies for all life forms, but is there a difference in the character of consciousness? We know that animals exhibit self-conscious behavior, clearly indicating that they are more than simply a collective mind, but there seems to be a difference in purpose.

The physical organism body consciousness is modeled in the Implicit Cosmology as being part of a collective, as well. The difference is that it is more of a collective as described in the Hypothesis of Formative Causation. (9) In that model, a morphic field has access to a worldview-like database that is a collection of *Nature's habits* for that family of closely related organisms. For the human, it represents how the human body has been made since its origin.

This is an area that needs much more consideration. It is important to know that there is nothing in the Trans-Survival Hypothesis that argues it only applies to people. How it might apply to a loving pet is beyond my speculation at this time; however, there is nothing in the cosmology that says other animals cannot be avatars for personalities.

Personal Style and Astrology

Just as people are born left or right-handed, people are born with basic personality traits that tend to influence their behavior. Personality styles are cataloged and studied in psychology as a means of understanding human behavior. They have also been adapted to teach sales people how to relate to customers. In 1981, David Merrill and Roger Reid published a book reporting their study of corporate personnel interactions which became something of an industry standard. (54) They noted four main personality styles:

Analytical: thinking, thorough, disciplined.

Amiable: supportive, patient, diplomatic.

Driver: independent, decisive, determined.

Expressive: good communicator, enthusiastic, imaginative.

Each basic style is typically further divided so that a person might be seen as a Driver-Analytical or a Driver-Expressive. The point of these styles is that people likely begin dealing with a situation from the perspective of one of these styles. The question is whether or not this inclination is inherited from personality or from avatar.

The idea of astrology is that people's behavior is influenced by the astrological conditions at the time of their birth. A year is divided into twelve signs based on the ancient zodiac and each indicates a different set of personality traits. The Personality Styles Diagram, below, shows a suggested relationship between astrological signs and the four personality styles.

It is important to note that personality styles and astrology are not being recommended here. They are used to demonstrate that we tend to display personality characteristics that are evident at a very young age and which tend to shape our lives.

There appears to be general agreement between the systems; however, the personality styles are more often described in terms of information acquisition or community, while the astrological signs tend to emphasize the same sort of characteristics usually associated with instincts. For instance, an Aries, who is a person born between March 21 and April 20, is described as (amongst other qualities) impulsive, physical and driven. Aries is related to a Driver in the Personality Styles Diagram, and a Driver is described in terms like independent, decisive and determined. A Capricorn (December 23 to January 20) is described as (amongst other qualities) instinctive, overreacting and moody. A Capricorn compares to an Analytical who is described in terms of thinking, thorough and disciplined.

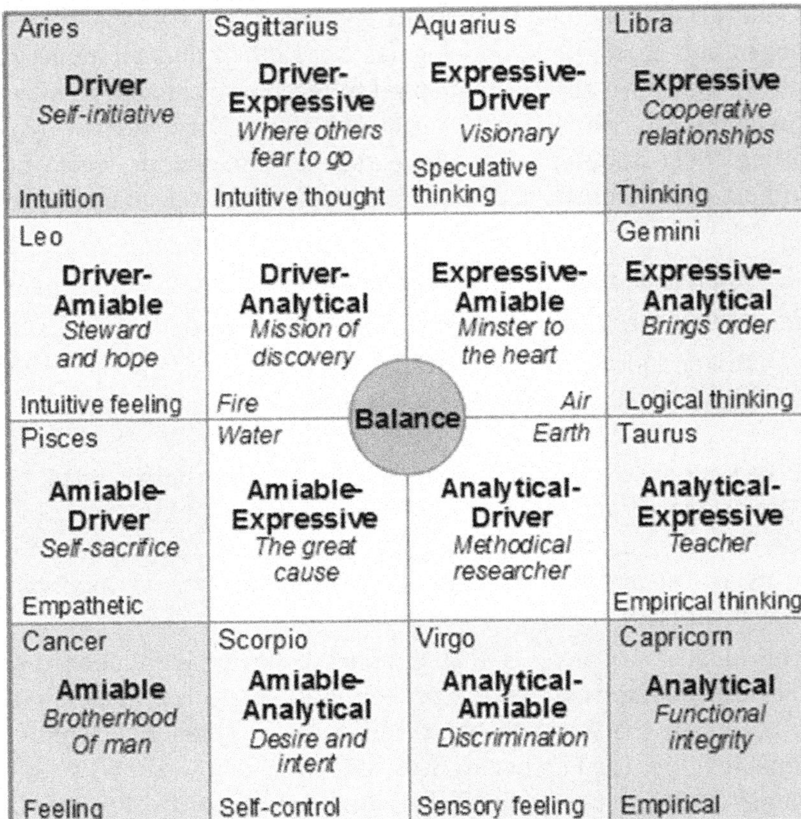

Personality Styles with Astrological Signs

Styles and signs are ways of modeling observed human behavior. This is not an endorsement of either model.

Since there is so little known about this from the personality-centric view, it must be left for us to take the initiative to study and self-analyze. The interpretation of astrological signs and personality styles offered here is just an opinion, and your experience may be different. Again, these are examples indicating the kind of cues you would look for if you conduct a self-appraisal.

There is more about this in the Temperament portion of the Point of View Commentary.)

Balance

Notice the center circle in the Personality Styles Diagram labeled *Balance*. As you come to better understand yourself and your avatar, and therefore gain in maturity, you will find yourself converging on the middle way. One of the most important secret lessons taught by ancient wisdom schools is often that balance is the middle way toward maturity. This does not mean a balanced person would never express extreme behavior, only that such a person would consciously do so, tactically for a purpose and then return to the center without attachment for the outcome of his or her actions See the Mindfulness Discourse.

Human Animal

The human brain can be considered in two parts: forebrain and brain stem. According to *Brain Structures and Their Functions,* the brain stem: *"is responsible for basic vital life functions such as breathing, heartbeat and blood pressure. Scientists say that this is the 'simplest' part of human brains because animals' entire brains, such as reptiles (who appear early on the evolutionary scale) resemble our brain stem."* (55)

The forebrain supports rational thought, so an assumption in the Implicit Cosmology is that the brain stem has evolved to support survival of the human body and the forebrain has evolved to support the entangled personality. In this view, it is reasonable to expect that, like other animals, humans would get along pretty well without an entangled personality.

Irrational Behavior

Do you have a pet, perhaps a dog or cat? If so, you may have noticed a lot of behavior that seems to be irrational. For instance, your pet might be fearful of an unexpected object in the backyard, or it might be unreasonably afraid of men, especially strangers. Sure, you would think it is just behaving like an animal. Animals often seem irrational. That is one of the distinguishing characteristics between humans and animals.

But what if you have irrational behaviors? Are you afraid of the dark? Do you go out of your way to avoid strangers? Is it difficult to communicate with people of the opposite sex? Are these behaviors rational? Do you have reasons for them which are more than just an excuse?

Everyone has fears that might be more exaggerated than circumstances would seem to require. The usual way of dealing with them is to talk the person out of the fear, either by appealing to logic or by showing that there is no need for fear, in effect, to wear out the fear response.

In the body-centric view, we naturally appeal to the rational mind: that part of us that is supposed to be logical, thinking and learning. However, in the personality-centric view, it can be seen that the human animal is the source of irrational fear. Because of the entangled avatar relationship, the human instincts are part of the shared worldview, which is in turn, the governing factor of the Perceptual Loop of a life field's Attention Complex. As such, too often, the strongest and first response to an external influence is the human animal's fight or flight reaction.

It is natural that a young person will more often respond to circumstance from the animal's instincts, but over time, it is expected that the person will more often have a rational response. That is the idea of maturity. Even so, people are plagued all of their lives with an unrecognized human influence. So long as they think with a body-centric point of view, it will be natural for them to seek to suppress the animal response, rather than understand the necessity to more directly manage the Perceptual Loop through mindfulness.

A better model for therapy might be to learn how to appeal to the animal in us, rather than the rational personality. We need a human whisperer more than we need a psychologist.

Degrading Avatar Relationship

There would be circumstances in which personality is forced to withdraw from the human avatar. Of course, physical death of the human is the usual reason, but personality might also withdraw if the body is physically healthy but is otherwise no longer able to support the expression of conscious self.

The inability of the brain to continue functioning as a physical-to-etheric transducer in cases of senility would be such a situation. This would not necessarily be a forced withdrawal but there may be little reason for conscious self to continue the avatar relationship if personality's intention to gain understanding through experience in the physical can no longer be supported.

Personality's withdrawal from the human may be temporary, as is seen in the case of physical injury to the brain or coma. The reports of a persona with advanced senility who becomes temporarily lucid is another example that might be explained by voluntary but impermanent withdrawal.

Keep in mind that this model recognizes that the human is a complete life form. As is illustrated in the Life Field Complex Diagram [xxv], the Body

Mind functional area includes human instincts which are filtered in the Attention Complex before presentation to conscious perception. This is the same set of functional processes experienced by organism when the human is entangled with an etheric personality as a person. Since conscious self is an aspect of etheric personality, the missing influence for the now independent human is the expression of intention informed by personality.

A human that has experienced a lifetime entangled with an etheric personality would not be well equipped to continue functioning as a person without the rational influence of conscious self. At the same time, the worldview function shared during entanglement would continue to be part of the human's perception and expression processes [125]. The missing influence would be intention expressed by conscious self.

Extrapolating what I know about human behavior, and relating that to the Implicit Cosmology, I would expect to see the human-minus conscious self to physically appear the same, but to exhibit changes in behavior which should be predictable from the model.

Worldview is memory, and the Perceptual Loop in the Attention Complex is moderated by intention. Without rational intention informed by personality, the perception-expression process would tend to present stream-of-consciousness thoughts to the human's conscious state. This would look like waking expression of the kind of mostly random dream imagery we experience in light sleep.

The human's instincts would no longer have the rational influence of conscious self, and would therefore, have greater influence on the perception-expression process. This would manifest as more animal-like responses and a tendency to fixate on or obsess about otherwise trivial things in the environment. This would likely include rapid, swings of emotion from passive happiness to aggressive behavior.

May I Introduce Myself?

With these considerations, then, how should we think of ourselves? How do we think with a personality-centric perspective? Each of us will have different takes on this but there are general points that should be considered.

To begin, try thinking of yourself as two people. For instance, I am Tom Butler. *Butler* is the family name of my body, and it serves to give you a sense of its lineage. Perhaps we can call my body Mr. Butler. *Tom* is a good name for my personality in this lifetime, so a proper introduction might be *"Hi, I am Tom and this is Mr. Butler."*

Mr. Butler was born in May, and according to astrology, should be stubborn, think habitat is very important, and have a strong dependence on tactile sense. When compared to characteristics of other signs, this is pretty accurate. According to the avatar model, these characteristics should provide hints about the nature of Mr. Butler.

My astrological chart always seemed to indicate more what I need to overcome than what I should be driven by, but I also fall into the Driver-Analytical side of the personality styles chart, which is where I feel comfortable. Again, according to the avatar model, these characteristics should provide hints about Tom's nature.

Implicit Cosmology

Discourse 9:
Perception and Expression

Introduction to *As a Man Thinketh* by James Allen, 1902 (56)

> The objective of "This little volume" is "to stimulate men and women to the discovery and perception of the truth that they themselves are makers of themselves by virtue of the thoughts which they choose and encourage; that mind is the master weaver, both of the inner garment of character and the outer garment of circumstance, and that, as they may have hitherto woven in ignorance and pain they may now weave in enlightenment and happiness."

Introduction

Our relationship with the environment is the product of the two characteristics of the life field described here as perception and expression. Both involve the same functional areas of the Attention Complex. It is only intended purpose that determines the output, be it perception to conscious self or expression to the environment.

Development of content for both perception and expression is normally an unconscious process, but it can be influenced by conscious intention. As with Mr. Allen's introduction to *As a Man Thinketh*, how we think is at the very foundation of who we are. This discourse addresses ways we might attempt to take control of the way our normally unconscious mind decides our conscious experiences.

The Mind is Etheric

Everything in the Functional Areas of Perception and Expression Diagram [xxiv] is nonphysical. That is, it is not in the physical space occupied by your physical body. Instead, it is in the greater reality of etheric space, of which the physical universe is just an aspect. The physical brain is best viewed as a transmitter/receiver for thought and the interface between your etheric personality (*I am this*) and your conscious self as you experience the physical (*I think I am this*). How aware your conscious self is of your etheric personality is referred to as lucidity and the objective of personal development is to become as lucid as possible. That, in turn, is accomplished by aligning worldview with the actual nature of reality.

Psi sensing is the normal way in which we sense the world. Being psychic is learning to understand how that unconscious psi sensing is translated by the mind into physical awareness. Suspended judgment is one technique, really a point of view, that will help you do this. Another is mindful living: make a habit of thinking of the consequences of your actions.

Understand the Unconscious Processes

It is essential to understand the thought process if we hope to control it. Parapsychologist James Carpenter has developed a theory based on current understanding about psi phenomena that models how we mentally process information. His hypothesis is detailed in the book, *First Sight: ESP and Parapsychology in Everyday Life.* (13) (49) (40)

Review the Perception and Expression Diagram [xxiv], and as I explain my understanding of First Sight Theory, keep in mind that I am talking about how information you sense from the world around you is unconsciously compared to all of your existing beliefs, emotions and experiences before you even become aware that you have sensed something.

First Sight Theory

In his book, Carpenter explains that people first sense the world psychically. As he puts it, "What if ESP is like subliminal perception? What if psychokinesis is like unconsciously but psychologically meaningful expressive behaviors?" He answers these questions by proposing two propositions based on the following assumption.[28]

Assumption

Everything and everyone, every action in the past, now or in the future, perturbs the reality field that connects all of us. Carpenter argues that

[28] Use discernment here. I am explaining this theory as I understand Carpenter's rather complex writing. The theory is presented here as a useful rule set for both the perceptual process and expression. For clarity, I have partially paraphrased the assumptions and corollaries from the perspective of the Implicit Cosmology. Be mindful that Carpenter did not necessarily expect this theory to be adopted for support of the Survival Hypothesis.

research indicates people psychically sense these changes in psi space just as they physically sense changes in physical space.[29]

If that is true, then

First, people sense their environment psychically as well as with their physical senses.

Second, people process this information unconsciously, and it is the conclusion of this processing that they are aware of and react to, not what has been psychically or physically sensed or unconsciously considered. A person might psychically sense someone near or far, a person's actions, and apparently their thoughts when they are expressed as intention.

Note 1: The Intention Channel shown in the Perception and Expression Diagram [xxiv] represents the mechanism by which the conscious self is able to influence the perceptual process.

Note 2: In this theory, the expression of intention is what produces a change in psi space which can be sensed by others. Simply thinking of something does not appear to produce information which is detectible by others.

Note 3: The result of the unconscious perceptual process is described by Carpenter as formatting of *"experience and action."* This is the Perception Functional area in the Perception and Expression Diagram. In this context, the person does not directly experience information from the environment. Instead, the person becomes aware of the information after it is formatted in a way that agrees with worldview.

Therefore

The following 12 corollaries address the perceptual process.

1. Personalness Corollary

The unconscious processes that constitute consciousness are personal and deliberate. (The more important it is to us, the more we unconsciously pay attention.)

2a. Ubiquity Corollary Part 1

---

[29] *"Research indicates"* is based on meta-analysis and personal research concerning hundreds of research reports, both mainstream and parapsychological.

Psi sensing is not limited by time or distance.

2b. Ubiquity Corollary Part 2

Psychokinesis contributes to the formation of experience by bringing intention to bear upon the physical processes of the nervous system. (The first response to a meaningful stimulus is transmission of a telekinetic *alert signal* to the body, and if useful, to the environment.)

3. Integration Corollary

Other preconscious processes are processed together with psi in a rapid, holistic, efficient, unconscious manner to format experience and action.

4. Anticipation Corollary

The mind seeks to anticipate events. In the Implicit Cosmology, this corollary is used as the basis of the curiosity concept. Source is curious about itself and that curiosity translates as the Anticipation Corollary in First Sight Theory (this is not necessarily what Carpenter intended).[30]

5. Weighting and Signing Corollary

The importance of sensory and extrasensory information is weighted as being more or less important before it is acted upon. (The yardstick for importance is based on the body's survival instincts and cultural conditioning stored in worldview.)

6. Summation Corollary

The content of conscious experience, emotional states, and behavioral choices are constituted in a summative way by unconscious thought.

7. Bidirectionality Corollary

In this summative process, the person may *turn toward information* (signed positively) to include it in the construction of experience, affect or action, or *turn away from information* (signed negatively) and exclude it.[31]

---

[30] The Anticipation Corollary provides reasonably good support for this idea of a personality having a natural tendency to want to understand its environment in order to anticipate changes. If this is a valid argument, then all else follows.

[31] In the Perception and Expression Diagram [xxiv], the Attention Limiter and the *Agree* or *Disagree* decision of the Perceptual Loop functions would facilitate this Corollary.

8. Intentionality Corollary

Including or excluding information is a function of unconscious intention in regard to an element of potential meaning. (While attention is the result of an *"Is this important to me?"* decision, information that results in action by the person is given force if there is an intention to act.)[32]

9. Switching Corollary

A person will be fairly consistent in how information is processed, (but) may switch in how information is weighted, the sign attributed to it, and therefore, whether or not it is included in behavior. This switching will occur rapidly or slowly depending on the consistency and purity (focus) of unconscious intention. This in turn is determined by the relative weight of the information over time, situational factors that promote or diminish critical analysis, changes of approach in a task, and mood.[33]

Persons who are disposed to rapid switching include those who

- tend to approach situations cognitively and analytically;
- lack consistent purpose and motivation;
- take a detached-observer posture toward most situations;
- are chronically ambivalent;
- are cognitively disorganized are highly distractible.

Persons who tend to switch slowly, conversely, tend to be those who

- approach situations globally and holistically;
- are strongly and consistently purposive;
- engage themselves wholeheartedly in situations;
- are not overly self-doubting or uncertain;
- are well-integrated cognitively;
- are prone to hold focus purposively and not become distracted;
- are dissociative (when in certain states).

Note 4: Highly dissociative people often experience altered ways of being that are more complex and enduring than simple moods.

---

[32] This is the basis for the Attention Limiter functional area.

[33] Switching, which is influenced by personality style, is not necessarily a good thing.

10. Extremity Corollary

The frequency of switching affects the relative density of accumulated additive or subtractive references to the meaning in question. Rapid switching renders potential meaning irrelevant to ongoing experience.

11. Inadvertency and Frustration Corollary

Information gathered via psi is not available to conscious experience but does contribute to the formation of conscious experience by the arousal of anticipatory networks of ideas and feelings (assuming that they are heavily weighted, afforded slow switching, and approached with the intention of assimilation). Because of this arousal, their action can be glimpsed consciously only by observing thoughts, feelings and behaviors that are inadvertent, that is, not intentional and not obviously caused by any current experiences. Someone who has become skillful in interpreting them is thought of as relatively psychic.

12. Liminality Corollary

The arousal of anticipatory networks of ideas and feelings resulting from unconscious psi information may be considered liminal ones, in terms of the boundary between conscious and unconscious thought. Habitual interest in liminal experiences facilitates expression of psi processes (openness), leading to unconscious reference to psi material (and other streams of unconscious material). A more positive, open, secure state of mind will tend to facilitate reference to a broader spectrum of contextual, potentially liminal experience. (In other words, habitually paying attention to subtle information emerging from your unconscious can lead to more direct awareness of what has been psychically sensed from psi space.)

This Corollary speaks to the idea that conscious intention can influence the perceptual process and evolve worldview. Mindful living is, in effect, a way of making the conscious expression of intention a lifestyle.

The Perceptual Process

The Perception and Expression Diagram [xxiv] shows the Attention Complex which translates external inputs into conscious perception, potentially resulting in an external expression. The external input is assumed to be either a psi source or a physical source which is preceded by a psi precursor. The functional areas for development of perception and expression are discussed in detail in the Life Fields Discourse.

Note that in the Implicit Cosmology, information from our five physical senses must be translated into some form of psi signal useful to the nonphysical, unconscious process of the Attention Complex. Thus, in a practical sense, all information is received as a psi signal.

The Worldview Function is like a database representing memory, understanding, and beliefs which define personal reality. That is, worldview represents the standard a person unconsciously uses to decide what is true and what is real.

The perceptual process constitutes an attempt to visualize the meaning of an external input based on a best match of content already present in the worldview database. It is an iterative process conducted in the Perceptual Loop with the following possible results:

- **Initial *No*:** If the *Agree?* decision produces a negative result on the first cycle, the process may be repeated many times in quick succession, depending on the perception of urgency and clarity of information being considered.

  If no solution is found, the information will likely be discarded. In this way, inattentional and incredulity blindness may cause a person not to see a novel object in an otherwise familiar scene.

- **Creative Solution May Modify Worldview:** Rupert Sheldrake (9) argues that organisms faced with environmental challenges sometimes arrive at a creative solution for survival which is integrated into Nature's memory and propagated throughout the species (probably by way of the species collective).

  As the perceptual model is defined here, the environmental information is responded to in the Visualization Function where it is integrated with the best fit from worldview. Here, a sort of *fuzzy logic* is called for. If the visualization is able to tell a likely story about the information, based on existing information in worldview, the story may pass the Perceptual Loop as a *Maybe* result and be used to modify worldview.

  If no acceptable match is found, the iterative cycle will eventually produce a *No* result causing the process to reject the information.

- **A *Yes* Solution:** Of the presumably huge amount of environmental information reaching the Attention Limiter, only a relatively small amount will enter the perceptual process. Much of that will be identified in the Perceptual Loop and eventually offered up to conscious perception.

But here is a very important point for your consideration. The storytelling capability of the Visualization Function can morph information into the familiar, even if it is not actually what was initially received or intended. This is probably the source of pareidolia, which is characterized as seeing patterns where there are none.

Implications of First Sight Theory

External stimuli are always psi related. If the input is physical, then it is assumed to be accompanied, or even preceded, by some kind of psi signature perturbing psi space (etheric space). This perturbation is psychically sensed by the organism as an unconscious stimulus which may or may not be consciously sensed, depending on the result of the perceptual process.

To illustrate, a person might psychically sense a developing threat before it becomes physically obvious. The perceptual process would give meaning to that threat based on past conditioning and the person would react, at first unconsciously with a slight tension of muscles and possibly increased adrenaline; however, if the threat manifests to the person's physical sensing, this unconscious preparation for the fight or flight response would be followed by conscious action. However, the selected action is normally determined unconsciously.

The idea of turning toward or turning away from information during the perceptual process evokes a good sense of what happens in our mind. (Refer to the First Sight Bidirectionality Corollary [128]) For instance, if you instinctively dislike a person or situation, it may be that you have unconsciously psychically sensed something and your Perceptual Loop has sent a *this is not for you* response to your conscious awareness.

In that way, we are naturally psychic because we are able to consciously sense the result of the perceptual process—not as a direct sensing of the psi influence, but as the conclusion the perceptual process has sent to our conscious mind. However, according to First Sight theory, you can train yourself to be more directly aware of that initial psi sensing by learning to encourage your conscious mind to pay attention to those subtle messages that emerge from your unconscious mind. (Liminality Corollary [130])

Perception Becomes Expression

The output of the Attention Complex is an expression. It may be directed toward another personality as a psychic link of rapport, it may be an ideoplastic structure, or it may be perception expressed to conscious self.

So you can see that expression utilizes the same functional areas as perception. The main difference is purpose. Perception involves sensing and translating environmental information into something useful to the conscious self. Expression involves the translation of intention expressed by the conscious self.

Intention Field

Expression is in the form of an intention field. A way to visualize this is to think of the field as a thought ball that is *sent out* by your mind to an intended destination. It contains the intended concept as a matrix for formation or influence, the organizing rules, information about the *target* life field or ideoplastic structure and the expected outcome. Consider information about the target life field as a unique psi field address.

The factors that determine the effect of an intention field are clarity of the visualized outcome, focus and strength of intention, and what is intended. For instance, a casual glance toward a person will likely not amount to a very influential intention field, but an admiring stare will likely be of sufficient influence that the object of the stare might at least unconsciously notice.

This concept is directly related to intended order, in that the product of the intention field is formation of a part of reality as intended by the personality/conscious self. The intention field can be very influential as expressed by an experienced healer-practitioner.

Informed Visualization

Expression begins with the influence of conscious self's intention on the Attention Complex, but it is informed by the person's worldview. That means expression is limited by the person's sense of personal reality. As an example, if the person has learned to fear the dark, then outward actions and expressions would reflect that fear, probably without the person's awareness.

The clarity of expression is a function of how well the person understands what is to be expressed. In this context, *informed* means how well the person has come to understand the subject.

People create their own reality but not in isolation. Society, the environment and even the physical body have an influence on what is created. Perception and expression are always based on current understanding (worldview as personal reality). In the concept of morphic fields, this is Nature's habit.[34]

On Being Informed

Probably the most important thing we can do to further our spiritual progression is to be as informed as possible about our world and especially about what we intend to create. It is important to be aware that being informed about the world can actually retard spiritual progression if it is based on erroneous assumptions. For that reason, *informed* should be in the context of the entire creative process.

Informed is used here to mean *based on or influenced by*. For instance, Objective in the creative process is *informed* by worldview and by the environment. In fact, the entire creative process is based on the usually subtle influence of worldview and external forces. By recognizing this influence, it is possible for a person to learn to manage how influential these forces are in the creative process. The more the influences are in agreement with the nature of reality, the more effective will be the creative process.

Changing Worldview

There is much research indicating that people who are open to more directly sensing psi stimuli have more psi-related experiences, while people who are not can be literally blind to the phenomena. In both cases, the people psychically sense the same information, but that sensing is not as easily translated into conscious awareness for nonbelievers. See the Point of View Commentary.

As such, James Carpenter suggests that a person wishing to be more psychic should maintain an open, non-analytical frame of mind while also maintaining a commitment to the goal of psi expression. Analysis of information, self-criticism, and doubt tend to cloud psi awareness.

Translate *open mind* as *suspended judgment*. Make a habit of not deciding whether or not you believe something. Instead, maintain a wait and see attitude. Consciously suspending judgment has at least an indirect effect

---

[34] Our life lessons and future experiences are not so much in what we create as they are in how we react to what is created.

on the perceptual process, which should in turn, exaggerate the ambiguity of the comparison between what is being visualized based on worldview to produce a beneficial change in worldview. More about worldview in the Point of View Discourse.

Suspended judgment is discussed in the Mindfulness Discourse.

Worldview and Expression

We are born with a degree of inherited understanding about the nature of reality which initially informs worldview. However, culture, parents, teachers, the media, and religion quickly fill our worldview with popular wisdom based on local beliefs, prejudices, and religious/academic dogma. It is for us to come to see reality as it is, rather than how it is taught.

We must do this by learning to align our worldview with the nature of reality based on gaining understanding through experiences. In doing so, we progress by way of increasing understanding and it is understanding which determines our experience in the etheric. In effect, after transition, we will gravitate toward the *heaven* with which we are most in perceptual agreement. As such, what we do in this lifetime has a profound and lasting effect on the rest of our existence.

Worldview has substantial influence on perception; in effect, it causes us to understand experiences as taught by the culture and perhaps not as they actually are. As we align our worldview with the nature of reality, our visualization becomes more in agreement with reality, making the resulting expression more consistent with what is supported by Organizing Principles [53].

Natural Law and Organizing Principles

The functional areas of Life Fields naturally occur in response to the influence of curiosity (intention and perception) and the state of understanding (worldview). This assumption is based on the expectation that such functional areas are necessary to produce known response patterns.

An Organizing Principle that naturally results from our perceptual processes is Perceptual agreement, which can be stated as *Personality must be in perceptual agreement with the aspect of reality with which it will associate.* The implication is that we are not able to experience parts of reality which do not agree with our expectations (personal reality). There is no need for an ethereal being to say we cannot go to heaven if our worldview will not allow our sensing of heaven to emerge to conscious self via the Attention Complex.

A second organizing principle is Cooperative Communities, which may be stated as *an effort to express understanding is necessary for progression.* This is the conscious expression of intention to compose a statement to be spoken to the listener. If this initial effort to explain a concept is followed by questions and new attempts to explain the concept, it can affect worldview in the same way as a *Maybe* outcome of the Perceptual Loop.

The Golden Rule [65] likely has at its foundation a more pragmatic *Teach me as I teach you*. The objective of a lifetime is to gain understanding. As it is modeled, the collective may not fulfill its purpose until expected understanding has been gained by every member life field. Cooperative communities are a natural response to that imperative.

Intended Order

Perception and expression are the outputs of the Attention Complex. Intended order is the effect expression has on the environment. It is seen in action in many forms of transcommunication. In that, order in the form of recognizable images, objects, or sounds *emerge* out of chaotic signals. In Instrumental TransCommunication, the physical process that enables this emergence of intended order appears to be stochastic amplification [276] acting on chaotic, broad-spectrum energy.[35]

Credibility and Incredulity

The Perception and Expression Diagram [xxiv] shows the decision-making process that helps determine perception. It can be seen at work whenever a person is asked to believe something. In this model, the person experiences

---

[35] A hint for how to identify the nature of psi phenomena is to look for the presence of a chaotic influence somewhere in the process. For instance, even though a magnetic field detector might indicate the presence of a psi influence as a change in magnetic field, it may not have anything to do with magnetism at all (probably not). The electronic circuit in detectors of about any physical influence has many components which provide the necessary chaotic resource to respond to the influence of psi as intended order.

In practice, the psi detection model used in the design of the Random Event Generator developed by Psyleron for consciousness research is applicable to virtually all forms of electronics that include active regions such as vacuum tubes and transistors. (57)

something such as a paranormal event and attempts to visualize what has been experienced in terms of worldview. If the experience is not supported by existent worldview, and the person has not learned to think beyond that cultural conditioning, then it is likely the experience would be rejected as impossible. In that case, the person would not perceive the experience, a response sometimes referred to as incredulity blindness.

Alternatively, the person might perceive the experience, but as partially modified by incredulity blindness or as *morphed* into something that does agree with worldview. For instance, an alien entity might register in the person's mind as a rabbit because it has big eyes.

It is common for a person to witness very convincing evidence but insist that it must be fraudulent. In this case, discounting the experience would be a way of morphing it into an acceptable form to agree with worldview.

An inverse effect of incredulity blindness can be seen when a person tends to find paranormal phenomena where (probably) none exists. This might be described as hyperlucidity [152], as people with this tendency insist they are experiencing phenomena, often against considerable testimony to the contrary by peers.

See also, the Villager-Explorer Effect [199] in the Point of View Commentary.

Venues for Learning

The creative process expresses a venue for learning when a person thinks of a possible new experience, for instance, to buy a new car. Typically, the person imagines being in the new car and what it would be like to own one. Perhaps friends would be imagined in the situation so that the person can have a sense of how having the car would affect self-esteem.

An important process is begun when a person imagines a, let us call it, *little me*. *Little me* is given a degree of self-determination. It is also given a directive to have a specific experience and an urge to understand certain aspects of the experience. Finally, it is given the directive to return a degree of understanding about the experience to the person, its local source personality.

The created venue for learning will remain a viable space as long as the person holds it in the imagination; however, the memory of *little me's* experience will remain part of the person's worldview. In effect, *little me* and the person become a group entity, and the accumulation of *little mes* produced in the person's imagining will accumulate to shape the person's etheric personality. (It is possible the same *little me* is *resurrected* for each experience.) The influence of each *little me* will be a function of the importance the person placed on the venue (intention) and the meaningfulness of what *little me* returns. Note also that the accumulated *little mes* represent a collective.

The creation of a venue for learning is exactly what appears to happen for individual lifetimes in this venue of learning we call Earth. The question is whether or not there is a *big me* holding this venue for learning in its mind with the intention of learning specific lessons. Since this venue appears to be a consensus or collective space, it is possible that the organizing field for physical reality is the product of many such individual efforts to learn functioning as a cooperative community.

Summary

The Great Work of Hermitic Wisdom is the transmuting of personality as immature lead into the gold of spiritual maturity. This is seen in the opening line of the *Emerald Tablet* [79], one of the few documents thought to actually come from Hermes: *"It is true and no lie, certain and to be depended upon, that which is above is as that which is below; and that which is below is as that which is above, for the performance of the one truly great work."* (58)

The metaphysical Cabala is all about the path toward spiritual maturity and the Tarot. (59) An example is Key 9 of the Major Arcana from the Rider-Waite deck: His beacon intimates that *"Where I am, you also may be."*

In John 14 of the Christian Bible, Jesus is shown to say to his disciples: *"I am the way, the truth, and the life: no man cometh unto the Father, but by me."* (John 14-6). Here, Jesus is showing himself to his followers not as a deity but as an example of the three aspects of the teacher: follow me that I am the path; follow me as the Spirit of Truth; and follow me as I have lived. (60)

Humankind has been told time and again that the path to wisdom, happiness and spiritual maturity is readily available to anyone who is willing to do the work. The creative process for perception and expression described in this discourse represents a knowable, doable path to agreement with nature. All one need do is realize the need to be informed, understand what *informed* means, learn to examine what is in worldview and accept responsibility for that which is created.

Happiness lies in how the person responds to the world much more than how well the person learns to manage the process. From this can be understood that there is no good or bad, only the degree to which a person is able to align perception with the true nature of reality. All is potential progression no matter the momentary pain or pleasure.

Interesting Speculation

In the Hypothesis of Formative Causation, Rupert Sheldrake proposes that the formation of an organism is organized by nested etheric fields. The simplest organic chemical process is managed by a field, which is in turn organized into more complex organisms by greater fields, and so on to the top field of, say, a human or a tree. There is good reason to believe that an etheric personality is the organizing field for the avatar's physical body, and is in turn part of a collective of other personalities which eventually brings us, in the hierarchy of life, to a source.

In this view, one might think that there is life and there is the product of life. That is, the Functional Areas of a Life Field Diagram [xxiii] needs only a slight modification to be a depiction of the process of expression, and as such, might be used as a model for life. Each organism, be it a single-cell amoeba or a human, involves this process of expression to one degree or another. Like life fractals, they are qualitatively the same but quantitatively different.

Objects of reality, such as a rock or a cup of water, are expressions of life. We see a primitive demonstration of this during some séances in which the etheric controls work through a physical medium to produce increasingly solid ideoplastic structures such as the ectoplasmic rods employed to suspend a trumpet or point of luminosity.

Implicit Cosmology

Discourse 10: Perceptual Agreement

Introduction

Cosmologies typically describe the greater reality in terms of physical energy. My first effort to describe a cosmology, *The Handbook of Metaphysics* (6), even has a chapter dedicated to energy, and includes an illustration of an etheric energy spectrum showing possible relationships of increasingly subtle expression of the psi field.

New Age and religious literature is full of references to energy as spirit energy, spirit, electromagnetism, and various predecessor names for psi such as *od*, *chi*, or *orgone energy*. Strong points of view are expressed about the nature of this energy with some arguing it is electromagnetic and others arguing it is not propagated at all. Parapsychologists have dealt with this naming problem by using *psi* as a placeholder term until a more useful term is found. They also solved the propagation problem by referring to psi as a field that is nonlocal and nontemporal.

In the study of Instrumental TransCommunication, a lot of time has been spent looking for the etheric-physical interface. As the argument goes, if the physical is an aspect of the greater reality, then the physical must be a product of the greater reality. This would logically make physical energy in some way a product of the etheric. In accord with that view, the ATransC website includes articles that explore the idea that etheric fields are precursors to electricity.

As I developed the Implicit Cosmology, it gradually dawned on me that the problem with the way I have been defining the greater reality is that I was using an earth-centric model. In fact, the more useful model is personality-centric in which the formation of reality is in response to intention acting on perception. That would be a Source toward the physical evolution of energy.

To say that in another way, the Implicit Cosmology is a foundation model. Specifics, such as how thought produces formative fields which order reality into ideoplastic structures, are left to further research. The cosmology is intended to provide a context from which to design further research.

Given that there is some form of source from which reality has evolved, then there is a substance of that source from which reality is formed. This substance is addressed here as a conceptual material which is compliant to the expression of intention to form an imagined aspect of reality.

Just as parapsychologists use *psi* as a placeholder, this substance from source is referred to here as *etheric*. Early efforts to discuss this substance usually involved use of the term *etheric energy*; however, while energy is conceptually the right term, it is more consistent with what is known about the greater reality to think of it as the etheric, etheric space, or etheric field.

With that said, the etheric is the matrix of reality. It is proposed in the Implicit Cosmology that it is differentiated according to intention to form various aspects of reality. It is also proposed that the greater reality can be thought of as the top organizing field. Further, *within* reality are nested fields produced as various parts of the collective of life which has sought to experience and understand the nature of reality. The physical universe is modeled as an aspect of reality and one of those nested fields.

While that is a lot to comprehend, the one point that is needed for this discussion is that between the source of reality and a personality which seeks to understand reality, there is a continuum of characteristics defining etheric space. This continuum ranges from its original state, which is the initial expression of Source, to whatever personality has expressed as personal reality. This change in character adheres to the rule of association described in this cosmology as the Principle of Perceptual Agreement: *Personality must be in perceptual agreement with the aspect of reality with which it will associate.*

Remember that this is a conceptual model in which the defining characteristic indicating the continuum of etheric from Source to aspect personality is understanding. This Principle holds that a personality is able to experience anything it wishes, so long as it is able to conceive of the experience.

As a physical world example, a New Yorker has a very different perspective of the world than a ranch hand in the West. In a conceptual sense, a New Yorker cannot be the ranch hand without changing perspective. Here, *perspective* is synonymous with perception, and it is correct to say that a New Yorker and a western ranch hand are probably not in perceptual agreement.

This is an important part of the cosmology because perception acts as a determining factor for how reality is organized. To experience an aspect of reality, it is necessary to be able to visualize that aspect. At the same time, to express an influence on an aspect of reality, it is first necessary to perceive that aspect.

Evolving Organizing Principles

The existence of Organizing Principles (some call them *Natural Law*) is a fundamental tenet of many systems of belief. The assumption is that there are rules which govern the behavior of everything in reality. Further, it is assumed these rules are everywhere the same (universal) and cannot be violated (immutable).

An equally long-held assumption of many belief systems is that personality (usually soul or spirit) experiences a progression in maturity from the relatively naive to the very mature personality. It is also thought that the purpose of progression is a response to an inherited imperative to experience, understand, and return this understanding to Source.

The two beliefs are contradictory, in that Natural Law requires that it never changes, and *returned understanding* requires that the source is still learning. If it is learning, then the nature of its emanations will change accordingly. Current research suggests that personalities do experience a progression and that this progression results in a change in personality's perception. This concept is supported by the emerging understanding of etheric fields and the concept of informed visualization [133].

In this cosmology, Organizing Principles are thought to evolve with Source's increased understanding of itself brought by the progression of its aspects. As such, there should be a gradual change in character of the etheric as personality progresses from local aspects of reality to Source.

With these basic Organizing Principles in mind, all else should be considered local organizing rules. For instance, the Natural Law referred to as balance is probably constant and immutable for the physical aspect of reality, but it may have no relevance in the greater reality. In fact, its validity beyond the physical should be considered indeterminate at this time.

Implicit Cosmology

Discourse 11:
Mindfulness

The Hermit from the Waite deck of the Tarot symbolizes the seeker who has achieved great understanding and has turned attention to the world to show the way for other seekers. The Hermit is both the seeker and the teacher.

Purpose

The phenomena of transcommunication appear to have a purpose beyond the reassurance it offers to loved ones. After examining mediumistic messages from the other side and *revelations* brought by past teachers, it is easy to imagine that our etheric communicators are trying to teach us about the reality of our immortality by showing us they exist. This discourse is written as an exploration of the idea that the EVP messages in our recorders, or the paranormal images we find in our photographs, are a new way of telling us that we are part of a larger community. Perhaps it is up to us to understand what that means.

Mindfulness

The terms *mindfulness* and *mindful living* have become catchphrases for right living, but not in a pretentious way or in an attempt to tell us what to do. People speak of mindfulness almost in a reverent tone, as if the concept relates more to God than to daily living. Always it is used to offer guidance in how to improve our life, how to be all that we can be.

Discussions about the phenomena of transcommunication are usually about technique and quality of examples. Who is talking may be discussed, especially if the information seems to come from a loved one, but the question of continuous life seldom comes up. But in fact, considered from the perspective of our immortality, transcommunication may actually be all about our immortality. If this is true, then learning to live mindfully may be the most important ability we can learn.

Teachers

With proper controls, Instrumental TransCommunication (ITC) can be a rich source of information about the other side. For instance, we have seen that we should expect a life review during our transition. We know they can see us, and we know our communicators sometimes *get together* with friends on the other side. We also know that there are changes in their ability or need to communicate so that some do not *report in* for years and some make contact right away, but seem to *move on* after a while.

Channeled material must be considered with reservation because we know cultural influences can color messages. Even so, consistency amongst communicators seems to add credibility to some messages. (26)

Perhaps two of the most important teachers of mindfulness have been the fabled Hermes Trismegistus and the biblical Jesus. The only document credited to Hermes that seems reliable is *The Emerald Tablet.* (47) In it, he speaks of *The One Thing,* which is the same as *The Great Work* of the Hermetic tradition. (61) The Great Work is all about the path followed by seekers to gain understanding. The lessons involved in this are virtually the same as those brought by many more contemporary teachers. The message is that a person benefits by learning to live in accordance with the true nature of reality.

A review of teachings attributed to Jesus, as found in Aramaic-to-English translations, shows that he taught that our I Am presence exists in the greater reality and that our transition out of this lifetime is toward our I Am presence: *"Where that I Am really is, there you already are, and you can be, consciously"* (from Luke 24:38-49). He also taught the unity of humankind, that one person's actions reflect on all people. (62)

A transcript of Hans Bender's words as conveyed by Kai Mügge during a séance can be read on the ATransC website. (43) To paraphrase, Bender explained that we are not alone and that how we view the other side has a lot to do with how we experienced it during our transition. He said that what we are doing here affects the other side and that we can project negativity into the greater reality which can cause problems for others.

Jane Roberts' Seth (63) material appears to be a reliable source of information about the other side. (26) Three important *instructions* from Seth are people create their own reality; people exist in more than one aspect of reality at once; and the only wrong act is to violate oneself or other life.

The common message from all of these sources is that who we really are, our *I am this* personality, always exists in the greater reality, that we are able to connect with our etheric aspect through *right thinking*, that how we think now affects us and others now and beyond this lifetime, and that it is for us to learn to live in accordance with the true nature of reality. This understanding is not based on one person or one organization teaching religious doctrine. Think of it as the handbook for *right living* given to us by our friends on the other side.

What We Do Now Matters

With close examination of ITC messages, a pattern begins to emerge that tells us much about the person. While the messages appear to be paranormal, it has been noted by many researchers that different practitioners are apt to record rather different kinds of messages from the same situation.

To illustrate, Lisa and another person went into a dark room of a reportedly haunted building and recorded for EVP. Lisa is a pragmatic, levelheaded witness and recorded EVP containing useful information. The other person delighted in being scared and expected scary EVP, and in fact, she recorded scary EVP. In both cases, the messages were clearly paranormal, but their character tended to agree with the practitioner's temperament and worldview.

As it turns out, it appears the person's expectations are projected onto experiences. This has been noted in what has become known as the sheep-goat effect [199]. In that, people who are more psi sensitive (psychic) tend to have more paranormal experiences. In his book *First Sight* (13), Dr. James Carpenter developed a hypothesis for the evidence currently being presented in parapsychology which holds that people are always informed about the world via their natural psychic sensing. Further, he argues that people are constantly psychokinetically influencing their world. See the Perception and Expression Discourse.

What all of this means is that we also see with our inner senses (first sight) and always have some influence on our world with our intention, which is based on what we think is true.

Worldview

As was discussed in the Perception and Expression Discourse, people imagine what they are experiencing, and the information for that imagining comes from our worldview database. If the incoming information agrees with the database, then it will actually be experienced by the person. If it does not match the database, then it will either be changed to agree with the database and experienced in that changed form or outright rejected.

The way we express ourselves involves the same processes. Something causes us to react and that initial stimulus is translated in our mostly unconscious Attention Complex based on worldview into an imagined reaction. It is that response based on worldview of which we become consciously aware.

Using this model, it becomes evident that worldview plays an important part in our lives. By all indications, we are born with a more-or-less empty worldview database. It appears reasonable to argue that we do begin with a degree of understanding so that one might say that a child is *an old soul* if born with more than average understanding about the world. It seems reasonable to say that the average person's worldview is full of what has been taught by teachers, parents, clergy and the media. Much of that is simply local custom or popular wisdom.

Personal Reality, Local Reality and the Greater Reality

As was discussed in the Nature of Reality Discourse, each of us experiences reality as our personal reality. Right or wrong, as far as we are concerned, our personal reality is the real reality, and again, that is determined by your worldview. In effect, our worldview defines our reality field.

In mindful living, we are taught to examine our worldview to see if what we believe is true makes sense. The idea is to align personal reality with local reality; the true nature of reality and not what we have been taught to think is true.

Suspended Judgment

Rethinking what you believe to be true may seem paradoxical. If you believe something to be true, how can you tell if you should change your mind or even examine the belief? In practical application, mindful living is a life-long process, a path to be followed one-step at a time, so how does one begin in the middle of a lifetime?

An effective way to begin mindful living is to make a conscious decision to have an open mind. Take conscious control of the process your mind uses to consider new information. As I have described before, we visualize what we are experiencing in a very fast, mostly unconscious reaction to information from our environment. This visualization is based on what we have been taught, which is in our worldview. Environmental information that does not match what is in our worldview may be outright ignored; however, if it produces a *maybe* result in the Perceptual Loop [112], it may be integrated into worldview as it is allowed to emerge into our conscious perception.

While large differences between environmental information and worldview are rejected, ambiguous information may evolve worldview in relatively small increments. Conscious intention provides the motive influence for the unconscious processes controlling perception. Therefore, an unfamiliar experience that the conscious self finds interesting (turns toward in First Sight Theory) will produce feedback that modifies worldview. The next time a same or similar environmental information is received, it will more easily pass the Perceptual Loop as an *Agree* decision.

People have a tendency to automatically reject things they do not understand. The idea of suspended judgment is that we seek to just experience and not automatically decide if we accept what is experienced as true. Instead, time is allowed for future experiences to better inform the perceptual process. This is a conscious decision that can eventually affect the unconscious Perceptual Loop to provide a way of consciously modifying our mostly unconscious memory. Also see the Principle of Suspended Judgment [73].

Self-determination

We have to decide ... everything. If not what we experience, then we must at least decide how to react. Self-determination also means that we create our world. Again, not necessarily the brick and mortar places and things we live in, but for sure how we react to these things. Two people might have essentially the same experience, but each will remember it in a different way. A person who is in the habit of thinking things always go wrong will likely remember it as a bad experience; however, a person who is generally optimistic about life is likely to remember it as a good experience or at least as a learning experience. It is all about attitude and that is a learned thing.

Here too, suspended judgment can help. Whatever we think the world is like, we can learn to consciously intercept that *"oh, it's awful"* response with either a *"wait and see"* or a *"it has a good side"* response. This may seem idealistic but it works. Once it becomes a habit to intercept those internal decisions, there is more room for alternative explanations for what we experience. An *awful* reaction tends to stop further consideration of alternative explanations.

We are always psychically interacting with your environment. How we think of incoming information also has a lot to do with how that information continues to develop. It is likely that a positive or at least neutral response will encourage a more beneficial effect in our environment.

A Personal Code of Ethics for Developing Mindfulness

A recent talk to our local Spiritualist Society concerned ethics and personal responsibility. It was based on Principle 7 of the NSAC Declaration of Principles which states that "We affirm the moral responsibility of individuals and that we make our own happiness or unhappiness as we obey or disobey Nature's physical and spiritual laws." (2)

Three important concepts concerned with mindfulness are expressed in Principle 7. First, there is such a thing as ordering principles in nature. In many systems of thought, these are referred to as Natural Laws. In the Implicit Cosmology, I refer to the concept as Organizing Principles.

The second concept is that understanding and living in agreement with such ordering principles can be beneficial. This is the foundation concept for Mindfulness.

The third concept is that we have a personal responsibility to do so. It is in this concept that we see a blending of religion and pragmatism. In fact, understanding and living in agreement with the organizing principles of reality is the purpose proposed in this book as our reason for existing.

As I developed my talk (64), I came to realize that personal responsibility, morality (more correctly, ethics) and Organizing Principles are all part of living mindfully. From that came a possible technique that you might find useful for learning to live mindfully.

The idea is to develop a personal code of ethics that can help you remember your purpose. Below is one I put together for the talk and which you might use as a starting point for your own code. It is divided into three tiers so that Tier 1 sets the tone for the code. Seth's *Do not violate* (63) is an excellent foundation concept.

Tier 2 is concerned with reasonably basic principles which complement or further define *Do not violate*. These should be intuitively obvious in the context of progression. Tier 3 is Expression which includes phrases intended to guide in how to live the principles. Catchy phrases are useful here, as they make it easier to remembering the principles.

Ethical Understanding	**Ethical Principles**	**Ethical Expressions**
Do not violate *Seth*	• Respect • Kindness • Do no harm • Justice • Fairness • Suspended judgment • Courage • Discernment	• Just because I can, doesn't mean I should • Mindfulness is a way of life • Citizenship means cooperation • How will my actions affect me and others? • Is it a belief or a supportable understanding? • I will not impose my will on others • Lessons come from new experiences • Contemplation not meditation

Possible Mindfulness Personal Code of Ethics

Lucidity

As I explained in the Perception and Expression Discourse, how we experience reality is largely determined by our worldview. The perceptual process is part of our normally unconscious mind, but it is possible to train our unconscious mind to be more receptive to our conscious control. Through mindful living, it is possible to cull undesired beliefs from our unconscious, and in the process, become more aware of our personality.

According to the *Gale Encyclopedia of Occultism & Parapsychology* (55), *Lucidity* is a term that was used in the early 1800s to describe the human abilities to sense subtle energy. This form of sensing has been more recently known as clairvoyance, psychometry and premonitions; however, these terms have been replaced by Extra Sensory Perception or ESP, which itself is now being replaced by psi functioning.

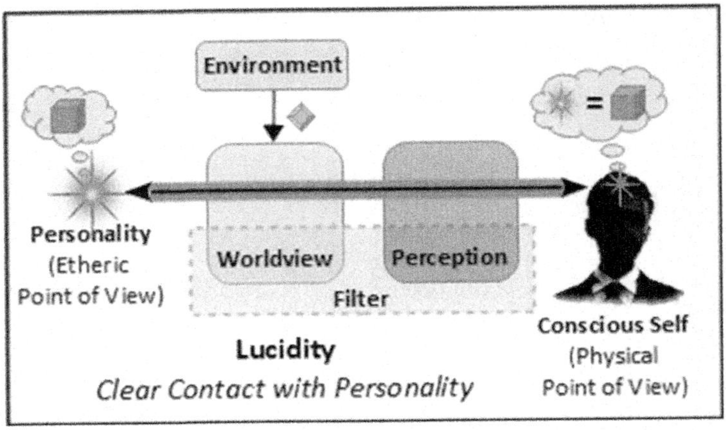

In the context of the Implicit Cosmology, *lucidity* is used to describe the clarity between a person's conscious self and unconscious personality. The term is used in much the same way to describe the extent to which a person is aware of activity during a dream. During a very lucid dream, the dreamer would be able to direct activity as if awake.

Hyperlucidity

A form of false lucidity is frequently encountered amongst new ITC practitioners. This is referred to here as *hyperlucidity*. In one example of this, a person claims the ability to produce astounding communication from the other side as if he or she is completely lucid. On close examination, it becomes evident the person is mistaking noise or imagination as communication.

This perceptual disorder is marked by the tendency to find phenomena *everywhere* despite considerable testimony to the contrary by peers. The concept comes from occasional reports of odd behavior exhibited by some people when they visit an emotionally charged location. It is known by such names as Paris Syndrome and Jerusalem Syndrome (65).

In regard to paranormal phenomena, the effect is the experiencer's mistaken belief in experiencing paranormal phenomena, such as reporting voices in recordings when there are none (66), seeing features in otherwise featureless visual media and misattributing thoughts as transcommunication.

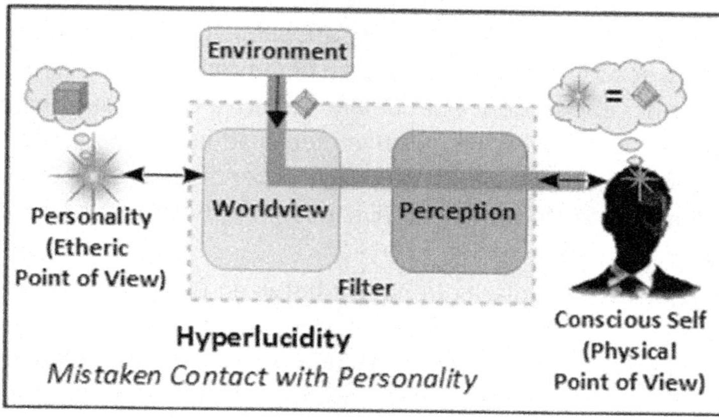

This is a difficult concept because there may be a tendency to use it as an argument against the existence of transcommunication. Instead, hyperlucidity should be looked at as an enthusiastic approach to the study of things paranormal that is not balanced by education about the subject and the necessary discernment to see things as they are. The symptoms eventually moderate as the person become better informed.

Inspirational writer, William Arthur Ward, told us that, *"If you can imagine it, you can achieve it. If you can dream it, you can become it."* That empowering advice pretty much sums up the point of view taught to people studying psi phenomena. Imagining something is true gives the mind a kind of permission to look beyond old beliefs ingrained in worldview. This is an effective tool for personal development; however, a distinction must be made between thinking one *can do* what is visualized and believing one *has done* what is visualized even though that is not the case. This is one of the reasons peer monitoring is so important. (67)

It is noted in the Perception and Expression Discourse that informed visualization is followed by expression. That is to say the work must be done to embody the imagined outcome. So the key to understanding hyperlucidity is to look for hints of magical thinking. If a person thinks ability comes upon them as a gift, rather than by way of study and practice, then an observer should first look for the objective evidence. If that is available, the next question concerns the validity of that evidence. Can it be independently experienced?

Good and Evil and Other Cultural Artifacts

Usually, the dichotomy between good and evil is cast in a biblical light. This should forewarn you that the concepts are subjects of belief. Of course, in conversation, *good* is understood to mean beneficial while *bad* is understood to mean not beneficial. Conversationally, *negative* means subtracting from the middle in some way, while *positive* means adding to the middle. These are literal terms that gain special meaning, depending on their context. Evil, however, is a term loaded with negative meaning which comes no matter the context.

There is, in fact, a polarity in nature between extremes, but the secret wisdom is that there is always the mean or middle way. For instance, with hot and cold comes warm; there is a seldom-discussed middle way for the four personality traits of driver, analytical, amiable and empathic; the middle way between belief and acceptance is discernment. See Personal Style and Astrology [119] in the Personality-Centric Perspective Discourse.

The definition of *evil* is *morally objectionable* and *doing harm.* By itself, that makes it a good word, but say *evil* to someone and they will predictably hear something more like *demonic* or *a threat to one's soul*. Some words are pretty much owned by special connotation and evil is one of the most powerful of those.

As a personal practice, I avoid the use of such terms as *evil, sin* and *hate* because it has been my experience that people never hear the intended context. We receive thousands of questions from website visitors and past ATransC members. When someone asks about the danger of attracting an evil entity while recording for EVP, for instance, I always make an effort to explain that we have found no evidence for the existence of the biblical form of evil.

Some people, on either side of the veil, behave in ways that might seem evil in the sense of being demon inspired. However, on closer examination, they virtually always indicate that they felt their actions were justified by circumstance.

An important point to remember is that people do many possibly antisocial things in an effort to protect their perceived rights or to protect something important to them. Also, we as etheric personalities share a body with our human host during this lifetime. Our body is not just influenced, but outright driven, by its instincts. Killing to protect or for food is natural, as is procreating to assure survival of the species. Such acts can easily become antisocial, even evil in the sense of being demonically inspired.

Demons

To be clear, we have found no evidence of demons in our study of things paranormal. I would not even know how to incorporate the biblical evil into the Implicit Cosmology. The Bible is a faith-based text written for readers who were modern two-thousand years ago. Back then, the concepts of demons and evil were likely useful exaggerations to help the community understand good and bad. We also have considerably more understanding of human nature today.

In fact, the sense of being justified to act and the influence of human instincts cause behaviors that are expected to be moderated by the influence of social norms. And yes, that may include religious teaching.

Perception of Evil

If a person has been taught to believe in evil, then that will be part of their worldview. When confronted by a negative influence, such as another driver yelling at you, it might be natural to assume the person is angry. If you are very religious, this might translate into evil. If there was an opportunity to sit down with the other driver, it might be learned that the other driver thought you had nearly hit his or her car when you changed lanes. The driver was not evil at all, only reacting to a perceived affront that might have caused a serious accident.

An antidote to the sense of the presence of evil is understanding the way we think and mindful living.

Sins, Suicide and Other Taboos

Jane Roberts' Seth (63), told us that *"Thou shalt not violate"* is perhaps the only commandment. I will add that all else is social norm. By that I mean choosing to commit suicide, for instance, may violate a social norm, and certainly it seems on the surface to be a waste of a lifetime. However, an important lesson might only come from suicide for the individual or for the people affected by the unexpected act.

If our objective is to gain understanding through experience, then there should be no disallowed behavior. After Seth's advice, I would add that failing to take an opportunity to learn is contrary to our Prime Imperative. To understand this, first consider what you have allowed into your worldview, and then learn to practice discernment. See the *On Being a Good Witness* Discourse on EthericStudies.org. (68)

Mindful Living

This is an abbreviated discussion about mindful living. The main message is that what you do now will follow you for the rest of your existence–here and hereafter. The more your personal reality agrees with the actual nature of reality, the more progress you will make in your evolution toward a spiritually mature personality; understanding begets understanding.

The key is to stop and think before you react. This also applies to things you do. Stop and think about how your actions affect others. You are a citizen of your community, the world and the greater reality. You psychically interact with it so that your feelings about another person in some way affect that person.

The only right you have is to decide what you think of your world and how you will react to what you decide. You are the only judge as to how well you are doing and that is not based on what you have been taught but on the understanding you have gathered during your existence.

In an ideal world, people would just naturally be mindful of how they are doing as citizens. Laws to enforce behavior that should be considered common decency would be unnecessary because people would be mindful of how their actions might affect others. Of course, we do not live in an ideal world, but that is the point. We are also a society of people whose personal reality is very different than the actual nature of reality. The ideal of mindful living is to evolve a society of people who understand they are part of a community.

Implicit Cosmology

Discourse 12:
Progression, Teaching and Community

The Way of Progression

Through community comes knowledge
Through teaching comes understanding
It takes a collective

Spirituality

It is in our spirituality that we find our true meaning: who we really are and our purpose for existence. One of the best explanations for the meaning of spirituality I have read is offered by Deepak Chopra in his Huffington Port blog (69):

> *Spirituality is the experience of that domain of awareness where we experience our universality. This domain of awareness is a core consciousness that is beyond our mind, intellect and ego. In religious traditions this core consciousness is referred to as the soul which is part of a collective soul or collective consciousness, which in turn is part of a more universal domain of consciousness referred to in religions as God.*

Since we are an aspect of Source in the Implicit Cosmology, *"our universality"* is the fact that our entanglement with Source makes us a citizen of the Source life field (reality, reality field). "Soul" is described as personality which is who we really are, our immortal *I am this*. This is the "core consciousness" of Chopra's definition.

The *"collective Soul"* in Chopra's definition is referred to as the collective in the Implicit Cosmology, and it is modeled as the multitude of personalities related to this realm of reality. The *"universal domain of consciousness"* is modeled as the multitude of aspect which inherits their existence and purpose from Source. God then is Source, Infinite Intelligence, the One, or whatever name you use for the first cause of reality in which we are all one. This is the *"domain of awareness."*

Spirituality, then, is our sense of this connectedness from the perspective of our conscious self (*I think I am this*).

In the Implicit Cosmology, life is modeled as the building block of reality, and expressions of life are the venues for experiencing the nature of Source. We express the world around us, usually as a collective view because we share this venue with others. There are likely many realms such as ours, each with many venues for learning. Together, they could be understood as *"universal consciousness."* This may be Jesus's *"many mansions"* from John 14:2.

Reality is modeled as a nested hierarchy of personalities so that the first round of aspectation is Source's effort to understand its nature. Our personality is an aspect of other personalities, probably many rounds removed from Source, which are further exploring Source's question, *who am I and what is my nature?* [36]

Spirituality, then, is our sense of this connectedness from the perspective of our conscious self (*I think I am this*).

Cooperation

Achieving understanding is not a solitary process. Just as our collective partners cooperate to help us while we are in this lifetime, so are we expected to help others who are in this lifetime with us. This does not mean we are our brother's keeper, but it does mean that we are unavoidably part of our brother's experiences. The part we play is our life experience, and so, we are unavoidably linked in the greater community and especially in our local community. See the Cooperative Communities Principle.

Friends as Teachers

With few exceptions, I have looked to a teacher to show me the way. I don't mean this in any mystical sense; it is just that I have managed to have someone I trust show me how to do extraordinary things before venturing out on my own. For instance, Lisa introduced me to Electronic Voice Phenomena; a mutual friend taught us how to mine gold; a work friend introduced me to the Tarot. Another showed me how to navigate the

[36] These assumptions are made to provide a beginning point for the model in much the same way the Big Bang is modeled as the beginning for the physical universe in physical sciences. The importance of the Implicit Cosmology is in the relationship of our *I am this* personality which is our unconscious core intelligence with our human avatar and our conscious self as *I think I am this.*

Sacramental Delta waters at night. They did not just say, *"Here it is."* They took me through everything I needed to know to get started. It was always a natural sharing.

Selfless People

There is value in contemplating the meaning of the selfless concept. Dictionary.com defines selfless as *"Having little or no concern for oneself, especially with regard to fame, position, money, etc.; unselfish."* There are many ways in which selflessness is expressed: the family that takes in a foster child, a person who stops to help an animal, individuals who feed the poor and inspire selflessness in others, or the activists who dedicate themselves to a cause which probably should be important to everyone.

What causes people to be selfless? In some obscure way, there may be an evolutionary benefit. Evolutionists argue that there is a selfless gene which induces people to selfless acts. If they do not survive, their selflessness tends to benefit their family which would by inheritance have some of the selfless gene. In that way, they argue, selflessness helps survival of the gene by helping the survival of family. (70)

While the selfless gene theory sounds reasonable, it is based on pretty tenuous evidence. However, when viewed from the perspective of human as avatar, the concept of cooperation amongst personalities seeking mutual progression is perhaps a more reasonable concept. The Prime Imperative would seem to invoke an urge in every personality to help one another, because by doing so, they are helping themselves and their collective. Ultimately, one could argue that selfless people are those who respond to the urge to answer Source's question. From a religious perspective, *"God desires that I do this."*

A Vision About Collectives

Just minutes after a meditation session at the Monroe Institute (53), I had the waking vision of the face of a clock. It was suspended in the air, face-up, but tilted toward me a little so that I could see that there were many black specks scurrying about on the white surface. My impression was that they were little stick people like those I would draw in a hurry. The space between three and four o'clock was an open hole and a few stick people had apparently fallen through.

The people on the face of the clock were somehow helping the people who had fallen through the hole. As the hour hand made a complete circuit, the ones in the hole came to the surface and a few of the others jumped into the hole.

My sense was that the stick people were all part of a collective of personalities, a soul group if you want, and they were doing all they could to help their fellows who had entered into a lifetime, symbolized by falling through the hole. I knew that they were helping one another to progress by gaining in understanding, and that none of them would be able to move on until all had made sufficient progress. The hour hand represented a lifetime.

Probably not all of them, but many of my helpers, friends, and guides, both on the other side and in the physical, are part of my collective. I am never very far from them, and just as I am a student, in turn, I am teacher, for we must all move as one.

First a Student Then a Teacher

You have probably heard the phrase, *"Our lot is to learn and having learned, our lot is to teach."* The student becoming the teacher is the natural order of life. Civilization survives because lessons learned by elders are passed on to offspring. We cannot be around other people, especially children, without influencing them in some way as accidental role models. However, in modern society, education is left to schools and hired teachers. There is a business of educating our young, leaving little more than role model duty to elders. Yet there are many examples in which individuals have taken the time to preserve knowledge for others.

The Internet has countless websites in which information for the sake of information is available at no charge. Looking past all of those commercial sites, one can find that there are many ordinary people who have spent time developing a personal website just to share what he or she has learned. An important point to this discourse is that many of these people probably know they will not receive a thank-you. Yes, there is a human compulsion to leave a legacy for the next generation, but virtually all of these websites will vanish when the server fees are not paid. So discounting legacy, the most probable reason is that people build them out of the desire to help others.

Consider the model offered by the ancient wisdom schools. There is typically a path one must follow to gain masterhood. The first step on the path is for an individual to seek understanding by deliberately asking for a teacher. What follows is a series of lessons, tests and initiations as the seeker progresses from neophyte to initiate and finally to teacher and master.

Here, it is important to make a distinction between information, knowledge and understanding. The lessons provide the seeker with pertinent information. Tests are intended to determine whether or not the seeker has assimilated the information in a way that can be expressed in novel ways. If so, then the seeker has demonstrated that the information has been assimilated as knowledge.

But here is the important part. During the initiation, the seeker is asked questions which may only indirectly relate to the lessons. If the seeker is able to provide reasonable answers that indicate understanding, then the seeker will be passed to the next level.

A selfless response to the urge to teach is difficult to ignore; however, the student is drawn to the lessons with usually only vague reasons for pursing understanding. It is this teacher-seeker relationship I wish to address here.

The Nature of Understanding

It may be a good idea to explain how *understanding* is used here. There is a distinction between information, knowledge and understanding:

Information

Information can be compared to raw data. In the context of mind, it is sensory input from the environment brought by the physical senses and countless psi signals. As it is received, information is undifferentiated and is only sensible when organized into some context. In a practical sense, much of what you have read in this discourse thus far is just information. Think of the bits and pieces of information as data points which must eventually be integrate into a sensible concept. Much of what people think they know is just information.

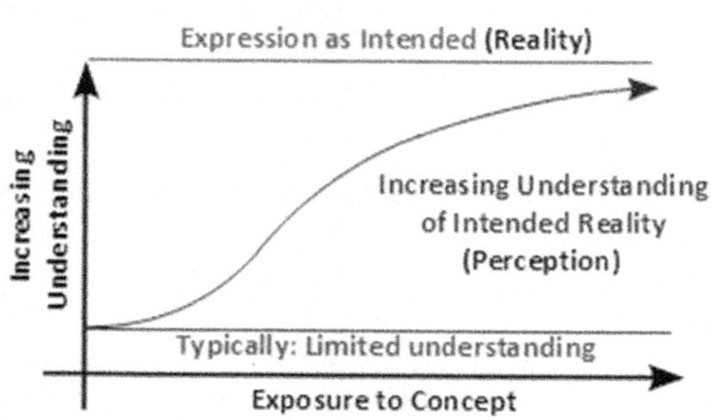

Limits of Understanding

Knowledge

When information is considered and the related concepts are understood, the resulting comprehension is considered knowledge. To be knowledgeable about something is to comprehend related concepts so that they can be reasonably well visualized. This is the key to the nature of knowledge. It must be comprehensible enough to be visualized as a concept. Here, *visualize* is used in the general sense of being imaged in some way in the mind. That might be sufficient recognition of an odor to name its source or formulation of many seemingly unrelated concepts into a form that can be described to others.

For instance, consider the old story of blind men examining an elephant: if each man represents a concept (leg, tail and such) then knowledge would be the ability of the men as a group to be able to describe an elephant. Knowledge is a combination of information, understanding concepts and comprehension of a global sense of meaning.

Understanding

In the Implicit Cosmology, perception is modeled as the outcome of a process that begins with encountering information that is external to the mind. We, as etheric mind and physical body, are immersed in an array of environmental signals. Some are physically sensed by the body, but most are psychically sensed. All of these signals are processed in our Attention Complex [110] where they are first filtered according to whether or not we are interested. (13)

The degree to which understanding agrees with reality (intended order) is also a function of the person's worldview. For instance, a person who has been taught to be prejudiced about something, and who retains that prejudice, will likely take what emerges from the unconscious perception process without question. Here, the exercises of mindfulness as a conscious effort to see things as they are intended become important in helping us see the world as it is, rather than as we are taught to think it is.

Understanding is not an absolute. As is shown in the Limits of Understanding Diagram [161], correct perception (understanding) is typically limited upon first exposure to a concept. As a mindful person seeks to better understand the concept, understanding approaches perception that agrees with reality. This is an obvious effect when, say, a new principle in science is introduced, but it is less obvious is when the process of overcoming prejudices learned in youth and fears foisted upon the avatar relationship by human instincts are considered.

As a person moves through a lifetime from youth to old age, things held to be true populate worldview and tend to reinforce future perception. A conscious effort is usually required to break this cycle. Perhaps even more distressing, a person carries these *truths* into transition from this lifetime. The result can be a complex, reportedly emotionally painful period of *getting well* while worldview is better aligned with reality.

The Process of Gaining Understanding

As a person gains in understanding via the accumulated lessons of many lifetimes (our own and/or those shared by other personalities in the collective), the person begins to understand the need for more mindful living and the desirability of having a worldview that is in agreement with reality.

A magical thing happens when a person begins to consciously seek understanding. You may have heard the phrase: *"When the student is ready, the teacher will appear."* A similar saying was offered by the famous psychologist Carl Jung: *"Synchronicity is an ever present reality for those who have eyes to see."*

"For those who have eyes to see" is also a saying used in the Bible and secret wisdom schools. In fact, the whole idea of the occult or secret wisdom is not so much the secret but the need for the seeker to have sufficient understanding to recognize the lessons. An example is the concept of polarity: *Everything has an opposite.* The seeker is expected to realize that the middle way is the hidden knowledge to be derived from such concepts. In practice, this realization translates as perceptual understanding based on worldview which is in alignment with reality.

At least one reason mindfulness works is that consciously seeking understanding effectively tells the unconscious part of mind to pay attention to external information that might help, rather than ignoring it as irrelevant. In many ways, the unconscious mind is an obedient servant that shields us from a world that does not support what we think is true. Questioning assumed truths moves the horizon of conscious self deeper into the unconscious process.

Worldview effectively prevents us from experiencing or believing anything that is not in agreement with worldview. That is how it shapes our personal reality, even when it conflicts with actual reality. It is also how a person can obsessively believe something even though the facts seem overwhelmingly to the contrary. See the Perceptual Agreement Discourse.

Change must come incrementally. That is, if incoming information is clearly not acceptable to worldview, it is ignored. However, information that is reasonably similar to what is in worldview will sometimes be accepted via the *Maybe* outcome of the Perceptual Loop. The *Maybe* outcome is consciously experienced and incrementally changes worldview. In that way, progression in understanding is a gradual thing. Engineers will recognize the curve in the Limits of Understanding Diagram [161] as a naturally occurring rate of change found in many physical processes. Change is usually evolutionary and seldom catastrophic.

Collective

The collective concept is not restricted to a personality and its many entangled aspects. It also applies to social communities of like-minded people cooperating to accomplish a common purpose.

We Exist to Learn

Given the concepts described above, it is arguable that each of us is here to gain understanding. However, there is an old saying that *"When you're up to your neck in alligators, it's easy to forget you came to drain the swamp."* While we are in this lifetime and immersed in daily living, it is easy to forget what we came for. In a practical sense, there are few signs to remind us of our Prime Imperative.

Progression may seem to be a solitary path. Perhaps that is why they call such paths *personal improvement.* In fact, the old saying that we are all connected is more truth than platitude, and the path we may seem to be walking alone is more a march of the multitude of personalities who await our progression while doing what they can to assist us.

We who are seeking understanding are a community, a collective of like-minded people that exists in the physical as personality entangled avatars and in the etheric as immortal beings.

It is worth repeating the opening Understanding:

The Way of Progression

Through community comes knowledge

Through teaching comes understanding

It takes a collective

Spiritualist Society of Reno

While it is certainly possible to gain progression in other ways, it is considerably easier to work as a member of a community. Using the Spiritualist Society of Reno as an example, the Sunday talks, offered by a different person each week, typically provide insightful messages that help members understand the nature of progression. The speaker may serve as a role model via a personal story and sometimes via focused lessons.

Such a gathering of like-minded people generates a contact field [100] which grows in strength with each meeting. Participants can sometimes sense the field during meetings and even as it extends into their daily lives. The thread of rapport strengthens with each shared memory. It is this field that facilitates communion across the veil, making it easier for loved ones on the other side to assist our seeking. In a multitude of ways, the community fosters a culture of seeking and mutual assistance.

The community offers many ways for individuals to teach others besides giving the Sunday talk. Members have a unique opportunity to ask questions of others who have similar interests, and to join into conversations that provide a powerful means of sorting out ideas and bringing clarity to both the one asking and the one answering. It is by teaching that the necessary ordering of thoughts and clarity of thinking brings understanding. It is very difficult to make progress without a forum in which to explore ideas.

The Prime Imperative as a Spiritual Obligation

While we all share an urge to gain understanding, in a very real sense, we have a sort of spiritual obligation to respond to that urge. While it may seem to be a sacrifice to teach or help others learn, every such effort is rewarded with new understanding and resulting progression that likely would not be possible otherwise.

It is much too complex of a concept to say that each of us needs to serve in some way in the physical. As I have noted above, there are many ways we teach. Perhaps the majority of us are too early on the path to be more than inspiration to others. However, if we have reached a point in our evolution at which we are accustomed to self-evaluation, it is time to begin finding ways to serve by teaching.

In a very real sense, progression can go only so far without teaching and that requires a community.

The Transition Experience

As I understand it, all of us experience a short period of disorientation as we begin the process of transition out of this lifetime. The first milestone of transition appears to be the often reported period of *getting well*. That is thought to be followed by a period of self-assessment and finally the completion of transition as we enter into a new venue for learning. The new venue could be in the physical, but it could as well be in some other aspect of reality.

Personal reality is a product of worldview, and we take worldview with us into transition. That is probably why, during the initial shock of transition, it seems to be relatively easy for a personality to remain *close* to familiar surroundings. It is during this time that most communication with people in the physical seems to occur. A common report is the departing visit to say goodbye to a good friend.

The *getting well* period is needed to allow the personality time to realize that old handicaps no longer apply. This may also be when we undergo a self-evaluation of our just finished lifetime. This is that dreaded judgment we are always warned about. However, instead of our being judged by some authority on high, it is a personal process in which we sense how our actions affected others from the perspective of those whom we affected. Perhaps it is from this self-evaluation that we determine our future lessons.

The period of adjustment apparently begins as the sense that we are in a world very much as we experienced in the physical; again, clinging to the familiar. However, if we expect a heaven from the religious point of view, we will likely find ourselves in that heaven. As we understand it, over time, we find that these are only constructs of our worldview, which itself is slowly realigned to better agree with the greater reality.

As we understand the process, we are born into a venue for learning at the end of this transition (a new lifetime), with a point of view, a degree of maturity in our understanding of the nature of reality gained from past experiences and an urge to learn specific lessons.

We are not all-wise when not in a lifetime. Instead, we have an imperative to learn and a degree of perceptual maturity which helps to form our point of view. It is this maturity and resulting point of view which determine what aspect of reality we can be in agreement with. In effect, the greater our understanding, the wider range of venues and opportunities for learning we have available to our personality.[37]

An important point is that there are no enforcement officers making sure we do what we should. We are self-governing. Our understanding manifests as perception and that determines what aspects of reality we can visit (Principle of Agreement).

It appears that transcommunication is possible during transition, but once a personality transitions into a new venue of learning, it appears that all communication into the physical stops unless it is assisted by other, more mature personalities.

We have noted in ITC that our etheric communicators behave as if they are living in a world much as they experienced in the physical. This is true even to the extent that they might refer to grandma's house at which members of the family in the etheric gather for a family dinner. We speculate that this familiar world eventually changes to a more conceptual view as the person's worldview is aligned with the local reality. We are told that our natural form is as a luminous being existing in a very luminous landscape.

Self-Realization

This discourse can be summarized as *Understanding Self-Realization*. The reason for metaphysics is not just to map reality. It is a process of discovery intended to help us learn how to gain personal understanding so as to fulfill our purpose. It is okay to ignore the details if you find the guidance.

Here, I do not mean purpose in a mystical sense. An acorn fulfills its purpose by growing into an oak tree. If the acorn tries to be a willow, it will likely end up being a pretty sad-looking oak tree.

Metaphysics tells us that our purpose is to align our personal reality with the actual nature of reality. If we ignore that Prime Imperative [61], we will likely end up being less than we could be. There is no need for me to explain this point. As you live your life as a person, you must surely see the benefits of living in accordance with ordering principles and the results if you do not.

---

[37] To emphasize this point, the relationship between personality, Attention Complex and conscious self remains during transition. The only real difference is that the human influence is no longer as strong. Part of transition may well be the process of sorting through worldview to flush out that human influence. Meanwhile, to put it in the vernacular, *if a person was an idiot in life, he is likely going to be an idiot in spirit.*

Fulfilling your purpose may or may not be the same as what our society considers being successful. Our society is dominated by people trying to be willows. Only a few are sufficiently self-aware to recognize there is a purpose that underlies the dynamics of our society. It is for you to live in this society as a useful citizen without allowing its less aware point of view to contaminate your worldview.

Remember that it is not what happens to you. It is how you react to what happens to you.

Implicit Cosmology

Discourse 13:
The Cosmology of Imaginary Space

Introduction

Reality is modeled in the Implicit Cosmology as conceptual (etheric space) while the physical aspect of reality is modeled as objective (physical space). The assumption is that we are taught what is physical and what is not. We are taught to assign physicality to objects of physical space, but intangibility to things of etheric space. This is a convention.

As modeled in this cosmology, a personality is only able to associate with aspects of reality with which it is in perceptual agreement [67]. This agreement is a function of worldview.

The agreeability factor of worldview is a function of how well the individual personality understands the nature of local reality. The degree of understanding is referred to as a measure of progression.

Other Names for Reality

Each system of belief has had a cosmology describing the relationship of what is considered *here* and where God and Heaven are thought to be. Most have shown reality as a series of layers, with the *here* on the bottom and God on top. The difference between the various layers is usually described as a difference in vibration or coarseness, and travel from one layer to the other is described in terms of *going there, being over there* or *the other side*. Lacking a better set of terms, these words are useful for conversation so long as it is understood that they are just metaphors.

In practice, these terms are too often taken literally. Perhaps this is because reality is virtually always considered from a body-centric perspective [115]. When information is revealed to us by our discarnate friends, it is necessarily in terms we understand: up, higher, over there.

An interesting evolution in terminology comes from what one might think of as mystical physics. Rather than many venues for learning, the fad now is to refer to the greater reality in terms of multiverses. As I remember, the concept originally came from speculation that particles rapidly vibrate in and out of physical space. If I remember correctly, this idea originates from the wave-particle states of photons.

This certainly is not my area of expertise, but the argument for multiverses is based on pretty tenuous logic. There is considerably more objective evidence for the reality described in this book. It is also pretty obvious that most of the people throwing in references to the sexier aspect of physics, including multiverse and quantum mechanics, are appealing to perceived authority more than speaking from actual understanding.

When you hear someone refer to multiverse in the context of things paranormal, at least as far as the Implicit Cosmology is concerned, think aspect of reality.

Imaginary Model of Reality

Given the idea that one gains access to an aspect of reality by gaining understanding rather than by traveling to it, a model of reality is needed that can accommodate conceptual relationships between its parts. The closest physical analog I could find to model the conceptual nature of etheric space is the imaginary space of the Mandelbrot Set.

Imaginary Space

In mathematics, a complex number is one that includes an ordinary number and an imaginary number, for instance, X + Y*i*. Y*i* is the imaginary number where *i* is written as *i* equals the square root of minus one, which is a disallowed state in ordinary math since a number times itself is always a positive number.

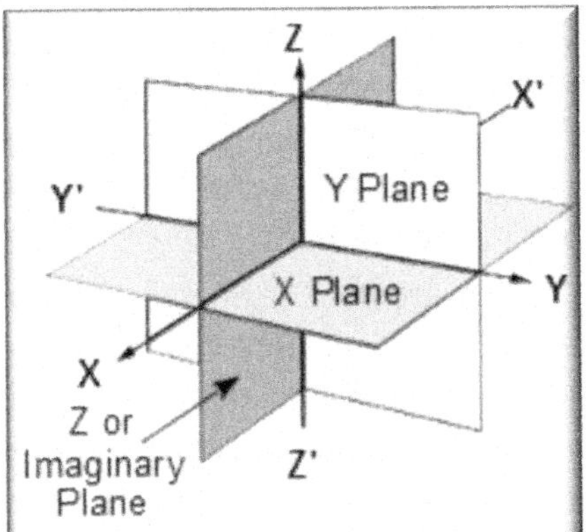

Complex Number Space
(X + Y*i*)

The complex number space (imaginary space) we are concerned with here is shown in the diagram above. It is a two-dimensional complex plane, extending from -2.0 to +1.0 on the real axis and -1.5 to +1.5 on the imaginary axis.

The Mandelbrot Set Equation

The equation for the Mandelbrot Set is:

$$Z_{n+1} = Z_n^2 + C \qquad \text{where } Z_0 = C$$

Values of **C** that are in the Mandelbrot Set are those for which the value of $\mathbf{Z_n}$ does not go to infinity. That is, the formula is calculated with an initial value of **C**, and then calculated again with that first result used for the new value of **C**. The equation is calculated in this way a predetermined number of times or until the result either rapidly approaches infinity or settles in a narrow range of numbers.

Each point in the Mandelbrot Set is used as **C** in a series of calculations to produce the plot of all points **C** in the set. For the plot, resulting values of the calculations may be assigned a color; usually black to represents the stable regions, and the lightest color represents values of **C** that very quickly take the calculation to infinity.

Although the area of this number set is a few units on a side, it is important to note that this imaginary space is infinitely small. That is, there is no limit to the number of zeros to the right of the decimal point for **C**. You can see this in some of the figures below where the coordinates and magnification factor (zoom) are shown.

How the set is viewed is a function of the resolution of the display technology and colors used to display the plot. Because color values are assigned to number ranges, the color display tends to be stepped. The Top Fractal plot below shows a step-wise color assignment. Shading in the plot Showing Navigation Diagram [173] has been blended.

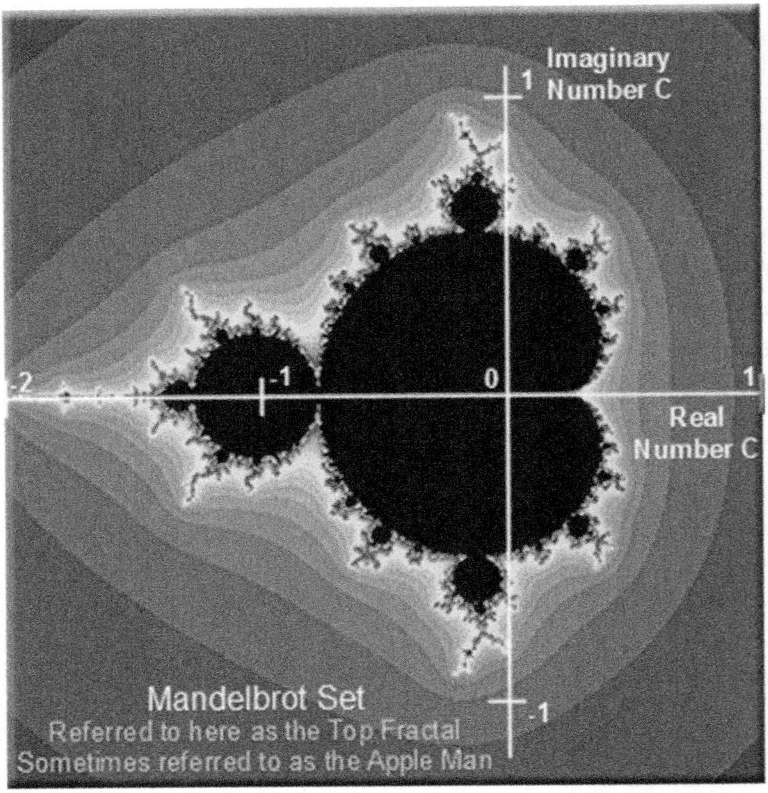

Fractals as Attractors

An important feature of chaos theory (71) is what mathematicians refer to as an attractor. (72) This is a conceptual region of stability toward which elements of a set gravitate. For our discussion, elements of a set are unstable but gain in stability as they become more closely associated with the core concept (attractor).

The top fractal in the Mandelbrot Set is an attractor for all of the subsequent fractals in the set. The black regions are fractals based on the top fractal. They are regions of relative stability and the coordinates outside of the fractals are considered in their orbit and are more or less stable, depending on their relationship with a fractal.

Comparing the Mandelbrot Set to Reality

As a person *travels through* the imaginary space of the Mandelbrot Set into ever smaller scales, an infinite number of Apple Man fractals are encountered. In the plot above, think of the top fractal as Source and the other fractals as aspects of Source. The entire set is comparable to the reality field.

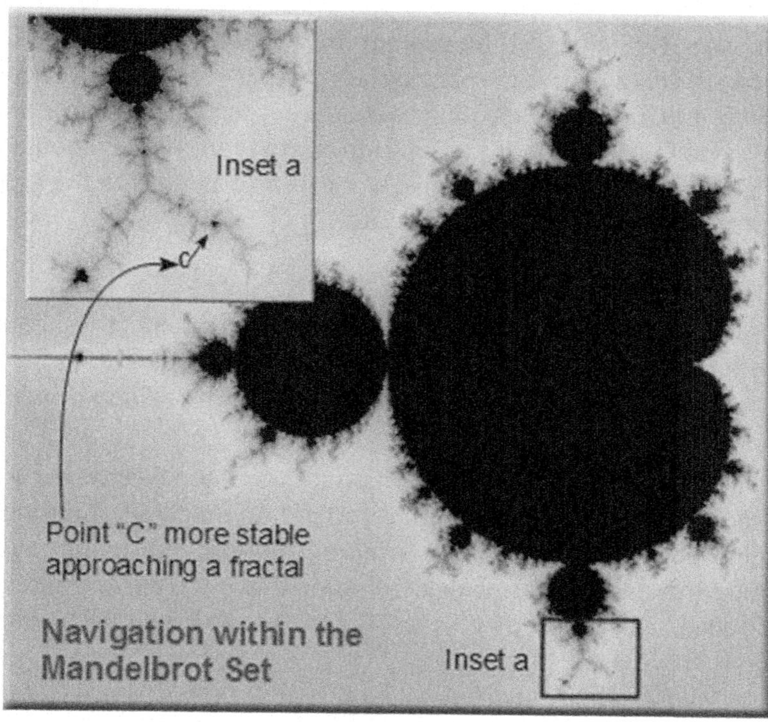

Point **C** in the plot is comparable to a person's perception of local reality. The result of the calculation discussed above, $\mathbf{Z_n}$, will approach infinity for all coordinates **C** except for those within the black region of a fractal. Think of $\mathbf{Z_n}$ as worldview. In this comparison, $\mathbf{Z_n}$ (worldview) *going to infinity* means the personality not understanding the nature of reality (Perceptual agreement). The more worldview is in agreement with reality, the longer it takes $\mathbf{Z_n}$ to go to infinity. The point on the plot, **C**, moves toward a stable region (aspect of the reality field) as $\mathbf{Z_n}$ is slower to go to infinity.

As a person (point **C**) gains in understanding, he or she moves closer to a stable place in reality. The stable area represented as Apple Man fractals which are comparable to collectives in the Implicit Cosmology.

Every point **C** is under the influence of a fractal (in its orbit). Every fractal is associated with the top fractal. The threads of structure made visible in the plot are artifacts of the technology, but they serve to represent lines of entanglement between the top fractal (Source) and the other fractals representing aspects of Source.

Navigating Imaginary Space

To complete the analogy, movement in the imaginary space of the Mandelbrot Set is a matter of changing worldview. Reversing the calculation and solving for **C** rather than $\mathbf{Z}_n$, one navigates imaginary space by developing a stable value for $\mathbf{Z}_n$. That is the same as saying that one moves about, not by traveling up or down, but by changing worldview to agree with the aspect of reality with which one wishes to associate.

A Useful Model

If you are familiar with the more traditional models of reality, the uniqueness of a model based on the Mandelbrot Set should be apparent. In the same way, the Implicit Cosmology is very different from traditional models of reality.

One reason for this is the precondition that the model explains currently understood phenomena without resorting to magical thinking and groundless appeal to mainstream science. There is substantial evidence supporting the Implicit Cosmology. Some of this evidence is produced by good science, and a lot of it is produced by replicable techniques for transcommunication.

By its very nature, the Mandelbrot Set is a conceptual model. It technically does not exist without resorting to the device of imagining a complex number plane based on the normally mathematically disallowed square root of minus one. Yet it exists.

Section II
Community

Introduction to Community

As it Began

The first book I wrote was the *Handbook of Metaphysics* in 1994. My objective was to give people an overview of metaphysical concepts that would provide a context in which to understand new ones as they came along. The book was complex and I suspect the publisher that picked it up did so because it looked like a textbook.

This paragraph originally included a long rant about publishers and the general lack of ethics in that industry as I experienced it with four books, three publishers and one agent. I am sparing you the negativity, but I do feel better having written it. It is enough to say that the *Handbook* was published but never marketed. Now I am reluctant to show it around because it needs proofreading and is out of date.

The thing about metaphysics is that one must live with the concepts for a while to understand their nature. I was fifty when I wrote the *Handbook*. My wife, Lisa, and I have taken many classes since then. And then Electronic Voice Phenomena (EVP) came along.

Lisa began working with EVP around 1987, and in 2000, we assumed leadership of the Association TransCommunication (ATransC). (It was the American Association of Electronic Voice Phenomena or AA-EVP back then.) It was a small organization with an average of 250 members representing around twenty countries.

Universal Studios asked us to help them market the 2005 movie *White Noise,* and for a while, we were *"all over the Internet,"* as one person explained when we asked how he found us. The *White Noise* DVD has a couple of very good bonus features that show us doing *walkabouts* in interesting places while we recorded for EVP and answered questions. As the producers wanted, EVP became a household term as the movie was released.

There are many reports of early efforts to record direct voice [434], but the discovery of EVP is generally credited to Fredrick Jürgenson in 1959. When Sarah Estep founded the organization in 1982, people had known about EVP for only twenty-three years. There was little access to the Internet, and most of what people knew about it was from the printed media. The first year of our leadership, we set up a website to provide background information about EVP and discussion board on which we spent a lot of time answering questions.

Everyone is an Expert

This is discussed in some detail in the Community Commentary, but here I will say that, without a culture of recognizable subject matter specialists as the source of authoritative information, it is common for people to assume that authority for themselves. A second factor here is the influence of religion. Of course, religion is all about faith and belief. It is concerned with heaven and people's relationship with God and has a lot to say about people going to heaven. EVP has a lot to do with discarnate people, so by association, EVP is often treated as a matter of faith. When it comes to matters of faith, many people—probably paranormalists in particular—are reluctant to let others tell them what to believe.

In the early days of EVP, virtually everyone who came to us came with preconceptions which we had to overcome before the concepts could be taught. Probably 90% of the thousands of questions we answered right after *White Noise* were concerned with fear of demons, fear that loved ones were stuck, and fear of bringing a demon home when they record for EVP.

A second idiosyncrasy of human nature became obvious around 2007. People were coming to us with preconceptions about EVP that were based on pretty silly assumptions. There was a growing number of ghost hunting clubs, and by then, each had one or two people who recorded for EVP. Individual clubs, and sometimes groups of clubs, developed unique theories of what EVP was, how to record for it and how it worked.

The rest of that story is that Lisa and I are not all that charismatic. We approach these phenomena with a pragmatic point of view which leaves little room for faith and unfounded belief. Our explanations were technical, and I am afraid that we did not do a very good job of communicating the concepts.

Probably the most damaging change in the community was the introduction of radio-sweep [296] as a technique for recording EVP. It was like a *silver bullet* because all people had to do was repeatedly move the tuning dial through the radio spectrum in about 2.5 second sweeps. The resulting audio stream of music and voice fragments was thought to contain meaningful messages. It did not help that *White Noise* used a similar technique as a prop.

Our study of radio-sweep (73) (74) has shown that it probably does not produce EVP, even though good stories can be concocted about the sounds to make it seem like there are messages. The *silver bullet* proved too tempting for most people, and as we began refusing to review radio-sweep examples, the ATransC became less and less popular as people looked for places to share their radio-sweep messages.

As the ghost hunting clubs became more popular, and as social media drew people from discussion boards to FaceBook-like outlets to satisfy the need for sharing, membership in the ATransC began to dwindle. At one time, I was able to gather ten or fifteen competent practitioners for an experiment. But because we will only use transform EVP [292] for research, we can no longer attract the necessary practitioners.

Wikipedia

In 2007, I registered as a Wikipedia editor. This is discussed in more detail in the Wikipedia Commentary. As that experience applies to this discussion, it became painfully obvious that there are people in the world who have the same kind of faith in science as others do in religion. This faith in the infallibility of science is referred to as scientism, and from personal experience, I can say that it is as damaging as religion in so far as the exchange of ideas is concerned.

The Wikipedia experience gave me a good sense of how people, in what one must think of as mainstream society, deal with metaphysical concepts. This is important because the skeptical editors, who control what is written in articles concerned with things paranormal, learn their point of view from opinion setters in the skeptical community.

As witness that the skeptic's message is being heard, virtually every parapsychologist who publishes articles in the media has at one time or another railed at the injustice of skeptical attacks. Funding for parapsychological research is nearly nonexistent, and the claim that parapsychological phenomena are pseudoscience and therefore a danger to society has turned the U.S. government and others against their study. It is said that a scientist can ruin his or her career by being associated with anything paranormal, making it almost a certainty that the mainstream status quo will continue for the foreseeable future.

Parapsychology

The majority of people in the paranormalist community are hobbyists. Some may have an advanced degree, but not necessarily one related to paranormal phenomena, and they do not usually attempt to work with transcommunication under the authority of their degree. Some people in the community are citizen scientists but most are just having fun. Those who are using EVP to help with their grief over losing a loved one tend to be a small minority, and from my experience, they seem to find more help from mental mediumship. (75)

With a few exceptions, parapsychologists are first psychologists, and interestingly, some of the most anti-survival people hold a degree in psychology. The study of Instrumental TransCommunication (ITC), which includes EVP, is technology intensive. By this, I mean that electronic devices are used and understanding physical principles is essential to understand transcommunication. So the study of ITC requires an understanding of human nature, audio and visual principles, electronics, and metaphysics. Psychologists are trained to understand human nature, and as I said, most are predisposed against the survival concept. See the Artifacts and Perceptual Error Study for a discussion about the complexities of transcommunication phenomena.

Most of the ITC research conducted by parapsychologists that I am aware of has been based on poorly conceived protocols, suggesting the authors had little or no operational understanding of the phenomena. Not surprisingly, the majority of what little ITC research has been conducted has been reported as *failure to replicate-style* articles. See the Anomalistic Psychology [218] in the Science Commentary.

Lack of a Workable Theory

The most common complaints I hear from skeptics about EVP are that it is fraud, mistaken noise (pareidolia), or unnoticed background conversations, in that order. I should say that here the skeptics are both mainstream science apologists and parapsychologists.

If the conversation does get past the initial accusation of fraud, pareidolia or mundane sounds, the next complaint is that there is no workable theory to explain ITC. If this complaint comes from an academic, never mind the field of study, the conversation pretty much ends. I am defeated for three reasons:

1. With only a four-year college degree, I am not an academic and have no intellectual standing amongst academics. (academic-layperson partition [214])
2. The alternative explanation favored by most parapsychologists for transcommunication is the Super-Psi Hypothesis which only allows for survived memory and not continued life.
3. There is little sense in arguing that there is the Survival Hypothesis because it exists as little more than a name.

In fact, most of the people we look to for such theories are not qualified to develop one. I know of one parapsychologist who has the academic credentials but he is poorly informed about ITC and predisposed to favor Super-Psi. The most damaging insult to the paranormalist community is the publication of *failure to replicate-style* articles by supposedly pro-paranormal organizations in their peer reviewed journals. The identity and qualifications of the *peer* reviewers are always hidden. At the same time, there has been little opportunity for nonacademic rebuttal. This is discussed in the Science Commentary.

The Implicit Cosmology

The Trans-Survival Hypothesis came first as my effort to provide a useful, well-supported argument for survival of personality after bodily death. It is my answer to the complaint that there is no theory, and is very different from other such theories because it includes what is known about transcommunication.

The Implicit Cosmology came next. It is based on the Trans-Survival Hypothesis and a wide range of metaphysical concepts. Some of these concepts are objectively supported and some inherit plausibility from other concepts. It is a cosmology because it begins with a first cause and extrapolates a model of reality that allows for known phenomena and their current explanations.

The intention of the Implicit Cosmology is to provide a structure in which to understand transcommunication. As it turns out, it also helps me understand my nature.

The hypothesis is also very different because it is written by an engineer rather than an academic. This is a point that needs to be kept front-most in your mind as you read this book. I do not have an advanced degree, and I am only familiar with human nature based on experience and not training.

The thing about engineers is that we are trained to account for all known influences on a system when we develop a design. To miss one might cause a product to fail, perhaps costing lives. When I develop a model, I am almost compelled to account for every factor I can imagine. At the risk of seeming snide, academics seem to be able to pick and choose what they are going to pay attention to in a study, perhaps because they are only working in a theoretical sense. That is a statement of academic policy in the study of anomalistic psychology.

Above all, my objective in writing this book is to foster more informed conversations about things paranormal that at least include consideration of survival.

PS:

Some people have what is considered a good moral compass which guides them toward being a good citizen. However, some people only obey the law to the extent that they think they might be punished for violations.

The Implicit Cosmology was also written to provide a reason why people should live mindfully. They should, not because they must, but because they eventually must, and doing so now will make life very much better in the long view.

The Organizing Principles are naturally self-enforcing. I fantasize that, if people come to understand the concepts, they will just naturally become better citizens.

Commentary 1: Community

A Community Divided

It may help you better understand the Implicit Cosmology if you have a sense of the environment in which it was developed. First, let me clarify a few terms. For lack of a better term, I refer to people interested in things paranormal as paranormalists and their community as the paranormalist community. Here, paranormal is used to mean subjects that are concerned with mind-body duality (nonphysical personality), the psi field, continuation of consciousness beyond bodily death (survival), the natural environment of mind in which personality survives (the etheric), and communication across the mind-body interface (transcommunication).

Such subjects as reincarnation, out-of-body experiences, and near-death experiences are not purposefully excluded. It is just that those communities of interest seldom overlap into what I want to talk about, so there is little reference to them here. To some extent, the same is true of the many people who experience spontaneous after-death communication.

Historical Perspective

A good place to begin discussing our community is with the New Age community according to Tom. I grew up in Susanville, which is a small mountain town in Northern California. We had a typical small town radio station with a lot of country music, but on a good night, I was able to pick up a really cool jazz station out of Salt Lake City.

Becoming the New Age Community

I fancied myself as a beatnik, and that self-identity followed me to Eureka, California where I attended high school. As I remember, Eureka bills itself as the *Heart of the Redwoods* and *Rhododendron Capital of the World*. That sense of beauty and art fed my beatnik mentality. Yes, I had bongo drums! In fact, I was distracted enough by art and fear of nuclear war that I would not have graduated from high school in 1961 had it not been for architecture and commercial art classes.[38]

The Hippy generation was in full bloom when I was discharged from the Air Force in 1966. Had I not been sequestered in that military world for four years, it is probable that I would have been seen as a hippy. But as I returned to civilian life still conditioned by military discipline, the beatnik in me was suspicious of the apparent lack of responsibility the hippy subculture exhibited.

All of this is important to this account, because I was able to watch the hippy subculture evolve into the New Age culture in the 1970s from the perspective of a far simpler worldview. By the time people began talking about the New Age, I had already been involved in the study of metaphysics for many years. This gave me a pretty good understanding of the fundamental concepts.

I like to think of New Agers as a community of like-minded people with a culture of seeking, rather than the more individual-centric culture of the hippy generation. As with paranormalists, a New Ager is apt to come across as a naïve idealist, but the New Age community still cultivates important ideas which represent our finest qualities.

The New Age Culture, then, represents the basic ideals that nature is a living organism (Gaia), that we are all connected as citizens of the greater community, that our individual potential is unrealized, and that it is important to find ways to access that potential. I think of this point of view as human potential. In comparison, the paranormalist community is a blend of human potential (psi functioning), survival, and transcommunication.

Enter Spiritualism

Okay, yes I know, spiritualism started way before beatniks or any of these modern subcultures. It is reasonable to say that the spiritualism I am concerned with here began in the last half of the 1800s. Most of the culture, religion and precepts were established prior to 1900, but shortly after 1900,

[38] Here is an example as to why Wikipedia is so bad for our culture. At the time of this writing (2015), the Wikipedia article for *Beatnik* was written by someone who probably detested the whole concept. (76) While I expect the subculture had different meaning for others, this little boy from a logging town saw it as a culture that evolved out of Emerson's Transcendentalism. Jazz was a conduit to soul and mainstream culture was not to be trusted.

spiritualism as a religion began to withdraw, and today, it represents a relatively small community.

Although organized spiritualism has diminished, well over fifty percent of people in our Western culture are spiritualists at heart. (77) Think of spiritualism as the belief in survival after bodily death and communication across the veil and then think of the New Age community as being focused on human potential. That is similar to the distinction I make in the Science Commentary between parapsychology being mostly focused on extraordinary human ability, and those of us who are focused more on survival-related phenomena.

Organized spiritualism claims quite a long list of phenomena as the *phenomena of spiritualism*, but from a more contemporary perspective, many of the phenomena are better seen as psi functioning or human potential. As a practical matter, any form of trans-etheric influence thought to be initiated by a discarnate personality is a phenomenon of spiritualism.

For this discussion, the important point is that the combined influences of spiritualism and the New Age culture have produced our present paranormalist community. It is characterized by the belief in human potential and survival, but as I will explain later, communities of interest within the paranormalist community lean more toward one or the other.

Paranormalist Community

The paranormalist community includes five main communities of interest. It is important to note that a person might be involved in all five:

1. **Parapsychology:** Academically trained, usually with doctorates, concerned with theorizing and conducting research, primarily considering the human nature-based response to phenomena, especially psi field-related phenomena.
2. **Citizen Scientists:** People who are involved in the study of these phenomena on the layperson side of the academic-layperson partition.
3. **Mediumistic Practitioners:** People who apply techniques to induce transcommunication, such as mental and physical mediumship, automatic writing and ITC. Healing intention is included here.
4. **Hobbyists:** This community of interest is primarily composed of ghost hunters. There are hundreds of ghost hunter clubs which are usually operated as social clubs with haunted house investigations

as outings. They usually employ quite a lot of technology to collect *evidence*, including recording for EVP.

5. **Seekers:** People who are interested in understanding these phenomena as it relates to their true nature and relationship with reality. Such concepts as personal improvement and human potential apply to this community of interest.

Each group represents a very different point of view, and while each group is relatively self-sufficient, all must be considered for the proper study of these phenomena.

Parapsychology

A parapsychologist holds a degree from an accredited college or university, usually a doctorate. Accreditation is an important criterion because there are organizations that teach classes in parapsychology and even offer degrees or certificates, but which are not recognized within mainstream academia.

Parapsychologists usually hold a degree in psychology; however, some do hold technical degrees. This is an important consideration which is discussed later in the Science Commentary.

While parapsychologists are technically paranormalist and part of the community, there is an academic-layperson partition that effectively isolates them from the rest of the community. Despite this isolation, and probably because people are culturally indoctrinated to respect scientists, it is common for people in the paranormalist community to look to parapsychologists for learned direction in all things paranormal. At the same time, the focus of parapsychology is typically on extraordinary human ability, specifically as related to psi functioning. In research, phenomena thought to be related to Trans-Survival are typically studied by parapsychologists as psi functioning or the normal mistaken as paranormal.

The two contending explanations for paranormal phenomena are the Super-Psi and Survival Hypotheses. As a general rule, parapsychologists can be expected to select the Super-Psi Hypothesis over the Survival Hypothesis if psi functioning or a normal human explanation cannot be found.

Because of its academic roots, the field of parapsychology has inherited academic standards of peer review, collaboration, scientific methodology, and a culture of respect for academic authority. However, as is shown in the Range of Interest Diagram, below, mainstream science does not generally accept the findings of parapsychology, and in fact, any involvement in paranormal subjects can harm a scientist's mainstream career.

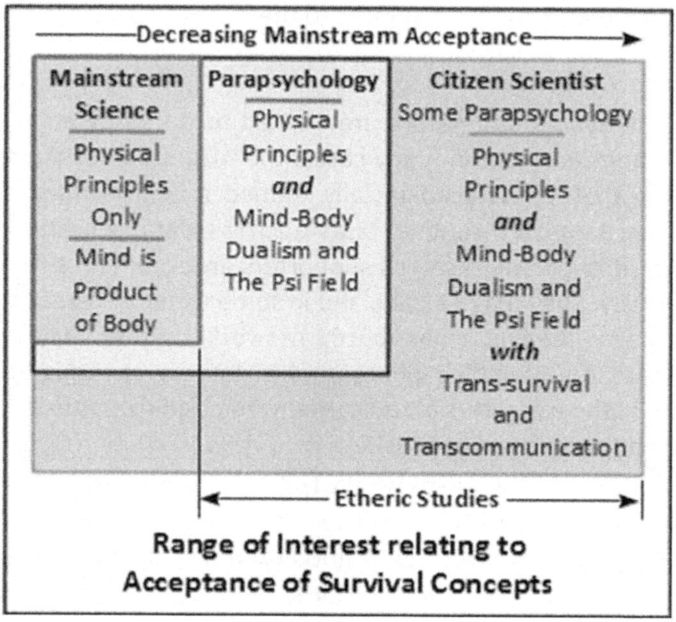

The mainstream academic system generally does not teach about paranormal subjects other than about the evils of pseudoscience and the many ways people can be delusional. As previously noted, most parapsychologists hold a psychology degree. I can only name one who has a second degree in technology, which I consider necessary for the study of ITC. The consequence is that most parapsychology research is restricted to human behavior else the researcher is working outside of his or her training.

While possibly not qualified to study the hard science side of paranormal phenomena, parapsychologists typically do have the necessary training in methodology to conduct research that is generally lacking on the layperson side of the partition. With that said, paranormalists are frequently the best informed people in the community when it comes to working with these phenomena. Even when compared to parapsychologists, people who have studied things paranormal for many years and have participated in numerous paranormal related activities are often the most knowledgeable about the specifics.

It should also be noted that a paranormalist may hold an advanced degree in an unrelated field and approach the study of things paranormal as an avocation. Coupling corporate savvy and probably unrelated technical skills with first-hand field experience often produces very qualified investigators, even if they do not refer to themselves as parapsychologists.

Citizen Scientist

Ideally, people who study nature are trained to do so by an academically certified organization, but in many cases, the magnitude of a research task may be such that a few academically trained scientists must depend on volunteer assistants for routine tasks such as data collection and field support. These layperson research supporters are sometimes referred to as citizen scientists. They may be paid, and in some forms of research, it is their computer that is used in a distributed network, or they may provide the necessary eyes for searching photographic data or counting species. For comparison, in the early days of radio, many ham radio operators could have been referred to as citizen scientists.

There are usually few academically trained scientists in frontier fields, in which case, virtually all of the organized study is conducted by enthusiasts. Hauntings investigation groups are a good example, and of course, the study of ITC has historically been shunned by academics, making it necessary for citizen scientists to fill the resulting void.

A characteristic of psi and transcommunication phenomena is that they cannot be studied without an able practitioner. As is noted below, this means that most research protocols depend on the use of citizen scientists as qualified practitioners.[39]

Mediumistic Practitioners

A person who has translated interest in a subject into some form of commercial or personal practice is considered a practitioner. Practitioners in the paranormalist community are mainly concerned with various forms of

---

[39] A distinction needs to be made between ghost hunters as hobbyists and hauntings investigators as citizen scientists. For the hundreds of ghost hunting clubs, an evening looking for evidence in a presumed haunted location is a fun outing. While some valuable information about the phenomena may come from such outings, such information is seldom reported in a form and venue suitable for collaboration.

Some investigations involve a protocol and a well-written final report. The investigators would be considered citizen scientists if the final report is made available to parapsychologists and other citizen scientists in a form which permits critical examination and in a forum which permits questions and answers. Admittedly these are my criteria.

mediumship and psi functioning, but this group also includes people working with ITC and healing intention. The main consideration for inclusion in this community of interest is that the practice is based on the person's desire to work with objective phenomena.

These phenomena are considered forms of transcommunication in that they suppose the influence of intentionality on the psi field and trans-etheric information access. Practitioners may begin with a natural talent, but study and work developing a personal methodology is typically necessary. Without such people, these phenomena cannot be studied under controlled conditions.[40]

Hobbyists

There is a progression of interest in things paranormal that often begins with a casual interest in something like ghosts which leads to membership in a local group. In this way, local groups can be a person's first introduction to the reality of things paranormal. For a few, this becomes a life-long study, and it is this progression that the community can potentially nurture.

Ghost hunters are like-minded people who gather together in local clubs to *investigate* a reportedly haunted place in an effort to collect evidence about the cause of the haunting phenomena. Some groups approach the *hunt* with formal protocols and produce meaningful reports, while others mainly just seek to have fun.

Even though there are substantial differences between recreational ghost hunting and serious investigation of supposedly haunted places, both groups are generally discounted by the more academically trained members of our community. This has produced something of a vacuum of leadership which has forced the less well-trained paranormalists to their own devices. The result is less than academic-quality research and reporting, and the isolation of a very valuable research capability from researchers who are trained, but lack access to the necessary practitioners.

In a few rare instances, parapsychologists have embedded themselves with paranormal investigation groups to produce important research.

[40] Without a doubt, there is a natural distribution in native ability. With the limitations of current teaching methods, some people are simply better able to develop into confident practitioners than are others. With that said, with study and a lot of work, most people should be able to achieve some success in working with these phenomena.

Seeker

A seeker is defined here as one who consciously endeavors to gain understanding that will optimize wellbeing and that might improve the ability to live in harmony with nature. Concepts such as *mindfulness, being a good citizen, we are all connected*, and *thoughts are things* typify the seeker point of view.

To be clear, we are all seekers. It is just that most of us do not consciously seek better understanding, wellbeing, and harmony. The seeker community of interest within the paranormalist community is not normally associated with a relatively large group of casual seekers [189] who are possibly just entering the paranormalist community, but have yet to develop the necessary knowledge base for discernment.

Peripheral Influences

There are a number of social groups which are not considered paranormalists but which have an important influence on the evolution of Etheric Studies. Surprisingly, the least organized is probably the most influential, so I will begin there.

People Who Believe

A popular survey question has to do with whether or not a person believes in some aspect of the paranormal. In a 2005 Gallup article, "Three in Four Americans Believe in Paranormal" (77), it is noted that *"about three in four Americans profess at least one paranormal belief."* At first glance, one would think this implies strong support for Etheric Studies, but this belief is conditional.

People are very good at compartmentalizing their worldview so that beliefs that contradict objective understanding can coexist with a person's objective view of reality. For instance, it is common for a usually pragmatic person who does not believe in paranormal phenomena to still be a religious person who believes in God and Divine influences in life. With this in mind, it is reasonable to think that many survey respondents answered from the *religious compartment* of their worldview and not from their objective perspective. For the purpose of this discussion, consider this casual belief.

A second form of casual belief is whether or not respondents have made a conscious association between belief in an aspect of the paranormal and the implications of that belief. For instance, it may not occur to a person who believes in ghosts that the ghost might be a visiting loved one. In a similar way, belief in intuition may not equate with belief in the possibility of psi functioning.

As opposed to casual belief, belief as conviction translates into willingness to study the subject or in some way support the study. We are not aware of any surveys designed to differentiate between casual belief and conviction, but from our experience, the number of people who are convinced some aspect of the paranormal is real is considerably smaller than seventy-five percent. When referring to the paranormalist community, I am including people who are willing to put their beliefs into action and not casual believers.

Both casual and conviction believers have a substantial influence on the media because of marketing potential and on the greater community when opinion is translated into funding of projects. As is noted in the discussion about pseudoscience, it is this group of people which stands between paranormalists and potential legislation designed to limit research. See the Psi Research Commentary.

Casual Seekers

It can be argued that all of us are seekers of understanding, but in this book, I am identifying seekers as a community of interest in the paranormalist community. They are interested in understanding things paranormal so as to better understand their nature and relationship with reality. Such concepts as personal improvement and human potential apply to this community of interest.

A way of generalizing this group is to think of what one must understand and do to be a good citizen in society. A basic understanding for seekers is that we are all connected on both sides of the veil. This often translates for seekers as being a good citizen in the greater society.

Paranormalist seekers tend to focus on a particular path of learning. For instance, one might be primarily involved in the study of and working with transcommunication. To be *on a path* generally means to be learning to be mindful and to practice discernment.

Seekers that I do not consider part of the paranormalist community are those who, for the lack of a better word, *dabble* in the study of things paranormal. As a longtime observer, I speculate that these casual seekers involve themselves in paranormal subjects because of curiosity and/or while seeking personal improvement without regard to their citizenship in the community. In time, many casual seekers do settle on a path of learning and become dedicated seekers.

There are no numbers I am aware of for the relative size of this group, but if mainstream society represents 100% and 75% of those who believe in some form of paranormal phenomena, then casual seekers are probably 2-3% and paranormalists are probably less than 0.5%.

Religion

The most fundamental criterion for a paranormalist community of interest is that it is not intentionally associated with faith or magical thinking. With that said, it is very difficult to completely transform the way traditionally religious concepts are described into new terminology that does not imply the usual religious concepts. The term *spirit* is probably the most difficult to replace, but others such as *death* and *angels* are also concepts carrying substantial religious baggage.[41]

The traditional influence religion has on paranormal phenomena has had and continues to have a substantial influence on paranormalists. For instance, it is common to hear a paranormalist speak in objective terms about a mediumistic ability, but then in the same sentence, describe it as a gift.

Skeptics

While many people have healthy skepticism about paranormalist claims, there is a small segment of mainstream society who have taken the name of *skeptics* and who are aggressively anti-paranormal. After a few years of attempting to balance paranormal articles in Wikipedia, and being more or less constantly abused by skeptics, I have come to understand that there is a point of view in mainstream society that has assumed the moral authority to combat the spread of pseudoscience, AKA, belief in all things paranormal.

The skeptic point of view is that, if science does not specifically account for a phenomenon, then it cannot be, and therefore it cannot exist. Of course this is scientism.

---

[41] Interestingly, the same problem occurs with terms used in science. For instance, the science-based meaning of *energy* is not applicable in purely conceptual etheric space, yet it is the term one reaches for when trying to talk about potential. With energy not having the usual meaning in etheric terms, concepts such as electromagnetic and frequency also do not apply. One knows what is intended, but using them leads people off on dead-end ways of thinking.

The nature of the skeptic threat to the paranormalist community is illustrated with this quote from the "Health and Education Implications" section of the Pseudoscience article in Wikipedia:

> *Pseudoscientific beliefs are irrational and impossible to combat with rational arguments, and even agreeing to talk about pseudoscience indicates acceptance as a credible discipline. Pseudoscience harbors a continuous and an increasing threat to our society. It is impossible to determine the irreversible harm that will happen in the long term. In a time when the public science literacy has declined and the danger of pseudoscience has increased.... (78) (current as of January 2016).*

When this point of view is translated into action, it could potentially lead to limitations on paranormalist ability to work with these phenomena. See the Skeptics and Wikipedia commentaries.

A More Pragmatic View of the Community

Community is parsed here as a functional view which I have learned through experience. Each part of the community is ultimately dependent on the others. Each part has a role to play if the reality of psi functioning, survival, and transcommunication are to become part of mainstream thought.

While I speak of community support for research, the actual support is coming from a few people who are interested in the science and want to support the work. These are generally successful, well-educated senior citizens who make it their business to be knowledgeable about these phenomena. Yes, of course, there are many more generally successful, well-educated senior citizens who can be found on the non-academic discussion boards, but most have not exhibited the discernment, understanding, or ability and inclination to help fund research.

Look around the places you frequent on the Internet. Are you seeing pragmatic discussion amongst discerning, knowledgeable people? Can you get a sense of informed participants? If not, perhaps it is time to reconsider how you spend your time.

The point of these comments is that the intellectual minority is supporting research and giving parapsychologists a reason to do their work. The next part of these commentaries is titled *Science*, and while much of my commentary there is in the form of complaining about parapsychology and the academic-layperson partition, it is intended to focus on what needs to be improved.

Personal Agendas

The New Age community was probably at its greatest in the late 1970s. Today, *New Ager* is something of a condescending term reserved for people who are willing to believe in *anything* that might unleash their hidden potential. At the same time, there are many people in the New Age community who will sell anything from blessed rocks to special treatments guaranteed to open chakras.

Care must be taken here not to be too general. Certainly some techniques, devices and beliefs are effective, if only to condition expectations so that the client can create the desired reality. However, there is much about the New Ager selling to New Ager culture that gives the rest of us a bad name in the eyes of mainstream society.

In the first draft of this book, I argued in this commentary that the commercial influence had not yet affected our community; however, it is clear now that I was ignoring the evidence. When Lisa and I assumed leadership of the ATransC, my focus was on EVP, and later on ITC. Sure, there were a few courses offered for a fee, but the devices used for ITC were mostly repurposed commercially available equipment. While most of what was offered for sale was information, most of that was freely available on organization websites such as atranc.org.

Between 2000 and about 2007, it was easy to gather people who were eager to experiment with ITC. Most of the conversations found on the still young Internet were about comparing experiences, sharing examples, and helping one another work with techniques. After 2007, our ATransC membership began to diminish, as did participation on paranormalist discussion boards around the Internet.

Facebook was introduced in 2005 and it was really catching on by 2007. Social media has had a profound influence on how people share ideas and satisfy their need for community. This has led to diminished quality of communication as here now and gone tomorrow posts provide the illusion of community while reducing it to a new kind of inane babble.

A strong supporter of the ATransC recently used Kickstarter to help fund a new television series on survival. His target was $500K. While I thought it was a little high, I also thought it might be possible considering the thousands of paranormalists in our community. As it turned out, 35 people pledged a total of $10,277. (79)

Considering the number of paranormalists that I am sure knew about the fund drive, such a poor turnout was unexpected. Was it just that people were lost in social media or was there a more practical explanation? Why does ATransC.org have six or seven hundred website visitors every day but find it impossible to attract participation on the discussion board or help with research? Why are many other discussion boards virtually abandoned?

It is clear that many people seek information about things paranormal. But few of these same people make an effort to further the understanding of others, or even to seek better understanding by asking questions. I understand that better educated science buffs tend to stay out of the public eye and that parapsychologists do not fraternize with laypeople, but what about the other ninety-nine percent of paranormalists?

If you pay close attention to who is posting in the paranormal oriented Facebook pages, you will notice that many, probably most, people posting are mediums, authors, or people teaching a paranormal oriented course or selling products. Some are also focused on spreading the word about a favorite theory. In short, much of the participation in online paranormalist circles is by people who have an agenda other than seeking understanding.

Here, I need to make it clear that I am one of the people teaching and promoting an organization. Probably like most other people promoting something, I am involved with these things because of a strong interest in all things paranormal and a desire to see the community mature. There is nothing wrong with being active on the Internet to make a point. I am the last person who wants someone to do something for free. The effort required to serve others warrants a reasonable fee, and certainly time and effort should be compensated.

It is also important to note that it is the people seeking to promote something who are building the infrastructure for the paranormalist community. The problem is that most of what is being built today are islands of information that seldom connect with or even relate with the larger community.

My point is that, while the Internet looks like a very busy place, many of the active people are seeking to do or teach and not seeking to learn. The seekers and intellectuals are mostly absent. While many people are fans of my friend's efforts to produce a television series we would all benefit from, hardly anyone is willing to pitch in a dollar to help out. While everyone is happy to read the four hundred plus pages on ATransC.org, virtually no one is willing to help out or even ask a question.

The only conclusion that seems to answer the evidence is that the paranormalist community is following the way of the New Age community as paranormalists seek to sell to paranormalists, rather than do what it takes to move understanding of these phenomena out of the frontier and into mainstream society.

In a very real sense, ours is not a cooperative community [64].

Blind Men and an Elephant

In the Blind Men and an Elephant parable, blind men were tasked with describing an elephant based on the part near which they stood. As you might expect, the elephant was variously described as a tree, wall, spear, even a snake. The point of the parable is that one cannot know what is being studied by only studying a part.

In a very practical sense, mainstream science, parapsychologists, practitioners and people involved in survival research are as blind men examining the elephant-paranormal. This blind examination of the phenomena is exasperated by the academic-layperson partition but it is compounded by a tendency of laypeople to assume understanding they do not have.

Natural Cycles

An important concept for seekers is the idea that progression tends to be cyclic. The same can be said for communities. It may take a new generation for meaningful progress to be made, but it is a reasonable certainty that future history will describe today's paranormalist community as a short-lived transition between superstition and mainstream understanding of the greater reality.

The Implicit Cosmology will not be the last such attempt to facilitate this evolution, but for now, it is my vision that you will take what you learn here and help develop the paranormalist community. Progression takes a community. Help others to help yourself.

Community

Commentary 2:
Point of View

Introduction

Point of view is our *I think the world looks like this* perspective. While I talk about the etheric personality perspective in the Personality-Centric Perspective Discourse, this commentary addresses point of view from the physical body or body-centric perspective: that is, from the perspective of our conscious self as it is informed by our immortal personality, cultural education, and human body instincts. This is the *I think I am this* identity we are accustomed to thinking of as who we are. In the Implicit Cosmology, the immortal personality is modeled as our mostly unconscious *I am this* identity.

Our point of view determines the community we gravitate toward and the kind of citizen we become within that community. Point of view is determined by worldview. Worldview represents what we believe to be true, and because it has such a profound impact on our lives, it is important to know what we believe and why. This commentary is intended to provide a sense of how we become who we are in terms of attitude about life, how beliefs affect a lifetime and how we might take charge of these very important influences.

The second aspect of this discussion concerns understanding the point of view of other people. Since organizations and communities tend to develop a collective personality, understanding point of view will also help you understand how to relate to these entities.

Models for Reality

There are three well-established models or theories designed to explain who people are and their relationship to the world. (80) These models are:

Personality and body are one: This is the mainstream view in which the mind is believed to be the product of the brain. The mind exists entirely within the brain and all functions of mind can be attributed to processes within the brain.

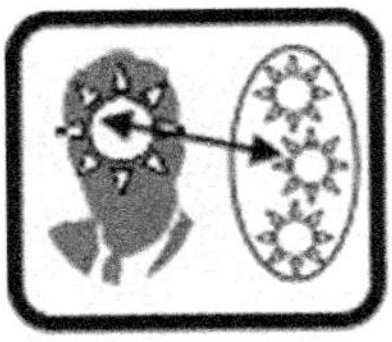

Personality and body are one and the psi field exists: This is the same as *mind and body are one* with the exception that it allows for the existence of a psi field which permeates the entire physical universe. Each person is thought to be connected to the whole by this field, and through psi functioning, is able to sense it.

In this view, personality is a product of brain, but memory of that personality remains in the psi field and is accessible via psi functioning.

Personality and body are independent: This view requires that there is a nonphysical aspect of reality, referred to here as the etheric. This view suggests that the physical body evolved in the physical and hosts the etheric personality which evolved in the etheric.

The *personality and body are independent* model best answers current evidence and is used throughout the Implicit Cosmology. In general, the physical body is considered an avatar for the etheric personality. In this view, personality requires the physical body in order to directly experience the physical. While the etheric personality is associated with the physical body during a lifetime, its perspective is as conscious self peering at the world from within the avatar's head. Here, a person is defined as a personality entangled with an avatar.

What We Are Born With

According to the Implicit Cosmology, we came into this lifetime with the urge to gain understanding about the nature of reality through life experiences. Our etheric personality has probably experienced many different venues for learning, and so we come into this lifetime with a certain amount of acquired understanding which prepares us to better understand the nature of new experiences. In effect, we are born with a hunger to learn and a preexisting inclination toward what we will learn. This urge guides us in life choices which might lead to beneficial or at least meaningful experiences and helps us make sense of them.

The Personality-Centric Perspective Discourse includes a description about how the environment, body consciousness, worldview, and experiences influence perception. Because point of view is usually physical body-centric, the ability of our etheric personality to influence behavior is very limited. As is illustrated in the Perception Diagram [xxiii], the path of learning for personality is via the Perceptual Loop as it is limited by worldview in the Attention Complex which supports perception and expression.

Worldview

The Attention Complex in the Perception Diagram [xxiii] is an unconscious process. It represents how environmental information is translated into conscious awareness. Note that the Perceptual Loop is moderated by worldview and functions as a filter that stands between a person's *I think I am this* perspective of the conscious self and the *I am this* perspective of the person's etheric personality. What we think of sensed information is what emerges into our awareness after it is moderated by worldview. As such, worldview represents personal reality because it determines how we understand experiences.

A person is born into this lifetime with a degree of understanding inherited from its collective, and perhaps gained through prior experiences in other venues, plus an urge to gain further understanding about the nature of reality. However, at birth, worldview is dominated by the human instincts of its avatar.

Worldview is quickly populated with information happily provided by our parents, teachers, religious leaders and community. We are taught what is right according to our culture. We are not born with a moral guide, and therefore, might not have the presence of mind to recognize attitudes taught about other people as prejudices. Conscious self's ability to influence worldview is also limited by the Perceptual Loop, meaning that cultural norms may be all we know.

As the perceptual process is modeled, if someone tells us something like *"New York City is a great place to live,"* the information is unconsciously visualized and compared with worldview in the Attention Complex. The unconscious process would likely relate the statement with how we currently live and how we would like to live. Because this is an unconscious process, our reaction to the statement might emerge into our conscious awareness with agreement or perhaps an alternative view without our realizing how we arrived at that response. Even more importantly, if the statement is reasonably close to what we previously believed, it will likely go toward modifying our worldview. Perhaps this would be experienced something like *"So and so thinks that...."*

Temperament

As is discussed in the Personal Style and Astrology [119] portion of the Personality-Centric Perspective Discourse, the concepts of personality styles and astrological influences may help us understand our temperament. The idea is that we are born with a temperament which stays with us through the lifetime.

Personality styles we appear to be born with are typically describe din terms we would use for rational thinking such as analytical or amiable. Astrological traits are described in terms more appropriate for animal instincts such as need for shelter or compatible with another sign.

You may remember that a basic question in psychology is whether temperament is inherited from a family member (nature) or learned while a child (nurture). Considering Sheldrake's Hypothesis of Formative Causation, temperament can be at least partially modeled as Nature's habit established by preceding generations. A refinement of Sheldrake's hypothesis offered in the Trans-Survival Hypothesis is that the entangled personality contributes the urge to gain understanding to a person's temperament.

Point of View as Influenced by Temperament

If we come into a lifetime with an urge to learn and a predisposition to prefer certain experiences, then the temperament models may be an indicator of the nature of that predisposition. Our worldview helps to shape attitudes, but that influence is moderated by our predisposition or the temperament with which we were born.

For instance, a community may teach its citizens to be deeply afraid of demons, but if the temperament of a member of that society is to question authority, it may be natural for that person to turn that fearfulness into a curiosity as to why demons are fearful. Upon discovering there is nothing to fear, temperament might help to modify worldview's influence, permitting the person to be more accepting of the unknown.

So it is temperament or predisposition to select learning opportunities that determines point of view. In turn, being aware of point of view helps us to better understand who we are and why we do what we do. This in turn changes automatic reactions to situations into more deliberate responses. We will experience and learn from those experiences without deliberate control, but being truly self-aware means knowing why. In that way, it is possible to align worldview and point of view with nature.

Villager-Explorer Effect

This is more commonly known as the Sheep-Goat Effect, but I do not much like being called a sheep. In 1942, Professor Gertrude Schmeidler identified a correlation between people scoring high on a belief in the paranormal survey with their psi functioning scores. Conversely, a poor *belief* score correlated with a lower than chance psi function score. She referred to this as the *Sheep-Goat Effect* with *believers* as the sheep. (81) As noted, for my self-esteem, I am referring to the *believers* as *explorers* because of their willingness to explore new ideas. Conversely, I am referring to disbelievers as *villagers* because of the more conservative attitude of people who follow cultural norms and depend on maintaining the status quo.

This concept implies that people who allow for the possibility of new ideas are more likely to experience new ideas. The inverse of this can be seen in the concept of incredulity blindness, a term I use to describe how some people are inexplicably unable to see or hear examples of paranormal phenomena. See the Perceptual Agreement Discourse.

Even if a person is skeptical, suspended judgment as an openness to the possibility of the unknown offers far more opportunities to experience phenomena. The Attention Complex which supports perception is designed to make decisions about what is real and what is not. It is also designed to characterize what is real in terms familiar to the person. In this same model, intention expressed by the conscious self is the means by which the perceptual process can be evolved. As such, following the Villager-Explorer Effect, the intention to withhold judgement about experiences enables more experiences. See the Perception and Expression Discourse.

Engineering Imperative

At least in fields of study concerned with the physical sciences, there is a hierarchy of skill in which scientists expand knowledge, engineers apply knowledge and technicians implement plans. This is a generalization, but one which is of practical importance to this discussion.

In a particular subject or skill set, neither a scientist nor an engineer can be expected to have the practical expertise held by a journeyman technician. In frontier subjects, technician-level skills might be equivalent to psi, medium and ITC practitioners. In frontier subjects, practitioners are often the most knowledgeable about the subject even though they may not be prepared to assess the paranormality of their experience.

For the most part, an engineer holding a Bachelor's of Science degree has been trained in the same subjects as a doctorate in a physical science. Doctoral programs are generally more detailed and some include training in many tasks related to research design, data collection, analysis and research documentation.

Engineering curriculums are designed to introduce the physical concepts necessary to understand involved principles with emphasis on application. From personal experience, this means that an electronics engineer should be able to identify known physical principles that might be involved in research data.

Of course there are many life circumstances that help to determine how far a person progresses in college, but experience has shown that there is a substantial difference in point of view between people who choose to be an engineer with a four-year degree and those who go on to earn a doctorate. In my experience, this difference becomes a factor when it comes to characterizing phenomena. Doctorates depend on precedence and the guidance of existing theory, and while engineers do depend on precedence, they are also obliged to account for observed circumstances without regard to their theoretical nature.

Put in terms important to the study of frontier subjects, mainstream scientists are taught that everything must be defined or at least predicted by existing theory. If an observed incident is not covered by known principles, it is likely to be discounted as either an illusion or fraud. For instance, statistical outliers are routinely discarded during data analysis. This dismissive point of view is demonstrated over and over again in the struggle for *truth* between skeptics and experiencers in frontier subjects with skeptics saying, *"It is impossible and therefore cannot be."*

This is difficult to characterize without over-generalizing. The point is that people with an engineer's point of view do not have the luxury of discounting phenomenal events just because they do not comply with known science. In this sense, there is something of an *engineering imperative* to account for the objective influences as well as the human factors constituting the circumstances of a design. If people experience it, then it must become a factor in the design.

In response to this imperative, the engineer's approach to the study of frontier subjects can be expected to begin with the development of a hypothesis based on the assumption that the phenomena exist. In a practical sense, doctorates are not allowed to make that assumption. It is critical that people working on either side of the discussion remember that this difference in point of view is a huge hindrance to meaningful communication across the academic-layperson partition [214]. (Of course, an engineer with a Bachelor of Science degree is seen as a layperson.)

In the spirit of Thomas Harris who told us in his book *I'm OK, You're OK* that real communication between two people could only occur if both are in an *OK* place in life, communication can only occur if both sides agree that phenomena have occurred. (46)

Point of View of Organizations

People have a point of view, but organizations are composed of a community of people. While personality style (temperaments) is a way to model the point of view of people, a similar model might be used for organizations based on typical goals, objectives and assumptions. From the perspective of Etheric Studies, an organization might have one of the following points of view:

> **Human Psychology:** Usually parapsychology; specifically, an academic approach with emphasis on a physical-world perspective;

emphasis on research; and collaboration but guided by community norms. This is the parapsychology community of interest.

Emergent Science: Emphasis on an academic approach; deliberate openness to new ideas; and open to concepts of nonphysical and survived personality if they can be explained with mainstream science. This is the parapsychology community of interest.

Human Potential: Usually academic; open to nonphysical explanations; focus on personal growth and the human condition; supports research; and education by opinion setters. This is the parapsychology community of interest.

Social Paranormal: Desire for scientific approach; for many, often first experience in paranormal; strong community; emphasis on nonphysical and survived personality; little formal collaboration; and, education by emergent cultural norm. These are the Trans-Survival and transcommunication communities of interest.

Metaphysical: Study/research of phenomena related to trans-etheric influence; desire for scientific approach; community emphasizing education and application of organizing principles; and specifically open to nonphysical and survived personality. These are the trans-survival and transcommunication communities of interest. Parapsychology is marginally involved.

Naturally, the Metaphysical point of view best suits Etheric Studies; however, it is important to emphasize that this list is conceived from the ATransC perspective. Other points of view may be appropriate for the representative organization's goals, but it is important that people understand there are significant differences between organizations based on mainstream thought and those allowing for possible postmortem survival.

Litmus Test

As a general observation, members of the parapsychological community of interest are less than candid about their point of view. As a professional choice, some will do almost anything to avoid admitting a preference for Trans-Survival, Super-Psi or mental aberration, I suppose, so as to avoid seeming unscientific.

There is little doubt this book is written from the point of view that the Trans-Survival Hypothesis is valid. So it should not surprise you that I use a litmus test of sorts when I consider the usefulness of organizations, theories and research reports. If Trans-Survival is not even considered in the discussion, my first reaction is that the person will not present an honest assessment of these phenomena. Anomalistic psychology [218] is a good example of this.

Community

Commentary 3:
Etheric Studies

Introduction

Even a cursory review of the scientific literature will show that survival of the personality after physical death (Trans-Survival), and an aspect of reality in which personality survives (the etheric), is not well studied by the scientific community. The field of parapsychology typically stops short of the study of survival as a possibility, tending instead to study it as probable residual memory or illusion.

The subjects of near-death experiences, reincarnation, and out-of-body experiences are also studied in parapsychology, but little from that community of interest reaches those of us interested in survival.

There remains a substantial body of experiences and empirical evidence that is mostly ignored by the academic community. It is this systematically ignored information which is the focus of Etheric Studies.

Scope of Etheric Studies

The relative range of interest of physical, psi, and Etheric Studies is shown in the Scope of Etheric Studies Diagram below [206]. The physical aspect of reality would be just a subset of all of reality. Anything other than physical reality, or that which is not considered physical, is etheric.

There is some form of etheric-to-physical interface which is probably being experienced via the psi field. Psi studies is treated as a stand-alone field of study, but in this cosmology, it is treated as a subset of Etheric Studies.

In fact, the physical is also part of Etheric Studies because, in the Implicit Cosmology, reality is modeled as consisting of life fields and the expression of life fields. Reality itself is modeled as a life field called Source. As such, Source is the reality field and all of the elements in the reality field including the physical are either direct or indirect expressions of Source. See the Perception and Expression Discourse.

As life fields entangled in an avatar relationship with a human, we express reality as we understand it. Here, it is important to note that the physical is seen as a collective expression in which we are both one of the creator personalities and one of the experiencers.

Reality is modeled as conceptual in form and only experienced as objective because the mind is trained to experience it as such. Hints via various forms of transcommunication indicate that our sense of what is objective will evolve as we move on to other venues for learning.

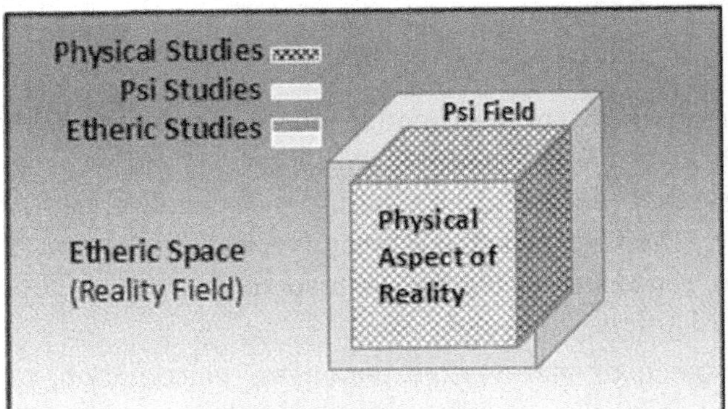

Scope of Etheric Studies

With these considerations, physical characteristics such as evolution and great planetary age are artifacts of how the physical was conceived. As it is modeled, worldview, which represents our personal reality, resists large changes, but is responsive to small, incremental changes. We create based on habit, so evolution of the physical is a reasonable artifact of how the collective learned to experience the physical.

Why the Name, *Etheric Studies*?

Most of the people involved in the study of survival from the perspective of ITC are citizen scientists. Because of this, the actual study of survival is from a very different point of view than typically expressed in parapsychology. This disconnect in points of view is the reason it is always important in report writing and conversation to clarify which point of view is being expressed. To facilitate conversation, a name was required to distinguish that part of the paranormalist community which is concerned with the study of Trans-Survival as possibly real from that part which studies it as probably not real.

One of the requirements for a name is that it is easy to use, not a new term, and recognized by the average person. For instance, when someone hears the term *things etheric*, he or she at least understands that the subject has something to do with the nonphysical.

Terms, such as *meta science* turn out to be too arcane, and anything that is specific to ITC or any other subject of the paranormal would not accommodate the other aspects of these phenomena. *Empirical metaphysics* was a good candidate at one time, since the study is primarily based on research and objective examples, but again, the term sent people scrambling for a dictionary. Simplicity is important for a household term.

Nonphysical was tried for a while. It works okay, but it is something of a physical-centric term while all of the models are etheric-centric. Surprisingly, fewer people have indicated that they have even an idea of what is intended by *nonphysical*, while *etheric* is easily related to the early efforts to define the greater reality. (82) Most people have been taught about aether in relationship with Greek philosophy in high school. (83)

It is interesting that even Albert Einstein speculated about the existence of a characteristic of space still best described as *ether*. (84) Even though the idea was set aside by later scientists, the need for an etheric characteristic of reality is often noted in quantum physics, although physicists will probably not refer to it as ether.[42]

Surely a better term for this field of study can be found, but by now, Etheric Studies is fast becoming established and changing it would cause much lost time. It seems best to settle on the current term for now and let circumstances lead us to whatever will become common usage as this field becomes established.

Who Studies the Etheric?

Anyone attempting to understand the nature of survival is by definition, involved in Etheric Studies. A mental medium or a person working with ITC is an Etheric Studies practitioner. A person who studies these subjects is an Etheric Studies researcher. A researcher may be a physicist or psychologist specializing in psi studies as a parapsychologist, but when studying the relationship between the medium and the etheric communicator, the researcher is working in the field of Etheric Studies.

The demarcation between physical, psi, and Etheric Studies is dependent on the subject. For instance, a theory proposed by Alexander MacRae (85) to explain the mechanism of etheric to physical influence is that the physical property first influenced by psi is the relative permeability of physical matter.

[42] What Einstein called the *Cosmological Constant* seems to have been resurrected as *Zero Point Energy* in quantum physics. This differential between absolute zero energy and what is actually measured in a vacuum at absolute zero temperature may represent the etheric-to-physical interface. This is a good illustration of how important mainstream science is to Etheric Studies.

(86) This hypothesis involves a conceptual influence organized by a person which produces a physical change. To evaluate this hypothesis, it is necessary to employ researchers with psychology, metaphysics, transcommunication, and physics skill sets.

In a different example, a proposal that there is a psi field which connects people is normally the domain of parapsychology (psi studies), yet when the question of how an etheric communicator influences that field is studied, Etheric Studies must be considered. The distinguishing assumptions here are the parapsychologist's preferred point of view that mediumistic messages are psi functioning only and the Etheric Studies point of view that they may be expressed by discarnate personalities.

Future

As a term, Etheric Studies is a placeholder until it has more popular acceptance or a more popular/useful name comes along. As a field of study, Etheric Studies and parapsychology should eventually merge, just as both will likely eventually merge into the mainstream with an even more appropriate name.

In the meantime, as citizen scientists try and fail to gain cooperation from parapsychologists, and as more one-sided research reports are published which include statements known to be wrong, naïve, and/or harmful to the community, the academic-layperson partition becomes a *"Do it yourself"* signpost for citizen scientists.

Community

Commentary 4:
Science

> As you read this commentary, please remember that it is written from the perspective of a non-academic person looking over the wall of the academic-layperson partition. My comments are a little strident because I look for meaningful direction from the academic community, but instead, daily deal with its absence. Your experience may be different—but is it reality?

Introduction

Hands down the most disruptive influence for the paranormalist community is science. When Lisa and I assumed leadership of the ATransC in 2000, one of our first goals was to engage science in our effort to understand Electronic Voice Phenomena (EVP). From all of our efforts since that naïve beginning, practically the only thing we learned from science has been why we should give up trying to work with the established scientific community.

Now to be clear, science as a philosophy is still king. It is only through disciplined study, codification of observations, theorizing, and hypothesis testing that humankind has been able to consolidate progress as civilization. The problem has been in how that noble philosophy has been applied on behalf of the paranormalist community. In short, too many people posing as parapsychologists under the cloak of authority of their doctorate degree have treated the paranormalist community as a plaything by ignoring best practices of science and ethics.

Cultural Landscape

In this commentary, I attempt to explain the point of view I have developed about science during my attempts to understand transcommunication. As a director of an organization that is chartered to support people who seek to understand and apply these phenomena, I see it as one of my duties to comment on the community dynamics which influences the paranormalist community.

Dominant Views of Reality

There are three primary subject areas which shape the paranormalist science landscape:

> **Psi Field:** Mind-body duality and psi field theory, as well as extraordinary human abilities such as psi functioning, which includes such abilities as clairsentience, intentionality, psychokinesis, remote viewing and healing intention. (Super-Psi Hypothesis)
>
> **Spontaneous Trans-Survival phenomena:** Near-death and out-of-body experiences, reincarnation and spontaneous trans-etheric influences which are generally referred to as After-Death Communication (ADC). (Survival Hypothesis)
>
> **Transcommunication:** This includes such induced forms of trans-etheric influence as induced ADC; mediumship, precipitation, apports, and both audio and visual forms of Instrumental TransCommunication (ITC). (Trans-Survival Hypothesis)

As I discussed in the Etheric Studies Commentary, the study of transcommunication is dependent on the study of psi. Conversely, the study of psi has a dependency on the study of transcommunication, in that it is unlikely the field will progress much further without including concepts associated with Trans-Survival.

Point of View

The dominant points of view are psychology, emergent or frontier science, human potential, social paranormal and metaphysical. Parapsychology supports all of these areas of interest but is more supportive of psi-related phenomena. The Society for Psychical Research (SPR) (87), Society for Scientific Exploration (SSE) (88), and Parapsychological Society (PA) (89) are the organizations I look to for support.

Organizations that anchor themselves with the concepts of transcommunication and survival include the NSAC Spiritualist (2) Association TransCommunication (ATransC) (90) and Academy for Spiritual and Consciousness Studies, Inc. (ASCSI). (3) These focus on phenomena that can be construed as the influence of a discarnate personality. These are represented above as Trans-Survival and Transcommunication communities of interest.[43]

Communities of Interest

The paranormalist community is fragmented into five communities of interest. These are Parapsychology, Citizen Scientists, Mediumistic Practitioners, and Seekers. They are discussed in detail in the Community Commentary. The focus of this commentary is science as it is practiced in parapsychology.

Names that Separate Science from Belief

It is natural for a person who wishes to be taken seriously by the academic community to rename phenomena so that what is being studied might appear less *woo woo*, to use a term favored by skeptics. An important example of this is the shift of perspective from studying spirit, which is a religious term, to studying psi, which is a more analytical term.

Such a change is more than cosmetic. Parapsychologists maintain that the term *psi* is a placeholder to be used until more is known about the apparently nonphysical, subtle field it represents. Spirit is a term that is often used to describe that subtle field, but the term's religious connotations unnecessarily complicates discussion of related phenomena. For instance, while spirit and psi may be different words for the same characteristics of reality, in a general sense, there is nothing about the way psi is defined to suggest it has any relationship with God, while spirit is commonly described as the stuff of God.

The use of *etheric* is also an effort to avoid *spirit* and to signify the greater reality, as opposed to the more local psi field. It is used as a characteristic of Source, but here I intend Source in a conceptual sense and not as a deity. Of course, the difference is probably nonexistent. It is just that I see support for none of the religious-related behaviors such as worshiping a father god. As defined, *Etheric* is also a superset of psi, or put another way, psi is just a local occurrence or aspectation of the more universal and less differentiated etheric space.

---

[43] **An encouraging change:** Mark Leary, Ph.D. (74) was asked to submit a section on EVP for the 2015 *Parapsychology: A Handbook for the 21st Century*. (91) Leary invited me to participate, which I happily did (thank you, Mark). This is perhaps the beginning of a new time for Etheric Studies which could bring changes negating much of what I complain about in this commentary.

The etheric verses psi choice of terms is a good example of point of view. As I have stated above, parapsychologists are reluctant to refer to heaven in any way. A religious person might use *etheric space* as a synonym for heaven. The *psi field* is often described as a characteristic of the physical. One might think of it as an aura of sorts. If so, then it would be a local aspect of heaven in a religious context. You can see the thin line scholars must walk to maintain an objective pose.

The shift in perspective from the terms used by experiencers to psi-oriented terms has changed the academic vocabulary so that academics no longer effectively communicate with experiencers. For instance, I have encountered the phrase *anomalous cognition* used when medium or clairvoyance was intended. *Biofield therapy* is a good one for spiritual healing, but here I am guilty as well because I use *healing intention* [369].

Normalizing Terms

One of the most important points to consider in the study of these phenomena is that two people may be discussing the same phenomena, even from the same point of view, yet be using very different terminology. On close examination, many systems of thought commonly assumed to be more religion than factual actually contain important insights that support modern objective observations. For instance, the idea of cyclic progression is expressed as Key 16, *The Tower*, in the Major Arcana of the Tarot (59), as *Dark Night of Soul* (92) in biblical text, and as the modern realization that people naturally learn in cycles. (93)

One reason for suspended judgment is that we understand what we experience based on our worldview. All three points of view mentioned above—Tarot, biblical, and modern education—are concerned with how we learn. Deciding that one or the other is wrong based on our perception of word meaning too easily robs us of an opportunity to gain insight. If we do not pause to normalize terms into the perspective of our point of view, we might well miss the point. In many ways, the path to understanding is by way of suspended judgment.

Theory of Everything

There are many Theories of Everything (TOE) that are specifically designed to explain these phenomena in terms of mainstream science. They are usually written in a way intended to convey academic rigor, but the result is too often incomprehensible to the casual reader. Few readers take the time necessary to understand what is intended. The result is that their often slick, usually persistent presentation may lead to wide acceptance of TOEs that contradict the beliefs of those who applaud the theory

The litmus test for a TOE is whether or not it allows for survival. Most do not and instead maintain that mind is a product of brain and ceases when the body dies. They may provide for some form of psychic functioning with retention of memory in some form of subtle energy field. These days, the subtle energy field will probably be described in terms of quantum theory. This form of survival is a fundamental concept in the Super-Psi Hypothesis and only provides for survival of memory, not living personality. To allow for Trans-Survival, a theory should provide for personality to exist prior to birth of the body and continue after death in an interactive, still evolving form.

To cut through the confusion, just ask if survived memory existed before the birth of the body. In one TOE, it is necessary to ask if it existed before physical world life began because the author argues that survived memory begins from the primordial soup.

A Close Look at the Science-Layperson Relationship

The definition of science is typically stated as the study of the natural world using theoretical models based on experimental data or observation. Scientists are people who have studied those theoretical models, methodically conducted experiments, and who have made observations to further refine those models.

In our society, a scientist is understood to hold a doctorate degree from an accredited institution. Certainly the quality and subject of education is important but immersion into the academic community introduces doctoral candidates to a culture in which critical thinking, careful attention to detail, literature reviews and collaboration is the norm. While this difference in culture is probably the reason for the academic-layperson partition, the discipline is necessary for a maturing community.

Some organizations offer a doctorate degree in exchange for a fee and perhaps a little academic effort. As I understand it, a degree purchased from an organization that does not have mainstream accreditation is useless. It is also likely that having such a degree means the person has not learned the disciplines needed for careful research.

A point of order to keep in mind is that a person holding a doctorate is not necessarily a scientist. A practicing psychologist, for instance, may not be a qualified researcher. He or she may turn toward that work, but just having a degree and practice by itself is not enough.

Academic-Layperson Partition

The academic-layperson partition is the functional separation of academia from laypeople. It is probably more cultural artifact than deliberate separation, but the effect is that people who have a doctorate degree from an accredited institution tend to shun people who do not.

More than once, Lisa and I have found ourselves ignored by doctorates. For instance, during a conference at which we presented, all of the doctorate presenters were on the stage during the question and answer session while the rest of the presenters were on the floor in front of the stage. Not a big deal in itself, but the message was clear.

This partition is very pervasive. We gave two parapsychologists free access to a world-class physical mediumship demonstration. I asked for feedback from them in return. Afterwards, one said that he was in an interesting discussion about physical mediumship in a parapsychologist's discussion group. Knowing better than to ask to participate, I again asked to be told what came of the discussion. We have never received an answer. By itself, this would not raise a red flag other than perhaps to note the inconsiderate behavior, but when considered in the light of an academic-layperson partition, it is infuriating!

Peer Review

In the academic view, science journals are not credible unless they follow a policy of peer review and maintain a high scholastic standard for authors. If a publication claims to be peer reviewed, it is saying that included articles have been read by advisors who are knowledgeable in the same or a closely related field with the intention of providing suggestions for improvement. In principle, readers can properly believe peer reviewed articles meet a high standard of report design, conceptual integrity, and adherence to the best practices of scientific methodology.

The three English-language organizations with peer review journals I see as having the most influence in transcommunication and survival studies are Society for Scientific Exploration (SSE) (88), Parapsychological Association (PA) (89), and Society for Psychical Research (SPR). (87) Who their peer reviewers are is kept secret. The reader is expected to assume the editor acts in good faith to team the most qualified reviewers with authors to assure published articles meet the highest standards possible. From my perspective, this is not the case for at least one of the organizations.

The SSE has published a number of *failure to replicate-style* research reports that appear to me to be based on ill-informed assumptions and poorly designed protocols. At the same time, the organization has been unfriendly toward letters to the editor from the lay public. It also has a history of favoring anti-survival reports. The result is to isolate the organization from much of the supporting community and create the kind of dissension I have expressed here.

Scientific Method

After truth, the most sacred article of faith in the scientific community is the scientific method. The first signpost of the scientific method is the understanding that, if something is to be researched, it must be done with the intention that a final report will be submitted to formal review leading to publication. Otherwise, it is just an informal study. According to the academic community, people holding a doctorate degree conduct research while laypeople can only study.

There are a number of fundamental elements that are reasonably expected in any research project:

- **Hypothesis:** The underlying assumptions should be clearly stated.
- **Research Question:** A research question based on the hypothesis should be able to be articulated.
- **Literature Review:** There should be a well-understood reason for the question which can be articulated in the final report. This should at least include reference to applicable existing theories and prior research.
- **Protocol:** A protocol should be prepared which states how the question will be studied. It should include a step-by-step procedure showing how data will be generated and collected and the form it is expected to take. The protocol must be followed.
- **Analysis:** How the resulting data is to be analyzed should be decided as part of protocol design. Analysis should be limited to data collected during execution of the protocol. That which is collected outside of the protocol is only information and should have little influence on the final report. (Of course, such information would be expected to help shape future protocols.)
- **Report:** The final report should be reviewed by subject matter specialists for both conceptual understanding and sound practices. It

is especially important that the data analysis phase of report development is evaluated by independent reviewers.

- **Modification of Hypothesis:** Results of the research should be used to modify the hypothesis, presumably leading to a new research question.

Best Practices

In the idea of best practices is that the basic scientific methodology will be adapted, based on current and most successful methodologies. That is one of the main reasons for the literature review. No one expects a single person to be an expert in all things, so an underlying assumption of science is that research is supported before, during and after by collaboration with subject-matter specialists in the same and related disciplines.

Legitimacy of Research Reports

A psychologist is not qualified to unilaterally conduct research on the nature of ITC. This is true in the same way that, with only a degree in electronics, I cannot credibly conduct research on a mental syndrome or efficacy of a drug.

It is something of a cultural norm that someone holding a doctorate can conduct research, but someone without one can only conduct studies. While this is something of a social rule, it is also a practical consideration of qualifications and a way to set the expectations of the public. I might say that I researched something, but it is important not to do so under cloak of the kind of authority people look for in people holding a doctorate degree. Researching something is not the same as presenting a research report.

By the same social rules, and I expect academic rules, it is not okay for a person to present work as science just because the person has an advanced degree. It is an abuse of academic authority and unethically biases the public's expectation of truth. The same can be said for a journal claiming to employ peer reviewers. Without disclosing the reviewer's qualifications, there can be no realistic expectation of good science. Conversely, good science is not established if the reviewers are not reasonably conversant in the subject.

In our too small paranormalist community, it is also unreasonable for a research report to be published without a mechanism for public comment. Here, I mean comment by the paranormalist community and not just fellow parapsychologists. The alternative is that people like me have no choice but to complain from outside of the system. While it is something of a social obligation to speak out about perceived problems, doing so only alienates segments of our community and displays internal confusion to our detractors.

One of the rules in science is that research must be published. The alternative is accusation of what some refer to as the *file drawer effect* in which data that does not agree with the researcher's expectations are discarded. It is also known as *selective reporting*.

The rule that research must be reported is reasonable, but it cannot apply to people working outside of their discipline. By that, I mean a psychologist studying ITC is not required to publish unless there is reason to think the research complies with best practices. In practical terms, most, if not all of the *failure to replicate-style* reports I have read should not have been published under the cloak of academic authority because the authors were working outside of the field in which they were trained.

Writing Style

Naturally, research report writing requires a good understanding of the subject, but it also requires good communication skills. It is reasonable for research to be published in membership supported journals. Having helped to publish the quarterly *ATransC NewsJournal* for many years, I understand that it takes organization, money, and qualified human resources to support journals.

Consistent with the science-layperson partition, research reports too often end up behind the paywall of academic libraries. Membership in these tends to be free to alumni and individual university libraries usually extend access to alumni from other universities. It might cost $60 or more to read a single report if the person is not in the library system. With this pay-to-read system, it is virtually impossible for laypeople to conduct a literature review, which is one of the prerequisites of proper research.

Public funds are allocated for supports of universities, and by extension their library system. An example of the very substantial infrastructure and public support of academia is the National Science Foundation Network (NSFNET) which *"was initiated in 1985 to support and promote advanced networking among U.S. research and education institutions."* (94)

The second part of this issue is the apparent need for parapsychologists to appear scientific. I say this because so many of the reports are clearly written more to impress than to communicate.

Another tell that parapsychologists are trying to look scientific is their inclination to frame of every point in statistical parameters. As I discuss in Decisive Determinism [223] in this commentary, most of these phenomena need to be studied with a *definitely there–definitely not there* standard for inclusion. There is a place for statistics, but by now the need to establish for proof is mostly past. It is time to begin establishing theoretical models.

Pseudoscience

A hypothesis must be able to be proven or disproven, a principle referred to as *falsifiability*. It is the falsifiability problem that leads most people to discount paranormal subjects. For instance, saying *"I believe in God"* is heard by a scientist as *"God is real."*

It is not considered possible to prove there is a God. You can present many example based *on "How can you doubt that God did this?"* but a cute little animal is not proof of divinity when it can be argued that the animal's evolution can be traced from primordial soup to its present form. Modern science does not predict the psi field, nonphysical beings, or mind-body duality. As such, hypotheses which depend on concepts that are not predicted by accepted theory are not considered falsifiable.

A scientist may believe in God, but it is typical for people to compartmentalize their religious beliefs from their perception of an otherwise objective world. From my experience, anyone trained in reductionist reasoning will tend to tune out when presumably unfalsifiable concepts are mentioned.

Thanks to the skeptic, mainstream society is being taught to think of our efforts to study these phenomena as pseudoscience. The *pseud-* prefix means *false*. Also review Pseudoscience [241] in the Skeptic Commentary.

Anomalistic Psychology

Later in this commentary, I talk about the concept of Trojan horse [226] paranormalists. Anomalistic psychology is a Trojan horse field of study. The author of "What is Anomalistic Psychology?" (95) defines paranormal as: *"Alleged phenomena that cannot be accounted for in terms of conventional scientific theories."* He explains that *"Anomalistic psychology may be defined as the study of extraordinary phenomena of behavior and experience, including (but not restricted to) those which are often labeled 'paranormal.' It is directed towards understanding bizarre experiences that many people have without assuming a priori that there is anything paranormal involved. It entails attempting to explain paranormal and related beliefs and ostensibly paranormal experiences in terms of known psychological and physical factors."*

At least some transcommunication phenomena such as ITC occur in an objective manner that can be experienced by others. ITC depends on psi functioning and an intelligent trans-communicator. Other phenomena have similar characteristics that can be studied by their effects on the physical. For instance, intentionality can be objectively studied by the influence it may have on living organisms. Since these effects are often studied as non-contact influences, even far across the world and as real-time exchanges between practitioners and sitters, nonlocality is also studied.

The concept of a psi field and intentionality influencing psi fields permeates all of these phenomena. This means that psi functioning is part of the study of ITC, mediumship, intentionality and healing intention. Transcommunication, especially induced ITC, and some aspects of mediumship are best explained with the Survival Hypothesis as living personality rather than residual memory.

Whichever hypothesis is considered, Super-Psi or Survival Hypothesis, the phenomena are modeled in such a way that they can be studied as a system of related experiences. Research that ignores these facts can only be seen as debunkery. In reference to the above quote from "What is Anomalistic Psychology?" it is insulting to suggest that I am ignorant enough to be *"assuming a priori that there is anything paranormal involved."*

The study of anomalistic psychology typically involves giving subjects a psychological test designed to determine their belief in things paranormal, and then showing them examples that might indicate something paranormal. If the person says the example is of something paranormal, then the response becomes part of the proof that people with a predilection toward things paranormal tend to identify things as paranormal even though *everyone knows there is no such thing.*

From the above quote, the emphasis is on *"It entails attempting to explain paranormal and related beliefs and ostensibly paranormal experiences in terms of known psychological and physical factors."* While it is apparently wrong to attribute something as paranormal, it is okay to assume a person is delusional while requiring only a good story to explain why.

This is a kind of circular referencing because the conclusion of witness bias is based on predisposition, frequently learned from actual experience. See the Villager-Explorer Effect [199] in the Point of View Commentary.

The study of anomalistic psychology is treated by the Society for Scientific Exploration (SSE) (88), Parapsychological Association (PA) (89), and Society for Psychical Research (SPR) (87) as a form of parapsychology. This is a problem for many paranormalists because they tend to assume that anything to do with psychology in paranormal research is going to be from the paranormalist's point of view. The result is that anomalistic psychology is having much more influence than it should because of false representation.

The Study of Paranormal Concepts

Experience has shown that research of paranormal phenomena may need to be conducted with a different scientific method than used for mainstream research. I say this because paranormal phenomena are all about conceptual influences causing apparently objective effects. In a very real sense, the meaning of perception is assigned by convention so that we are taught from birth to experience information associated with the physical as objective and information associated with the etheric as imaginary. Probably because of this, reductionist methodologies often fail to consider the conceptual influences and treat what we assign as objective as the only things that are real.

The Paradox of Objectivity

From a body-centric point of view, conceptual means all in the mind and objective means physically solid; however, the influence of intentionality is from *outside* of the physical system, so that paradoxically, intended expression is perceived by the conscious self as objective. It is assigned that characteristic by convention.

As it is used for mainstream research, the scientific method is designed to study such characteristics as action-reaction, mass, momentum, velocity and propagation. In this cosmology, these characteristics are results of intended expression, actually mental artifacts that are ideoplastic [441] structures.

In mainstream science, the conceptual-objective interface is not considered. Instead, theories are designed to explain a purely objective nature of reality. Parapsychologists have attempted to adapt this point of view for psi research with mixed results. However, there has been little success in adapting mainstream methodology to survival-related research.

Departures from the Science Methodology

The fundamentals of the scientific method are probably universal. However, the following are important points of departure in Etheric Studies from the mainstream approach to science:

- **Brain as transceiver:** It is assumed by mainstream academia that the brain's processing of sensed information can be monitored by monitoring biological activity. From that, it is assumed that disabling the part of the brain that is processing sensed information also destroys that information

  In Etheric Studies, the brain's processing of sensed information might be able to be monitored, but the information is modeled as existing outside of the brain. The mind-body duality model followed in Etheric Studies holds that disabling the brain only disables its ability to express associated information *into the physical*. The information itself is assumed to remain in the psi field. See the Perception and Expression Discourse.

- **Modeling perception:** A host of issues are involved with the idea of how people perceive reality. Agnosia, incredulity blindness, inattentional blindness, villager-explorer effect (sheep-goat), pareidolia, and gestalt association are just a few of the terms often used to describe perceptual error. The best model I have found to describe perception is First Sight Theory (13), which is used in the Implicit Cosmology as the rule set for the Perceptual Loop. See the Perception and Expression Discourse.

  Any research concerning perception is incomplete if it does not consider these influences in protocol design and data reduction.

- **Body-centric and personality-centric:** There are a number of ways in which the body influences perception that can bias research results. The Life Field Complex Diagram [xxv] illustrates the relationship between personality and body mind as it is modeled in the Implicit Cosmology. The diagram is useful for visualizing the differences between body-centric and personality-centric perception.

  To indicate this difference, a general statement can be made that the body reacts while the mind examines. For instance, the body fears while personality is cautious. Protocols should consider this difference between the influence of human instincts and the presumably more rational perception of the entangled personality. See the Point of View Commentary.

- **Inability to shield intention:** A common practitioner error is the assumption that, if the practitioner is not touching the equipment, then he or she is not involved. In fact, one must think in terms of influence from the practitioner or an interested observer. The practitioner and observer may be anywhere.
- **The influence of intentionality:** Electronic measuring devices are designed to display influences they are intended to detect. Properly designed, they suppress spurious physical influences. However, psi influence can cause changes in transistor circuitry that may produce a false positive reading. For instance, a magnetic field detector might signal the presence of a magnetic field, when in fact it is detecting a psi influence on the noise within its circuitry. The device is, by design, only able to indicate this influence as a change in magnetism.

  Evaluation of protocols and data should be a collaborative effort that includes people qualified to identify sources of false positives. See the Artifacts and Perceptual Error Study.
- **Peer review:** The qualifications of peer reviewers are probably as important as the qualifications of the researchers. Their qualifications should be known to the reader and they should represent multiple disciplines. If this is not reasonable, then the reader should be advised as to the limits of review.

  In our sparsely populated community, it is reasonable to publish a research report in a normally peer reviewed journal even if suitable peers are not available. It is more important that good studies are made available for collaboration. If such a report is published, it is the first duty of the editor to assure the lack of peer review is explained.
- **Statistical analysis:** Current best practices for statistical analysis apparently require that outliers be discarded. A Class A EVP [394], for instance, is by definition, an outlier. As such, statistical analysis may not be appropriate for EVP. The same can be said for most ITC and mediumship. Mediums who are able to reliably deliver verifiable information are outliers themselves. The average person is able to function as a medium, but the rare verifiable message would have to be considered a statistical outlier.

Decisive Determinism

The influence of intentionality as it is detected in controlled conditions is typically very small so that a study of psychokinesis involves detection of small deviations from the norm. It is also well established that experimenters are able to influence the outcome of experiments. This can be due to unconscious bias in procedural decisions and analysis of data, but there are indications experimenters might contaminate data collection by introducing a small psi influence during experiments.

There is no known way to shield the psi field from the influence of intentionality. That simple fact might turn out to be one of the best proofs of the existence of the influence of intentionality, but for now, it is a reason why protocols intended to detect small deviation should be avoided.

In transcommunication, the communication is either there or it is not. For instance, it is ATransC policy to not use Class C and Class B EVP for demonstration unless there is some compelling reason.

Requiring Class A examples can be construed an all or nothing approach to research. In controlled studies of mental mediumship, the medium produces verifiable information or not. Statistical analysis of whether or not a message is verifiable ignores the reality of the phenomena and how people translate psi information into messages. On the other hand, statistical analysis of the frequency of verifiable messages is useful for determining the confidence of the practitioner, but that is an operational issue and should not be a measure of proof.

Decisive determinism also helps with the issue of experimenter bias. Such bias usually has a very small effect, and the saturation state of all or nothing at all means ignoring small signal influences.

Qualified Practitioner

Another deviation from common practices in science is how data is acquired. A common practice is to hire college students to do the routine work required by research protocols. In a study of EVP reported in the Society for Scientific Exploration (96), recording EVP was considered routine work and was conducted by college students. As should be expected, other important aspects of this protocol designed and executed by a psychologist were fatally flawed, but the assumption that would usually doom the study was that any inexperienced person can reliably record and hear EVP.

This is an example of how wrong assumptions produced a useless protocol: the would-be researcher recorded around sixty hours of sound files and then listened to the results for EVP. Our practice is to record for about three minutes in a single session. It takes us an average of thirty minutes to examine the three-minute recording.

Based on ten minutes of analysis per minute of recording, it would have taken us some 600 hours to properly analyze the recordings for EVP. I am assuming he did not spend that amount of time, and in fact, used a procedure that assured he would overlook possible EVP. (96)

The best practice for such studies is to standardize recording sessions to about three minutes and then to consider the number of utterances and quality per session using multiple blind listeners.

The average person should be able to learn how to record EVP, but it takes a little focus and practice. Probably the biggest problem is learning how to find, and then understand, the recorded utterances. People have described it as learning a second language because the paranormally formed voices usually lack the usual audio cues people need to recognize words from noise. (21)

I am a good example. Try as I do, I can only occasionally record a useful EVP. Lisa can reliably record one almost every session. Most of mine are Class C and B, while many of hers are very exceptional Class A. Probably one in a hundred people who try will become a competent practitioner. Anyone can master the mechanics, but it appears to be more a question of lucidity [151] than one of determination.

So here is the bottom line. As a best practice, transcommunication research protocols must require proven practitioners and the use of qualified witness panels [401]. This applies to all forms of transcommunication including mediumship.

Research Grants Verses Contracts

Parapsychologists depend a great deal on research grants; however, as might be expected, there are very few granting organizations that support the study of anything having to do with things paranormal. One important reason for so few granting organizations is that they are virtually entirely supported by our relatively small paranormalist community.

As directors, Lisa and I have always considered the ATransC a research organization. (Here, I use *research* in the generic sense as *formal study*.) We have initiated a number of studies with both member and website visitor participation. Our expectation has always been that information would be shared throughout the project and that a final report would be issued.

Lacking realistic access to peer reviewed journals, we depended on publication in the quarterly *ATransC NewsJournal* and the online Journal section of ATransC.org.

We have also issued two research grants that were targeted toward specific research questions. It is the experience of these two projects that has produced much of what I have included in this Science Commentary. In the first project, our first shock was the researcher's demand for secrecy during the several years of the project. We also feel that the contracting organization's refusal to consider our input, inappropriate application of the scientific method, and failure to comprehend the concept of a *pilot study* contributed to the disaster that was delivered to us.

The second project was with a more mature organization, but it was abandoned by that organization before it was begun. My guess is that soberer minds prevailed when they realized the subject was EVP and that the ATransC is not an academic organization. An ATransC member and parapsychologist salvaged the project, and even completed it using some of his own funds. I am not mentioning the name of the organization because there remains hope for future professionalism.

The ATransC is still a publicly funded research organization, but thanks to our experience with supposed *real science*, and after witnessing the almost religious application of the scientific method in our first effort, our policy now is to keep research under a strict contract model. In effect, when we have a question, we pay to get it answered. There is no secrecy and the results are publicly accessible.

The administrator of one granting organization I am familiar with admits that all they can do is screen the candidate before issuing a grant, but after it is issued, they are not able to impose restrictions or even assure a report is published. This is a problem because, as enablers, granting organizations have a responsibility to assure they have not funded work that is contrary to their stated objectives. Stated another way, the granting organization has a degree of ethical responsibility for the contents of the resulting research report. As an example, see One Bad Apple: A case Study [227] below.

I recognize it is idealistic to expect granting organizations to manage work they have funded, but the alternative appears to be too many of the kind of debunking failure to replicate kind of reports that I am complaining about in this commentary.

Litmus Test

Having a yes-no test for something is intellectually risky, but there are tells. If I can know from the abstract that the author assumed mental aberration over the possibility of genuine phenomena, the journal editor should be able to know not to include it without appropriate warning to the reader. This is really a simple point. *Trojan horse* articles typically ignore current understanding. As such, the research is incomplete and the report should not pass peer review. Most scientists know that the discussion section should at least consider alternative hypotheses.

Trojan Horse Organizations

I am of two minds concerning naming names in this commentary. As an ATransC director, I have always assumed the responsibility to warn members about problems in the larger community. If we did not, then who would? The other side of the argument is that naming names is divisive for the community and skeptics too often use the tension as proof of our confusion.

For the most part, I have tried to generalize my comments here, but in some cases, the problem is just too acute to ignore and I feel compelled to accept a leadership responsibility to shine light where it is needed. As you read this book, please consider the need for awareness by other practitioners who might become the object of some armature scientist's attention.

My comments are limited here, but I am a little more aggressive on EthericStudies.org, should you wish to explore. Especially look under the Essays Tab.

Poorly Managed Expectations

Some organizations claim to represent the full range of paranormal phenomena, yet their actions suggest they are covertly skeptical about Trans-Survival. Since the rest of us want to take them at face value, the result is a form of entrapment as our expectations are not matched by their actions.

The same can be said for psi phenomena when a parapsychological organization publishes reports that are ostensibly honest studies of extraordinary human potential, but are in fact intended to debunk the paranormality of the experiences. Here, I am speaking of the growing number of anomalous psychology reports in the leading parapsychological journals.

The fault is really not with the organizations. The problem is more likely that we are culturally conditioned to respect academic organizations. I know that I began thinking parapsychologists are *our scientists*. It took a closer look and a few bad experiences to teach me that at least some people in the parapsychological community of interest are not actually scientists and that some are aggressively anti-survival.

It is in the failure to meet expectations that invokes the idea of a Trojan horse. We invite people posing as scientists to join our work and help us understand experiences we know are real. Our only recourse when that trust is betrayed is to complain and feel confused by those mixed signals.

I have seen this interaction too many times to think it is accidental. At the risk of overgeneralizing, it is the culture in the parapsychological community which fosters this deception. Take a close look. Most members of that community are retired academics, often professors who take on this study as a late-life hobby. They are playing at science and can do so with little or no peer oversight. The community s too professional for self-policing of member behavior, even though it hurts all of us.

One Bad Apple: A case Study

An example of a Trojan horse is seen in the treatment given the medium for the Felix Experimental Group (28). The short version is that the medium submitted to a study that involved a long series of séances during which the researchers had pretty much total control over the medium and the circumstances. This even included body and cavity searches.

Even though the medium succeeded in producing astounding phenomena under those controlled conditions, three members of the research teams focused on accusations of fraud that supposedly occurred prior to the study. The accusations are based on assumptions and innuendoes about the medium's behavior while ignoring alternative explanations.

Stephen Braude (PhD philosophy) was the lead researcher and is chief editor of the supposedly peer reviewed *Journal of Scientific Exploration* in which two of the articles were published (97) (98). A grant to fund the research was issued by the Parapsychological Association. The Society for

Psychical Research gave Braude their Myers Memorial Medal for "significant contributions to psychical research." This was at the same event in which he gave a talk on the Felix medium. (99)

The following *Paranormal Review* published by the SPR included an interview with Braude; an article by another of the research team, Michael Nahm (PhD biology); and an article titled "Fall of the House of Felix" by Peter Mulacz (PhD psychology), a third member of the research team.

I discuss this in some detail in the Essay section of EthericStudies.org (100) and will not go into more detail here. The entire incident is distasteful and diminishes all who identify with the paranormalist community. My point should be obvious: researchers with mostly unrelated training are doing everything they can to discredit physical mediumship, and by extension trans-survival, at the expense of a test subject who gave them his trust. Braude's article in the journal he edits amounts to self-publishing an article that should never have passed peer review if only because the study ignored minimal requirements of ethical treatment of a test subject. (101)

The takeaway message from all of this is that the three parapsychological organizations I look to for intellectual guidance in this field have turned out to be Trojan horse organizations. Members of the research team, especially Braude, have been taking every opportunity to further their attack on the FEG medium. Yet it seems that no one is bothered by this treatment of perhaps one of our most important practitioners demonstrating survival.

For the record, in 2010, Lisa and I attended a séance in the FEG medium's home (28), during which substantial phenomena were demonstrated under conditions that made fraud virtually impossible. I believe this was well before the supposed trickery Nahm complains about. The ATransC sponsored two visits to Reno for five séances by the medium, and again, astounding phenomena were demonstrated under circumstances over which I had complete control.

We are not privy to all of the circumstances leading to the complaints by the doctorates, and as such, our complaints must focus on the good of the community. In simple terms, my complaint is that academics have unethically attacked a layperson practitioner in a way that will diminish our community. They do so with impunity because their professional organizations applaud their work. Finally, this whole sordid affair will contribute to contraction of the paranormalist community and gives new condemning talking point to the skeptics.

Research Ethics

Ethical treatment of research subjects has become a regulatory policy imposed by organizations and governments to help assure humane treatment of test subjects. The Belmont Report published by the U.S. Department of Health and Human Services (101) identifies three foundation considerations recommended to university research ethics review boards.

Respect for persons:

> To show lack of respect for an autonomous agent is to repudiate that person's considered judgments, to deny an individual the freedom to act on those considered judgments, or to withhold information necessary to make a considered judgment, when there are no compelling reasons to do so.

Beneficence:

> Persons are treated in an ethical manner not only by respecting their decisions and protecting them from harm, but also by making efforts to secure their well-being.
>
> Beneficence can be understood as *do no harm*.

Justice:

> Who ought to receive the benefits of research and bear its burdens? This is a question of justice, in the sense of *fairness in distribution* or *what is deserved*. An injustice occurs when some benefit to which a person is entitled is denied without good reason or when some burden is imposed unduly.

To my knowledge, there is no code of ethics for parapsychological research, and certainly no review panel. The parapsychological community is guided by the mainstream academic culture, so I will guess this lack of concern for ethics is because many of our academics retired from the mainstream before the new policies were enacted and have not learned the need for ethical codes. For whatever reason, an ethical standard is currently missing, but clearly needed in the paranormalist community. Also see A Personal Code of Ethics for Developing Mindfulness [157] in the Mindfulness Discourse.

Omniscient Science Syndrome

This title was chosen with much deliberation. *Omniscient* means infinitely wise. *Science* is used here in the sense of the body of knowledge, community of scientists, and the act of conducting science. *Syndrome* usually refers to human nature as a complex of behaviors related to a common factor. So *Omniscient Science Syndrome* is used here to mean *a complex of beliefs that science is infinitely wise*. By extension, *scientists are believed to be infinitely wise*. This point of view is too often expressed in behavior and actions by laypeople and scientists, themselves. Yes, the academic-layperson partition again.[44]

Something like Omniscient Science Syndrome is intended by *under the cloak of academic authority*. We have been culturally taught to trust and respect science, scientists, and what scientists tell us. Because we have been taught to expect scientists to have a doctorate, virtually anyone with a doctorate degree inherits that respect. As I think I have made pretty clear above, this unthinking trust is seldom deserved and is too often abused.

As a point of order, scientists acquire new knowledge, but it is engineers who transform that knowledge into applications that serve society. In practice, a few scientists tend to be supported by many engineers, and those are supported by many more technicians as knowledge is transformed into products. The point is that the people who understand the practical sense of what scientists present are the engineers, and they typically have just a four-year bachelor's degree.

We need to respect doctorates but it is important that we understand why, and that should be on a case-by-case basis. If the subject is transistor radios, a medical doctor is just a citizen. If the subject is optical illusion, a psychologist with a doctorate has no more standing than an optical engineer with a bachelor's degree.

With that said, if the subject is how to research optical phenomena, then anyone with an advanced degree, the curriculum for which included classes in research techniques, probably has important insight for protocol design of any subject. Again, qualifications should be considered on a case-by-case basis.

---

[44] It is difficult to resist a little sarcasm. The third expression of Omniscient Science Syndrome is the apparent belief of the academics that they are infinitely wise. In reality though, it is more like the Wizard of Oz Syndrome in which the wizard expresses himself as this larger-than-life, fearsome-in-his-intelligence practitioner of the magical arts while the citizenry fears his awesomeness. To complete the picture, too many of our academics have a wizard complex!

The objective of my comments on this is to say that we should not blindly assume intellectual authority if a person has an advanced degree. A person with an advanced degree should not expect or encourage belief that he or she has that authority. At the same time, all of us should respect the training and experience some people with an advanced degree have gained in such subjects as application of the scientific method and critical thinking. Perhaps that respect can be extended to the layperson with a lifetime of study in the subject at hand.

The Syndrome

The assumptions about science and scientists are rather different, depending on the point of view of the people making them. The general assumptions are that scientists:

- Are well educated in appropriate subjects.
- Are knowledgeable about the subjects about which they comment.
- With integrity, strive to serve the public's interest.
- Search for understanding no matter where it leads.
- Are infallible: If science does not support the concept, then it cannot be (the skeptic's mantra).

Mainstream society seems to be most guilty of blind acceptance of culturally taught *truths*. The skeptic community aggressively attacks all things paranormal and anyone involved in paranormal study or application. As a reaction, people in the paranormalist community tend to be a little more suspicious of mainstream scientists. Interestingly, as parapsychologists protest the intellectual injustice of skeptics, they also teach us to distrust people claiming intellectual authority.

It is fair to say that the study of ITC has been harmed more by parapsychologists than by mainstream scientists. Mainstream scientists usually ignore ITC, while many parapsychologists have made uninformed (simply wrong) pronouncements about ITC under the cloak of their academic authority. More than one skeptic has referred me to a statement by a respected parapsychologist that contradicted a point I was trying to make in favor of paranormal phenomena.

Lay-Paranormalist Assumptions

In addition to the above list, we in the paranormalist community tend to assume that scientists:

- Are working on our behalf to understand these phenomena.
- Listen to our input and organize their theories accordingly.
- Are candidly honest when they engage us in discussion about the phenomena with which we are experienced.
- Tell us what they are concluding in a collaborative manner when they are studying our examples.

These assumptions are mostly unsupported by action. An academic policy of secrecy until publication usually extends to the person being studied and related phenomena. As such, the academic-layperson partition clearly turns conversations into a deception. From my experience, the thoughts of *"I'll listen but I know the person does not really understand the implications of what is said"* on the parapsychologist's part and *"I am so happy the scientist is considering what I have to say"* on the part of the layperson assures that no real communication occurs.

Academic-Layperson Partition as a Syndrome

It is likely that the academic-layperson partition is just one manifestation of a self-organizing behavior. Based on how I acted when I came out of engineering school, it is my guess that the universities inadvertently teach this behavior. It is also probably difficult to maintain a little humility when the general public so easily believes what scientists say and is so willing to fund research for almost anything a scientist claims is important—anything seen as mainstream, that is. (Think Large Hadron Collider.)

Again, I will say that this is a general assessment based on personal experience. There are many important exceptions, so please do not assume the worst of your academic/scientist friends. This discussion is intended to make our community aware of these issues but not to indict the entire community.

Of course, scientists are not going to easily admit fallibility. It is human nature to develop a point of view that is based on education. The problem is that it is necessary to evaluate those assumptions. That is a foundation concept for mindfulness. So here I will say that these are apparent assumptions demonstrated by John Q. Scientist:

- The public will trust what I say without requiring supporting information.
- The public will not understand how I arrive at my conclusions even if I explain.
- The public should not see processes such as peer review and data analysis. It will only confuse them.
- I know best what is important to the public.
- Scientists who study theories that are not supported by accepted evidence are bad players.
- People who attack me do so because my research did not support their beliefs.
- If it is not specifically supported by the evidence I accept, then it must not be real.

In defense of parapsychologists, I am adding one more assumption that seems to underlie much of what parapsychologists publish:

- It is important that I protect the academic integrity of my work.

There is not much that parapsychologists say and do that is not scrutinized by people like me (friendly fire) and the mainstream academic community. For instance, my guess is that the overreliance on statistical analysis is a response to mainstream criticism. But here is an example of obsessive use of statistical notation, from a survey of parapsychologists' view of their field:

> These changes in belief (belief verses skepticism) did not correlate with age (Spearman's rho=-0.08, df=112, p=0.42). (102)

There is much intellectual/academic verbiage like this in some reports that effectively overshadows intended communication.

Reason to Expect Leadership

Quite a lot of college and university operating costs are funded by the public. Yes, a college degree is expensive for the student, but it is subsidized by the public with the assumption that the people with advanced degrees further the greater good of the community. The following is from a 2009 article, "Higher Education Subsidies" on the CATO website:

The Department of Education spends about $30 billion a year on subsidies for higher education. The bulk of that funding goes toward student aid programs, with the balance going toward grants to educational institutions. In 2008, grants to institutions cost $2.3 billion and aid programs cost $27.6 billion, which included $17.4 billion for student grants, $9.6 billion for student loans, and $0.6 billion for administration. (103)

Perspective

Science as a disciplined search for understanding is an essential component of the paranormalist community. Realistically, it is the academic community that provides that science. However, it is the lay community that hosts it. This must be clear: parapsychology alone would have nothing to study without practitioners, experiencers and witnesses. Test subjects in parapsychology are practitioners in the larger paranormalist community. The two are partners in the search for understanding.

When I say that the parapsychology community of interest has abdicated its authority to explain the paranormal, I do so from the perspective of transcommunication. Psi has been identified and is being mapped, but by now that search for proof is reaching the point of diminishing returns. This will not change until transcommunication is added to the equation.

As a final note, I accept that it can be difficult to have an intelligent conversation with people who do not hold an advanced degree. But I experience a similar problem talking with people who are not well informed about transcommunication, and this includes many people with an advanced degree. The solution for me is education of the general public. An important concept in mindfulness is that *our lot is to learn, and having learned, our lot is to teach*. Shunning the unwashed is self-defeating!

Community

Commentary 5:
Psi Research

Also see the Transcommunication Research Study

Keep in Mind the Good

The preceding commentary was a long list of complaints about science. For balance, it is important to point out some of the ways our community is supported by parapsychologists. Many of the characteristics I complain about have also given us foundation science which has helped us evolve from faith-based to research-based understanding of these phenomena.

Some seekers still proceed with the belief that the lessons they learn from metaphysical teachers and ancient wisdom schools represent state of the art understanding about human nature and reality. In fact, parapsychologists are changing our long-standing assumptions about fundamental concepts such as mediumship and intentionality. As a long-time paranormalist, I can say with confidence that most of our fondly held assumptions about our nature and the nature of reality are simply out of date. It is the science community of interest that will shine a light on more contemporary understanding, and therefore on a better way to progression.

First Sight Theory

Dr. James Carpenter's First Sight Theory makes a clear case for the separation between our conscious self and our unconscious mind. (13) The theory includes a set of characteristics describing how the unconscious mind processes information. Three important points that come from this theory should cause us to reconsider what we know and what we believe to be true. See the Perception and Expression Discourse.

The first point is that everything appears to produce a psi signature which we unconsciously sense. In turn, part of our unconscious expression is a psi signal sent into the environment representing our expression.

The second point is that information we psychically and physically sense from the environment is unconsciously processed, and this processing is very much influenced by our personality type and worldview.

The third point is that we only become consciously aware of the results of that mostly unconscious processing. In fact, we only become aware of what our worldview thinks of the information.

Some of the consequences of First Sight Theory are that we begin to physically react to information before becoming conscious of it. This seems to explain presentiment, in which people seem to sense near-future events.

What is sensed from the other side is necessarily filtered in our unconscious process, what I refer to as the Perceptual Loop. This means that mental mediumship, even much of the information coming to us via deep-trance mediumship, is unavoidably colored by what the medium unconsciously thinks; that is, what is in the medium's worldview.

First Sight Theory is based on meta-analysis of a very large number of clinical studies, including those related to psi functioning, and current thought in psychology. Interestingly, a brief survey of the Internet will show that mainstream scientists are beginning to report that much of our decision-making occurs in the unconscious. (11) This seems to be a convergence of mainstream and parapsychological thought in support of psi phenomena.

Formative Causation

Darwinian Evolution requires accidental solutions to environmental challenges. That is, organisms mutate and some mutations allow the organism to survive better than others without the mutation. Before Darwin, Jean Lamarck (Lamarckian Evolution) (104) proposed that organisms learn how to cope with the environment, and that the learned behavior is inherited by offspring. Of course, Darwinian Evolution prevailed to become mainstream gospel in science. The problem is that Darwinism does not explain morphogenesis, specifically, how does a single cell at conception know to divide into cells which are differentiated for specific parts of the body such as skin and hair cells?

Dr. Rupert Sheldrake has proposed the Hypothesis of Formative Causation, in which he argues that a subtle field, referred to as a morphic field, organizes cell formation according to how that cell has always formed. (9) He refers to this memory as *Nature's habit,* and the process as morphic resonance. The top field of an organism is represented by a morphic field which organizes a hierarchy of nested fields right down to cell fields. Each field consists of a nonphysical (etheric) set of functions which support memory (Nature's habit or worldview) and decision making.

An important function in this hypothesis is the ability to consider changes for possible modification of Nature's habit. It is this functional area that is interpreted by me as the Perceptual Loop described in the Perception and Expression Discourse.

While this model still allows for evolution by survival of the fittest, it also considers the existence of a psi field and provides a way to change the momentum of Nature's habit with creative solutions to environmental challenges. Visualize a morphic field as a thoughtform that contains all of the information necessary to form an organism. This would be organized into functional areas such as memory, test of external influences for agreement with memory and a means of modeling and then expressing change.

Global Consciousness

The following is from the Global Consciousness Project website:

> The Global Consciousness Project is an international, multidisciplinary collaboration of scientists and engineers. We collect data continuously from a global network of physical random number generators located in up to 70 host sites around the world at any given time. The data are transmitted to a central archive which now contains more than 15 years of random data in parallel sequences of synchronized 200-bit trials generated every second. (51)

The Random Number Generators (RNG) are based on tunneling diode technology to generate a highly random string of zeros and ones. The array has consistently shown changes in average randomness moments before major events such as the 911 attack. These seem to indicate a global sense of dread before disasters occur. This is also an indication of nonlocal characteristics of reality.

The importance of this research for paranormalists is that the project produces repeatable indications of a collective consciousness, the influence of this consciousness on random processes and an indication of at least short-term nontemporal behavior of consciousness.[45]

Quantum Entanglement

Amongst his many contributions to our understanding of consciousness, Dean Radin, Chief Scientist for the Institute of Noetic Sciences (IONS) (105) , has conducted experiments to determine the effect of attention on the behavior of entangled photons. The following is from the Abstract:

---

[45] An RNG based on lessons learned from the Global Consciousness Project is available from Psyleron. (57) The device and software make it possible for individuals to study the effect of intentionality on the device. One such study approach involves looking for changes in the output of the REG as a group of people meditate.

> *A double-slit optical system was used to test the possible role of consciousness in the collapse of the quantum wavefunction. The ratio of the interference pattern's double-slit spectral power to its single-slit spectral power was predicted to decrease when attention was focused toward the double slit as compared to away from it.... factors associated with consciousness, such as meditation experience, electrocortical markers of focused attention, and psychological factors including openness and absorption, significantly correlated in predicted ways with perturbations in the double-slit interference pattern. The results appear to be consistent with a consciousness-related interpretation of the quantum measurement problem. (41)*

There are two points I would like to make here. One is that Radin holds a master's degree in electrical engineering and a doctorate in psychology. It has always been my contention that transcommunication research requires a multi-disciplinary approach. As I understand it, Radin works with a team at IONS and is himself qualified to understand the behavior of technology by understanding the capabilities designed into experimental devices. The combination of his awareness of technological idiosyncrasies and human behavior and the fact that he has access to team collaboration makes his work an important model for parapsychology.

Of course my second point is that the double-slit experiments provide an important benchmark for consciousness and intentionality research.

Citizen Scientist Research

From the perspective of the Implicit Cosmology, transcommunication is, by definition, communication between (etheric) personalities with an objective expression in the physical. As such, the study of transcommunication is primarily conducted with emphasis on the Survival Hypothesis. Of course, most parapsychological study of transcommunication is more focused on human nature and why people think there is such a thing as the paranormal.

Please review the Transcommunication Research Study for a discussion of that work.

Community

Commentary 6:
Skeptics

Introduction

Anyone who does not share our truth is apt to be a skeptic. Truth is maintained in worldview which is the arbiter of perception and the template of personal reality. Assuming two people do not share exactly the same worldview, it is reasonable to expect one will be skeptical about some aspect of the other's truth.

Years ago, I was told that older people's first reaction to a new idea is rejection. The argument was that older people are more set in their ways and take time to come around to new ideas. As I aged, I watched for that negative first reaction to new ideas, and sure enough, that has become something that I must now consciously moderate.

There are two reasons for this I can think of and both involve the same underlying process. The less familiar we are with a concept, the less likely our mostly unconscious perceptual processes will find a match in our worldview, and therefore, the more likely it will be rejected in the Perceptual Loop.

The other side is the more we know about something, the more confidence we have in our understanding. A new idea that challenges a dominant paradigm will almost certainly be rejected by our perceptual processes.

For our discussion, skepticism is in three forms:

Spontaneous Skepticism: Probably the most common source of a skeptical response to an idea is the mostly spontaneous rejection of an idea because it is unfamiliar or because it challenges what is familiar. This response is generated in the Perceptual Loop based on worldview. If this response is not balanced with suspended judgement, it may produce a skeptic's response.

Emotional Skepticism (a.k.a. skeptic): This is typically a determined attack on an idea that challenges a belief. This form defines what I am referring to as a skeptic, and in my experience, the system of belief is best described as scientism.

Objective Skepticism: When people speak of healthy skepticism, it is this form they intend. It is typically a reluctance to accept an idea based on an informed opinion which produces well-reasoned questions. If the person is able to recognize and accept well-reasoned responses, it remains healthy skepticism. If it becomes apparent that only an answer that agrees with the person's preconceptions is acceptable, it becomes a skeptic response.

Make a distinction between *skepticism*, which is usually expressed as reasoned questions that may, with further understanding, lead to acceptance of an idea, and *skeptic*, which is the name claimed by people who attack ideas that do not agree with their sense of proper science.

It is worth noting that a number of friends have urged me to use *pseudoskeptic* (false skeptic) as a name for the people I describe as skeptics; however, the people causing all of the trouble are found with Internet searches for *skeptic,* and for the most part, remain invisible if *pseudoskeptic* is used.

Also note my comment about Objective Skepticism. An insidious form of skeptic is the person who asks informed, reasonable questions but rejects all answers supporting the new idea. On close examination, it usually becomes evident that the person is an emotional skeptic and will only accept answers that disprove the idea.

Discernment as an Alternative Name for Skepticism

In our community, *skeptics* are pretty much the enemy of free enquiry. For those who wish to respect *healthy skepticism,* I suggest they adopt *discernment* as a better term. To be discerning means to express informed judgment.

One of the more important traits for a paranormalist is discernment, but as it happens, it is also a trait many do not exhibit. From my experience, this lack of informed judgment is more a problem of education than lazy thinking. One of the objectives of this book is to rectify that by providing a useful model on which informed judgment can be based.

Scientism and Uninformed Skepticism

Another way of looking at skeptics is in terms of the source of their objection. Scientism is the belief that science, the scientific method and research are the only way to validate reality. In practical terms, scientism holds that, if something is not recognized by mainstream science, then it is not real and is, therefore, impossible. When people under the cloak of authority of science advise the public about any subject without analysis of available information about that subject, they are effectively practicing scientism. (106)

Pseudoscience

I invite you to take a look at the Pseudoscience article in Wikipedia. (107) As I will discuss in the next commentary, the paranormal articles in that online encyclopedia are controlled by a cadre of skeptics. One of the cornerstone articles in their organized attack on all things paranormal is the Pseudoscience article.

As far as I can tell, the term *pseudoscience* was coined by a skeptic as a way of branding concepts that are not specifically supported by mainstream science. *Pseudoscience* is used as a derogatory term to brand virtually all paranormal subjects. The people paranormalists see as *skeptics* are seldom mainstream scientists. They are more like science apologists who behave as if it is their sacred duty to see to it that everyone adheres to pronounced truths of science as they see them.

The skeptics paranormalists are concerned with are the ones that seek to prevent funding for research and who would make our study illegal as a matter of the public good. They are all about science verses what they perceive as pseudoscience. Also see Pseudoscience [218] in the Science Commentary.

Skeptics Impact on Research

It is reasonable to speculate that a large percentage of our population is open to the possibility of these phenomena. Yet the National Science Foundation (NSF), the department of the U.S. Government chartered to advance the national health, prosperity and welfare, considers these phenomena pseudoscience. The NSF in the 2006 annual Science and Technology report includes:

> *A recent study of 20 years of survey data collected by NSF concluded that 'many Americans accept pseudoscientific beliefs,' such as astrology, lucky numbers, the existence of unidentified flying objects (UFOs), extrasensory perception (ESP) and magnetic therapy (Losh et al. 2003). Such beliefs indicate a lack of understanding of how science works and how evidence is investigated and subsequently determined to be either valid or not. (108)*

Losh et al. is a reference to an article in the *Skeptical Inquirer* magazine. A further comment about pseudoscientific beliefs is based on a reference from the Committee for Skeptical Inquiry (CSI) which states:

> *According to one group studying such phenomena, pseudoscientific topics include yogi flying, therapeutic touch, astrology, fire walking, voodoo, magical thinking, alternative medicine, channeling, psychic hotlines and detectives, near-death experiences, unidentified flying objects and alien abductions, the Bermuda Triangle, homeopathy, faith healing, and reincarnation.*

Note that the celebrity skeptic, James Randi, is one of the founders and the publisher of the Skeptical Inquirer Magazine.

The skeptical community is well organized and dedicated to eliminating pseudoscience in our culture. They have had an impact on government funding for research and have likely been successful in aligning many school systems with their beliefs. Consequently, being associated with the study of paranormal subjects can be damaging to a science career. We have also had many members in the ATransC who insisted that their involvement not be made public out of fear of harming their career in a conservative field such as law and medicine.

Tells of a Skeptic

Skeptics attempt to show the unacceptability of an idea by belittling it and associating it with obviously silly ideas, rather than with facts, evidence and sound logic. They describe ideas they disagree with, and people associated with them, in terms that would usually cause a fight if spoken face-to-face. They seem to assume *believers* are seen as second-class citizens by the mainstream public and thus do not enjoy the protection of social norms afforded mainstream citizens. So if you read something that calls someone a whacko or fraud, or describes a practice as fraudulent or woo-woo, then you know that the material was written by a skeptic and is probably based on belief rather than fact.

Name-calling is especially true of skeptics who focus their attacks on alternative or complementary health practices. They commonly refer to these with the derogatory term of *quackery* and practitioners as *quacks*. Of course any such practice that is not specifically approved by the government is considered pseudoscience, and even some that are approved, such as chiropractic, feel the wrath of those who think *complementary* is just another word for *fraud*.

Without research to support their accusations, skeptics seldom add knowledge to the subjects they attack but are only able to destroy knowledge. Here the Latin term *a priori* has special meaning. Skeptics routinely make statements about subjects for which they have no knowledge other than that their peer group is against them. In this context, *a priori* means without prior knowledge and is used to say that the person is judging without having become informed about the subject. The practical result of this book-burning mentality is that new ideas are suppressed and examination of new ideas by academically trained researchers has become probable professional suicide.

Comparing the View of Science with the View of Skeptics

Wikipedia editor Ludwigs2 expressed one of the better descriptions of the skeptic view:

> Science and skepticism are entirely different projects; they share the word skepticism, but it has different meanings for each group. For a scientist, skepticism means (roughly) *"I choose not to have any beliefs about a subject in the absence of evidence."* It's a philosophically conservative position designed to keep people from making a priori assertions about the world (except those dictated by logic or math). For skeptics, by contrast, skepticism means (roughly) *"I choose to believe that non-conventional ideas are wrong until they have met some burden of evidence."* This is an ideological position designed to advocate against certain kinds of viewpoints. See how these differ on (for example) acupuncture:
>
> > *Looking at something like acupuncture scientifically one would be forced to admit that there really isn't much evidence either way—there is no scientific reason to recommend its use, but no obvious reason to say that it's wrong, either. That is, acupuncture is morally neutral, like drinking tea with honey and lemon when you have a cold.*
> >
> > *Looking at something like acupuncture as a skeptic one would find oneself saying that acupuncture hasn't met the needed burden of evidence, and so acupuncture is wrong—and this will lead to ideological claims that people who take acupuncture are stupid, that people who do acupuncture are charlatans, and etc. That is*

acupuncture is morally bereft, like selling sugar pills as cure for cancer.

Scientists and skeptics overlap in the assertion that one should use practices that have been borne out by systematic experience. But that's where the similarity ends: skeptics go on to make moral judgments about practices that science can never make, and to engage in advocacy with respect to those moral claims. Consider the vast range of skeptical literature, almost none of which contains any actual research (aside from literature reviews of other people's published work), and which is almost entirely dedicated to critical declamations against one or another questionable activity. Skepticism is (frankly) scientific punditry, and while I won't deny its value in that consumer advocate sort of way, one needs to be cautious with it as an intellectual enterprise.

Skepticism is Relative

Unless practicing suspended judgement, anyone is likely a skeptic about something. In our study of transcommunication, we routinely have parapsychologists a priori say that something like EVP is probably stray radio waves or pareidolia. Ironically, these are the same people who rail against the injustice of skeptic attacks.

I had a friend who was a longtime Spiritualist and familiar with many of the related phenomena such as healing intention and mental mediumship. Yet he came out of a very impressive demonstration of physical mediumship completely doubting its authenticity. Granted the séance was held in a completely dark room, but every precaution was taken to assure no trickery. The medium was bound in a way from which it would be simply impossible to escape without much noise and shuffling about. Certainly he would never have been able to get back into the bindings before the lights were turned on. The room was new to the medium, and I had personally inspected every inch of it and assured that the medium and sitters had all been searched.

A number of evidential things occurred during the séance. If a sitter accepted that the medium was not able to physically cause movement or voice far from his body, then the experience should have at least been given conditional acceptance that something important occurred. There is also the precedence that other research has shown phenomena does occur, which is why the séance was called a demonstration and not a proof.

My friend insisted that the séance was completely unconvincing. At least concerning that séance, he was a skeptic and was not expressing healthy skepticism, but outright rejection. The task for paranormalists is to understand why people react in this way and how to better manage perception.

As a general rule, if a person does not have personal experience with a concept, he or she will likely reject it on first exposure. As it is modeled in the Implicit Cosmology, we learn incrementally, as opposed to readily accepting entirely new ideas. My friend was a veteran Spiritualist, but I suspect he had little experience with darkroom séances. Further exposure to such phenomena would likely have incrementally changed his mind.

The researchers involved in the case study [227] I described in the Science Commentary are applauded by other parapsychologists because they are dedicated works in the field. In fact, it is my opinion that people do the best they can, and that it is the limits of their physical capabilities and understanding that determines how they relate to concepts. Accepting the Super-Psi concept simply requires a smaller change in worldview than does Survival.

Paranormalist's Diminishing Influence

Paranormalists represent a small and mostly unorganized community. The skeptical community, on the other hand, is relatively well organized and fast growing. In *Skepticism: The New Religion*, Roy Stemman notes that Spiritualism's public outreach is contracting while the skeptical community is becoming more organized and much more effective in influencing the media. (109)

In fact, the paranormalist community in general is contracting. Yes, those who hunt for ghosts are increasing in number, and there seems to be more parapsychologists. The problem is in the cohesiveness of our community. We remain like the blind men and the elephant. Parapsychologists seem fixated on more proof, ghost hunters tend to prefer *silver bullets* tools for detecting ghosts, and serious ITC practitioners interested in research have become almost nonexistent. While many people accept the need for self-regulation, their solutions are almost always to develop their own, rather than seeking collaboration with like-minded people.

Skeptics, on the other hand, are thriving as a coherent community. A search of the Internet for *skeptic* will produce dozens of pages full of well-designed and often-visited skeptic websites. They have absolute control over Wikipedia and appear to be blessed by the mainstream science community. They have a shared objective to protect humankind from the degenerate influence of pseudoscience and those who would practice pseudoscience concepts.

A case study: Government Acting on Skeptical Views

I am not well informed about how governments have acted against people who have been accused of activity deemed by skeptics to be pseudoscience. Please be sure to examine this for yourself. The first point I would make is that the study of frontier subjects is not protected by law and can go away with the right social pressure. The pattern is simple:

- Establish that anything not specifically supported by mainstream science is pseudoscience (false science). A model for how to do this has been established in Wikipedia* (107) (110) See the Wikipedia Commentary.
- Define pseudoscience as a danger to the public welfare. This is usually accomplished concerning health dangers, but it is also a major focus of the National Science Federation's efforts to educate the public about what is real and not real science. (111) (108)
- Establish that paranormal subjects (psi studies, healing intention, mediumship) are pseudoscience. Again, the skeptic Wikipedia editors have established this as policy for the online encyclopedia. (107) (110) (112)
- Once it is socially okay to say that what paranormalists study contradicts science and may be harmful to the public, it becomes possible for governments and organizations to make examples of paranormalists *for the greater good of society*. We have seen this reenacted many times in our history with everything from witch hunts to internment during the Second World War.

Wilhelm Reich is a good example of such branding. (7) The short story is that he was put in jail for making claims about subtle energy which were deemed by the skeptics to be unsupportable and therefore pseudoscience. He also developed devices that might put the energy to work and claimed he could heal people of some diseases with the energy.

In fact, the actual jail time was because he ignored the charges, and apparently his partner transported some of their material across state lines, which was seen as criminal contempt of court. The following is from the Wilhelm Reich Museum website:

> *While Reich appealed his sentence, the government carried out the destruction of orgone accumulators and literature. In Maine, several boxes of literature were burned, and accumulators and accumulator materials either destroyed or dismantled. In New York City, on August 23, 1956, the FDA supervised the burning of several tons of Reich's publications in one of the city's garbage incinerators.... This destruction of literature constitutes one of the most heinous examples of censorship in United States history. (7)*

Reich died of heart failure while in prison, and I understand his research partner committed suicide shortly thereafter.

Reviewing "What is Orgone Energy?" by Charles R. Kelley, Ph.D. (113) will give you a sense of the nature of this subtle energy which Reich called Orgone. You will also see that Reich's discovery is really just one of many rediscoveries of the same influence field which is today studied as psi functioning and the psi field. Because of its apparent effect on living tissue, the influence intentionality has on this field to heal a person is a primary means of its study. See "An Unusual Form of Radiation has a Reproducible Effect in the Laboratory" by Robert A. Charman (114).

The U.S. Food and Drug Administration took its direction from public pressure brought by the skeptic community. Otherwise, Reich would probably have been seen as just one of many crackpots.

A Case Study: Skeptical Control of the Media

It is obvious from a simple search of the Internet that the skeptics dominate the media when it comes to public outreach about frontier subjects. Yes, there are thousands of ghost hunter websites and websites promoting the many forms of complementary medicine, but if you look for substantive support for the concepts, you run into a wall of skeptical websites supported by skeptic clubs, universities and mainstream science organizations.

An example of how skeptics have had a substantial influence on the media is the Technology, Entertainment and Design (TED) group which produces conferences. TED was formed to disseminate "Ideas worth spreading." According to some reports, videos of the talks are widely distributed via the Internet, and collectively have been viewed by more than a billion people.

TED grants licenses to third parties to organize independent TED-like events known as TEDx conferences. At a TEDx WestHollywood conference,

Rupert Sheldrake gave a talk in which he suggested that modern science is based on ten dogmas, and then made the case that none of them hold up to scrutiny.

The videos of his talk and that of co-presenter Graham Hancock were removed from the TED website when two TED science advisors complained that the talks were *unscientific*. The videos were later returned to the website, albeit in a hard to find section, after a strong response from the paranormalist community.

The struggle between skeptics and supporters over the Sheldrake and Hancock videos is documented in the well-researched book by Craig Weiler: *PSI WARS: TED,* Wikipedia *and the Battle for the Internet.* (115)

The TED altercation between mainstream defenders and paranormalists spilled over to the Rupert Sheldrake biography in *Wikipedia*. (116) There, a long and bitter battle was conducted between skeptic editors trying to use the article to discredit Sheldrake and more balanced editors who were trying to keep the biography of a living person from being used as a tool for skeptic propaganda. Several editors were banned from Wikipedia because of the battle. I was banned from editing that article for life. See the Wikipedia Commentary and Craig Weiler's blog, The Weiler Psi. (117)

The skeptics have complete control of the Sheldrake article, but it is much less biased than when the battle was on. Today, the *View History* and *Talk* tabs provide an important lesson in how skeptics defeat opposing views in all frontier subject articles. Be sure to read the archived talk pages, as the skeptics have been using a *bot tool* to quickly archive material in an effort to remove discussions from the public view.

Healthy Skepticism

There is a balance between *a priori* skepticism and open-minded gullibility. Some claims are not even reasonable judging by the more lenient standards necessary for frontier subjects. Many reported experiences are clearly some form of misperception or the ordinary mistaken as unusual. Every bump in the night is not a ghost and not every instance of improved health is because of healing intention.

On the other hand, some of these reported experiences are not explained by current principles of science and may point to new understanding of nature. It is not reasonable to accept some of the extreme explanations without substantiating research but it is also not reasonable to discount the reports because they are not currently part of known science.

My motto *Believe what you wish but understand the implications of what you believe* is based on the idea that we should practice critical thinking leading to discernment. That means taking time to examine the evidence before adopting a conditional opinion. I say *conditional opinion* because the rest of the story is that whatever is decided should be routinely reexamined to see if it still makes sense. If possible, the opinion should be tested. If there is not sufficient information to arrive at an informed opinion, then no opinion is the only appropriate response. To respond otherwise is to base opinion on faith derived from popular wisdom, superstition and/or the opinion of others who may be even less knowledgeable about the subject.

A fair portion of academia believes all major principles in nature have been discovered and that all we are doing now is filling in the details. Basically that is to say that there is only the physical universe, period! Anything outside of that, such as a psi field, etheric personality and survival of personality after bodily death, is not included in these major principles of nature and therefore cannot be. As I have said, such an *a priori* assumption is Scientism at its worst.

Each of us has a responsibility to practice discernment about what we believe. At the same time, anyone who makes a claim about these phenomena has the responsibility to clearly distinguish between what can be objectively proven and what is believed as a matter of faith. If a claimed study is not based on the applicable scientific methodology with the intention of publishing a report, it cannot be claimed as experimentally proven. Fruits of the study can be described as ongoing study, but be careful not to claim science unless reasonably well-considered methodologies have been applied. See A Personal Code of Ethics for Developing Mindfulness [150].

A second, equally important consideration when someone makes a claim of truth under the cloak of science is the qualifications of the people who conducted the research. Even a person with a doctorate in parapsychology must establish credibility to study the particular subject. As an example, a person claiming to be a parapsychologist and who is academically trained as a philosopher is not qualified to conduct research concerning transcommunication under the cloak of academic authority.

In the same way, a biologist who participates in a study of physical mediumship has no direct academic standing. Yes, the training in research methodology learned for the doctorate may be applicable, but claiming authority as a biologist is a far reach in the study of phenomena which appear to involve small signal processing and perception.

Our field is especially vulnerable to skeptical attacks based on the qualifications of our specialists. Such terms as *experience*, *study* and *research* have specific meaning in the eyes of mainstream academia. Anyone can claim experience with some credibility. Anyone can study phenomena and comment with some degree of credibility. But research reports have little or no credibility unless the work was conducted by an academically trained researcher in an applicable field, and has been reviewed by qualified, informed subject matter specialists.

Of course, citizen scientists are usually capable of conducting very good research, but it is important to be mindful of how mainstream society views this. Each of us is a representative of our field. Skeptics have influence because they are zealous, not because they are right. Yes, there is still the problem that things paranormal are outside of known and therefore acceptable science, but that can never be addressed so long as skeptics are able to make their ridicule of us believable by using our own words and actions. It is for us to show the world they are wrong.

Community

Commentary 7: Wikipedia

Introduction

Wikipedia is a wonderful idea and it includes a huge amount of information, but the fastest way to turn a scholarly dissertation into meaningless blog is to use Wikipedia as a reference. Because of the way the articles are developed, this condemnation is generally true of the entire online encyclopedia. It is especially true of articles concerned with paranormal subjects.

If you search the Internet for substantive support for paranormalist concepts, you run into a wall of skeptical websites supported by skeptic clubs, universities and mainstream science organizations. One of the most dominant of the skeptic's media is Wikipedia. In nearly every search subject, Wikipedia is the first or nearly the first website to come up. Perhaps even more confusing is the circular referencing that is common amongst skeptic websites and Wikipedia.

To see this for yourself, search for your favorite frontier subject using one of the main search tools, such as Google.com. It is probable that an article in Wikipedia will come up first or nearly first. For instance, in a February 6, 2015 Google search for *pseudoscience* and *mediumship*, the Wikipedia article was number one. For healing and *EVP*, it was one and two, while for *ghost*, it was three and four.

Statistics for how many times a Wikipedia article is visited are available for February 4, 2015 and they are Pseudoscience = 1085, mediumship = 404, healing = 182 and EVP = 191 visits. The ATransC website has over four hundred pages and for the same day, the entire site only had 673 unique website visitors. Without a doubt, Wikipedia has much more public reach than ATransC, and likely any website with a positive paranormalist message.

Anyone Can Edit Articles

Anyone can register to be an editor. It is reported that even school-age children are sometimes given an assignment to make changes in Wikipedia. New editors are advised to use a pseudonym instead of their real name and to keep their personal information private. The problem is that people might personally attack an editor for an entry. One of the fastest ways to be permanently banned is to deliberately *out* an editor by revealing the person's name.

The downside of using screen names is that editors can say pretty much whatever they want without fear of reprisal. Naturally, this means immunity to real-world social norms. There are Wiki norms which amount to whatever the dominant group of editors wants them to be.

This is an important reason why I have come to the conclusion that screen names are a hindrance to the exchange of meaningful information and a corrupting factor for all aspects of Internet society.

Wikipedia Rules of Behavior

Wikipedia has rules governing the interaction of editors such as the need to assume good faith and be civil toward other editors. There are also procedural rules, such as how often and why an editor might change another person's edits. It is considered *edit warring* when an editor reverts another editor's entry more than three times. This is an offense that can result in the offending editor being blocked from editing for a time.

Civility is policy, but the dominating skeptics found civility just got in the way, and so today it is okay to attack people outside of the skeptic group, but hell to pay for outsiders to attack skeptics. This is best seen by reading article talk pages recorded during more contentious periods.

An editor's failure to follow the rules is usually addressed by other editors, but if that does not work, then it is possible to bring an editor before a tribunal that has the power to ban an editor from making further contributions. I was banned from ever editing the Rupert Sheldrake article because several skeptical editors took me to the administrators claiming I was a one subject editor. That one subject was anything paranormal so that now I am one complaint away from being banned from ever editing any Wikipedia article in my field. In this way, the now dominant skeptics are able to drive away contending editors.

Subject Matter Specialists

The policy of Wikipedia, or at least the dominant editor's policy, is that subject matter specialists are discouraged from editing articles within their area of expertise. (118) (119) In fact, it is common for subject matter specialists to be so abused that they soon stop attempting to contribute content. One of the Wikipedia founders, Larry Sanger, has even written the article, "Why Wikipedia Must Jettison Its Anti-Elitism," explaining the pitfalls of editors not using their real names (no accountability) and not being knowledgeable about the subjects they edit. His response was to begin Citizendium.org, which is supposed to be a kinder and more dependable online encyclopedia.

I did attempt to organize paranormalists to work on paranormal articles in Citizendium. There is now a paranormal subgroup at en.citizendium.org/wiki/CZ:Paranormal_Subgroup. Included there is the structure for articles and indexes, but no editing contributions from others beyond what I put together.

The two contributions that came from existing Citizendium editors were deletions of some of the material I had added. Had there been other paranormalist editors, I could have reversed those, but instead, I have copied the material to the Collective (ATransC.org/bp/Collective) and given up on the publicly edited online encyclopedia.

Wikipedia Editing Rules

Perhaps most important are the rules governing what may be included in articles. For instance, articles are required to be written from a neutral point of view and everything in articles must be based on verifiable references. The references themselves must be from mainstream publications and must not be self-serving.

All of the parapsychological journals are considered fringe and their content is not acceptable as a reference. (112) This effectively assures that only the less informed, usually more anti-paranormal publications are acceptable. Of course, this means the skeptical publications are acceptable.

An interesting consequence of not allowing subject-matter specialists or frontier subject publications is that Wikipedia articles about paranormal subjects are usually out of date. For instance, the article on Spiritualism is virtually all historical, so that one would get the impression that Spiritualists are still living in the early 1900s. Virtually the entire EVP article is out of date, and there are no articles for Instrumental TransCommunication (ITC) or physical mediumship.

This past tense bias in articles also has the effect of slowing progress. A person seeking information about a paranormal subject likely ends up at an authoritative sounding Wikipedia article that provides an erroneous sense of the state of the art.

It is reasonable to expect that young people's understanding is likely based on what is in Wikipedia. In fact, it appears that most people depend on Wikipedia, ghost hunting television programs and club websites for their understanding of the paranormalist community. As a note to the paranormalist community, this array of misleading media must certainly serve the skeptic's objective of marginalizing our study of these phenomena.

Circular Referencing

One of the more blatant ways to bias an article is to first have someone say what you want in an article published in an acceptable publication. Sometimes a skeptic blog will do. The article is then used as a reference to make the editor's point.

An example of how this works can be found on the Pseudoscience talk page (107) in the section titled "Learning styles as pseudoscience." The section will probably be archived by the time this book is published, so look in the Talk Page archive for the section title and/or date. I have only included the final, January 15, 2015 comment here.

Wikipedia editor *User:Puedo82* wanted to have learning styles (120) added to the subjects listed in the pseudoscience article. *User:Jytdog* has a problem with *User:Puedo82's* references because none of them specifically name learning styles as pseudoscience. In the end, *User:Jytdog* advises:

> *Here is what I suggest. See if you can succeed in having the word "pseudoscience" included in the lead of the learning style article. [on Wikipedia] The word does not appear in that article, as of today. To have the word "pseudoscience" in the lead, you will need support for it in the body, per WP:LEAD. If you can do that, I think you will have no problem getting that listed here. But this is not the place to make the determination. Jytdog 15 January 2015* [46]

[46] Briefly, WP:LEAD says that the lead is just a summary of the body; nothing should be in the lead that is not in the body. *Pseudoscience* is now in the Learning Styles article.

The practice of tailoring articles in *acceptable* publications to specifically support the dominant group of editor's view is common and effectively defeats the rule requiring authoritative references to support statements.

Foundation Rules

The fundamental principles of Wikipedia are described as the Five Pillars: (121)

1. Wikipedia is an encyclopedia.
2. Wikipedia is written from a neutral point of view.
3. Wikipedia is free content that anyone can use, edit and distribute.
4. Editors should treat each other with respect and civility.
5. Wikipedia has no firm rules.

I say that skeptics control the paranormal articles in Wikipedia, but the cultural norm for the entire encyclopedia appears to be based on scientism. Three major arbitration cases, which I believe were moderated by arbitrators from all around Wikipedia, specifically concern paranormal subjects but likely have far-reaching implications for the entire wiki.

Information about these is accessible via the Viewpoint section of ethericstudies.org. Also note that arbitrators are senior editors elected by the dominant group of editors.

Collectively, these arbitration cases establish official Wikipedia approval for what is considered acceptable treatment of paranormal subjects:

Requests for arbitration/Fringe science (2009)

The arbitrators generally found in favor of the skeptics with a series of decisions. (112) The most important ones include:

Principle 9) Relevant comparisons: *The prominence of fringe views need to be put in perspective relative to the views of the entire encompassing field; limiting that relative perspective to a restricted subset of specialists or only amongst the proponents of that view is, necessarily, biased and unrepresentative.*

Principle 10) Advocacy: *Wikipedia is not for advocacy. The purpose of an encyclopedia is to state neutrally the current knowledge in a field, not to put forward arguments to promote or deride any particular view. In particular, conjectures that hold significant prominence must no more be suppressed than be promoted as factual.*

Principle 11) Citations: *Citations should not be used disproportionately to the prominence of the view they are citing or in a manner that conveys undue weight. Exceptional claims in Wikipedia require high-quality sources; if such sources are not available, the material should not be included.*

Principle 13) Scientific focus: *Wikipedia is an encyclopedia and its content on scientific and quasi-scientific topics will primarily reflect current mainstream scientific consensus.*[47]

Requests for arbitration/ Paranormal (2007): **(122)**

Principle 3) Basis for inclusion: *In addition to firmly established scientific truth, Wikipedia contains many other types of information. "The threshold for inclusion in Wikipedia is verifiability, not truth" (from Wikipedia:Verifiability).*[48]

Principle 6.1) Appropriate handling of epistemological status: *It is the responsibility of editors to appropriately handle any question regarding the epistemological status of a subject, that is, questions of whether something exists, is hypothesized to exist, general scientific consensus, etc. The goal is not arrival at the correct conclusion, but adequate treatment of any controversy.*

Principle 9.1) Editorial judgment regarding reliability: *Determining the reliability of sources is a matter of sound editorial judgment informed by expertise. Exceptional claims should be supported by strong sources. Sensationalist sources, when used at all, should not be the sole sources for an article. Topics for which no reliable source can be found are not suitable for inclusion in Wikipedia.*

Principle 10.1) Conflict of interest: ***Wikipedia:*** *Conflict of interest, strongly cautions but does not forbid an editor from working in subject areas where the editor is strongly invested. Such editing must be done responsibly. Other*

---

[47] This is my whole point. Skeptics know the power of Wikipedia to influence the public. For every word they fight to keep in an articles, such as is it pseudoscience or not, or is Rupert Sheldrake to be called a scientist or not, they are clearly editing to sway public opinion.

[48] Here, *scientific truth* translates as *"only what mainstream science says is true." Verifiability* is based on mainstream media, since the Fringe arbitration gave skeptics permission to reject virtually all of the frontier subject media unless it supports their point of view.

editors are expected to respond diplomatically even when they believe a conflict of interest may exist.[49]

Principle 11) Generally considered pseudoscience: *Theories which have a following, such as various manifestations of the paranormal, but which are generally considered pseudoscience by the scientific community may, with adequate sourcing, properly contain that information and may be categorized as pseudoscience.*

Finding of Fact 3) Status of parapsychology: *Parapsychology has an ambiguous status, engaging in scientific research, but strongly criticized for lack of rigor.*

Finding of Fact 7) External campaigning: *Activists, including a Tom Butler, have put up pages which campaign regarding the content of Wikipedia articles...*[50]

Value of Wikipedia

Because of the requirement that all substantive statements have a reliable source, virtually all material in Wikipedia articles is at least secondhand and often based on very outdated material. Because books are preferred over websites as being more academic, and it often takes years to publish a book, it is common to find book references that have long-since been outdated by new research published in journals and on websites. More importantly, references are often used that are unavailable to the reader, making it nearly impossible to verify that the included information is actually supported. Too often, it is not.

An article published in a science journal, which is written by a researcher about his or her research, is considered original research. This means that information about research must be written by someone else, which usually means secondhand and potentially inaccurate reports are preferred.

Small and/or non-mainstream publications are considered fringe, and are therefore easily discounted by an editor determined not to allow their use.

---

[49] This is a restatement of prior policy, but including it in the Pseudoscience arbitration gives it special meaning. I have been admonished a number of times with the *strongly cautions* of this principle being read as *definitely not allowed.*

[50] This is a good illustration of the embattled Wikipedia mentality. The complaint concerns an article I wrote for the *ATransC NewsJournal* about how paranormal subjects are treated. Of course I would write about the subject, but Wikipedia editors seem to believe they are above public criticism.

In most frontier subjects, there are only small publications because of the immaturity of the field and small readership. At the same time, mainstream publications will not venture to publish a positive report about a frontier subject. Also, major book publishers will not invest the resources to publish a book intended as a serious research report unless there is a large audience.

All of this means that collaboration in frontier subjects is accomplished via newsletters, self-published books and websites. The most current research information is too often on the same search engine page as hobbyists speculating about the subject from a point of view of *"how it helped me today,"* rather than whether or not it has any validity in fact.

Using Wikipedia

First and foremost, Wikipedia should never be used as a reference. There really are no exceptions. Since anyone can edit, there is no way of knowing if comments are reasonably correct or if the references supporting an article actually say what is needed to support the information.

The articles are changed as new people come along with different opinions so that what is in the article one day may not be there the next. It is possible that a useful comment in an article will be changed so that your reader might visit your reference and find a contradictory statement.

Since every substantive comment is supposed to be supported with a reliable reference, Wikipedia articles should be excellent research tools and study guides. If you want to know about a subject, read the Wikipedia article, but then use the references to see what the subject-matter specialists have to say. Be mindful though, that references from frontier subject journals will probably be unrepresented if there are any at all. In the Rupert Sheldrake article, references are something like ten to one in support of pro skeptic points of view. (116)

It is informative to follow the *View History* tab to see if there is an edit war underway. Also follow the *Talk* tab to see if there are any current controversies that might make the article unstable. A new trick of the skeptics is to hide comments on the talk page with a very quick archive bot. Comments on the talk page of the Rupert Sheldrake article, for instance, are almost all in the archive. Look for the Archive box.

Personal Attacks

It is common for skeptical editors to denounce anyone who studies frontier subjects as morons, idiots, deluded, charlatans and frauds. There are administrative-level editors and procedures to request help with such abuse, but in many instances, complaints are answered by a barrage of comments agreeing with the original insult and adding many more disparaging words to the list. In the end, it is an inescapable conclusion that Wikipedia intends to maintain a civil work environment, but is unable to apply existing rules to protect editors from other editors. The common term is *poisoned atmosphere.*

I was an active editor for a couple of years and suffered considerable abuse from the skeptic editors both in Wikipedia and elsewhere on the Internet. All were childish attacks on me and any point I might have been trying to make. The attacks were not the problem so much as the poisoned atmosphere that allowed them to seem to be okay and the arrogant attitude of the attackers.

Occasionally, there were individuals who stood with me. These editors did not necessarily understand paranormal phenomena, but did have a keen sense of fairness and saw the need for balance in articles. For instance, some did not like Rupert Sheldrake's ideas, but felt a proper encyclopedia needed to explain Sheldrake's hypotheses in a neutral way that would not endorse or condemn while letting the reader know what they included. Each balanced editor working on that article was eventually driven out of Wikipedia with lifelong bans or such severe personal attacks that they could not stay.

As an indicator of how difficult it can be for a frontier subject editor, Wikipedia at least did have a civility board on which I once posted a complaint about verbal abuse. Ironically, I was attacked more aggressively there than in any other part of Wikipedia.

EVP Article

It occurs to me that some readers might wonder why, if I am an editor and knowledgeable about EVP, the Electronic Voice Phenomena article is in such bad shape. Your first hint is the title. It was *phenomena* (plural as it is used in the community) but now it is *phenomenon* even though the opening sentence begins with phenomena. That was just one of the battles I lost.

My original reason for becoming an editor was to update a very poorly written EVP article. I began by suggesting a paper that has since evolved into the *Transcommunication White* Paper (123) which is available for download from ATransC.org. The other editors threw a fit because it is not done that way, and so, I learned my first lesson about Wikipedia culture.

One or two neutral editors and several hardcore skeptics joined the discussion. What followed was a long edit war in which I was officially charged with various wiki crimes such as conflict of interest, pushing original research and bias. In the end, the very fact that I was a director of the ATransC (then it was the AA-EVP) proved that I had a conflict of interest.[51]

All of the publications I cited to support EVP theory were classified as fringe, and therefore unreliable, as the result of the Fringe Science [255] arbitration. (112) They were not allowed while blogs, columns and hearsay articles written by skeptics were allowed as reliable. On the other hand, any of our *fringe* articles are allowed if it proves the skeptic's point.

Some of our best researchers were attacked as crackpots and I was routinely insulted off-wiki in skeptical forums. Again, the qualifications of a psychologist or magician skeptic to pronounce with authority about EVP was given respect while parapsychologists were deemed unqualified.

Many of the explanations for EVP in the article are a priori pronouncements that sound good but have no empirical foundation. Even meteor showers are given credence, for instance: *"there is a possibility of meteor reflection of the radio signal."* None of the more reasonable speculation is included because it depends on parapsychological research which is not allowed. Also, original research is not allowed, while reports from a wiki-reliable second sources are. Virtually all work in our little community is original research.

One of the more irritating surprises was that negative comments made by people in the paranormalist community were used as proof these phenomena are not real. It is especially damaging when one of our own denounces our work. Consider these quotes from the Wikipedia article:

- **Rorschach Audio:** *"And a broad class of phenomena referred to by author Joe Banks as Rorschach Audio has been described as a global explanation for all manifestations of EVP."*[52]

---

[51] Because of my scope of study, I can probably be banned from editing any paranormal article for potential conflict of interest. It is because of the nature of our community and how Wikipedia rules are enforced that virtually any active member of our community can be banned for one such Wiki rule or another. Again, it is at the convenience of the controlling skeptics.

[52] Banks used old, usually Class B and C examples from the beginning of EVP and did not seem to think the more contemporary and better authenticated Class A material worth using to balance his work.

- Rosemary Guiley has written, *"Within the parapsychology establishment, Rogo was often faulted for poor scholarship, which, critics said, led to erroneous conclusions."*[53]
- Imants Barušs: *"While we did replicate EVP in the weak sense of finding voices on audio tapes, none of the phenomena found in our study was clearly anomalous, let alone attributable to discarnate beings. Hence we have failed to replicate EVP in the strong sense."*[54]

In the EVP article, if you click on the reference number associated with a statement, the system will take you to the reference at the bottom of the article. At that reference, you can click on the *up* symbol at the beginning of the article to see where it is used. If there are also lowercase letters, they will take you to additional uses of the reference.

Note that many of the references are only allowed because they are negative and not because they are authoritative.

I am not allowed to edit the article because of potential conflict of interest. My failed attempt to support the development of an informative article within the bounds of Wikipedia policy has taught me a lot about the skeptical world. It has also caused me to *go it alone* in terms of trying to inform the public. My loss of patience for long, drawn out arguments over what is truth is collateral damage from that experience.

As you probably see by now, the paranormalist community has not stepped up to the task of reasonable public education. Unthinking comments as I excerpted above are quickly used to prove we are delusional. We are our worst enemy in this battle for respectability.

Passive Paranormalist

As I have previously stated, the reason paranormal articles in Wikipedia are controlled by skeptics is because they outnumber moderate editors so

---

[53] In fact, anything to do with EVP in Rogo's day was as taboo as it is today amongst parapsychologists. A common skeptic tactic is to deflect the reader from the facts by attacking procedures that produced the facts. By including that kind of aspersion in her writing, Guiley just gave the skeptics another talking point.

[54] I have discussed Barušs' "Failure to Replicate" article on my website. (96) In fact, his protocol was fatally flawed, yet his work is happily published by one of our parapsychological journals and widely quoted by the skeptics.

completely that they are able to decide how the Five Pillars are applied as practice. This has not always been the case.

When I first began editing Wikipedia, the arbitration cases had not been translated into practice and there was still an ongoing discussion about what was acceptable for inclusion in articles. I tried very hard to entice other paranormalists to join into the discussion. I know others in our community also tried. Hardly anyone joined in and virtually none stayed to help. The Wikipedia we have today is arguably the result of paranormalist complacency.

You will find many places in my writing in which I say that paranormalists are a gentle folk who prefer not to get involved in disagreements such as those that are all too common on Wikipedia. In fact, I have had a number of paranormalists privately tell me that they would never subject themselves to what I have experienced.

The problem, I think, is that we are still developing a sense of community and cooperative efforts, such as establishing a strong group of editors in Wikipedia, are simply not yet part of our cultural norm.

It is my hope that these commentaries might help to change that.

Alternative to Wikipedia

Wikipedia is an open source WikiMedia project: "The Wikimedia Foundation, Inc. is a nonprofit charitable organization dedicated to encouraging the growth, development and distribution of free, multilingual, educational content, and to providing the full content of these wiki-based projects to the public free of charge." (124)[55]

The WikiMedia platform is probably the most effective tool I have found on the Internet for collaborative article development. It continues to be maintained with frequent updates, improvements and security fixes. It also provides a means by which the public is able to see the interaction of editors and evolution of articles. Assuming editors use their real names and take the time to include a brief description of their qualifications on their user page, this extraordinary visibility would amount to an effective alternative to peer review for observant readers.

[55] All of the websites I manage use the Creative Commons "Attribution-ShareAlike 3.0 Unported license: Share—copy and redistribute the material in any medium or format; Adapt—remix, transform, and build upon the material for any purpose, even commercially." (125)

Because the arbitration cases noted above are now part of Wikipedia rule, it is unlikely that Wikipedia can be changed toward a more neutral encyclopedia; however, it is possible that an alternative wiki could be developed. With proper attribution, the entirety of Wikipedia could be ported to a new wiki under new management that is more agreeable to balanced reporting. In principle, Citizendium (126) is such an alternative wiki, but again, it would require a group of editors working together, as right now it is still pretty much dominated by editors leaning toward scientism.

The Collective (67) is a WikiMidia wiki dedicated to developing best practices [387]. It is sponsored by the nonprofit ATransC. We are always seeking qualified volunteers to help develop articles.

Wiki Aggregator

A number of discussion boards and alternative wikis are showing up on paranormalist sites. Of course, there is the ATransC Idea Exchange and Collective, but some other organizations also have their version. All are probably doomed to obscurity, though. The problem is that, rather than cooperating with other groups, each group, including the ones I manage, has preferred to develop its own, presumably because of special needs.

A natural result of these islands of information is that it is difficult for them to rise very high in search results, meaning that they are mostly lost to those who are not part of that community of interest. One alternative would be to encourage the development of such specialty knowledge bases, and then as a community, to develop an aggregator tool which allows people to conduct information searches specifically in those sites. I have set up a Google search tool at ATransC.org/links_community.htm that does this for some sites, but the need is for a much more sophisticated tool. (127) Such a tool would amount to a virtual alternative wiki.

On Being an Editor

Until recently, it had been my policy to encourage people to become a Wikipedia editor and help in the effort to balance articles. However, now it is clearly a lost cause. The only solution is a policy change and the managers have stated that they are happy with the current environment.

While I think Wikipedia serves an important service in the mainstream articles, the online encyclopedia is just a propaganda tool for skeptics. As such, it is taking up a niche on the Internet that would be better filled by a new effort.

Here are a few things to consider for community information sharing:

- Blogs and personal websites are a good place to develop your ideas because they tend to be stable for public assess.
- Public forums are too freeform for effective long-term idea development, and new comments quickly push old ones out of sight.
- Unless a wiki is managed by paranormalists, it will be taken over by skeptics. Forums such as the ATransC Collective wiki at ATransC.org/bp/Collective (4) are a safe place to help develop community practices and useful informational articles.
- Aggregator tools such as the dedicated search tool at ATransC.org/links_community.htm enable a person to access information from many included sites. Please let me know if you wish to have your website added to the ATransC search tool.

It is by sharing your unique perspective with others that we all gain understanding and by which we develop a cohesive cooperative community. Personal blogs and websites allow you to state your point without needing to consider other people's input, but you can probably expect the number of people reading your material to be very small.

Cooperative websites develop much greater audiences, but you will probably need to work with others to develop a shared view. We have seen with Wikipedia that such communal efforts are very effective. In most cases, I consider reaching a consensus view for an article worth sacrificing my opinion in favor of a shared opinion. In turn, I almost always learn.

Conclusion

Wikipedia is a very important tool for accessing information about noncontroversial subjects and as a study guide. However, be mindful that it has become a platform for social engineering. Because of its editing rules, articles are almost always out of date. This means that even the references are probably little better than suggestions for what you should be looking for to better understand the subject.

Whenever I see someone using Wikipedia as a reference in an article intended to be taken seriously, my first impression is that the person has not done the necessary work to understand the references, and therefore, the subject. If you want to be taken seriously as a subject-matter specialist, it is in your best interest not to use Wikipedia as a reference.

Community

Commentary 8: Perspective

My Reader

Being mindful means pausing to consider personal actions *as we act.* It also means taking time to see ourselves as others see us. We do this so that we can be aware of how our actions affect others, and whether we are communicating as we intend.

In the essay "Clarity of Communications," (128) I discussed how important it is to understand that there are different ways people relate to information, and therefore, how they *unconsciously* process our words and actions into *conscious* awareness. Of course, this is based on the model used in the Perception and Expression Discourse. In that, I explain that First Sight Theory (13) argues there is a difference between unconscious processing of information and conscious awareness of the results of that processing. Since mainstream science is beginning to support that model, it is reasonable to argue that a person (conscious self as *"I think I am this"*) is the product of unconscious mental processes which are based on prior conditioning of which we are hardly aware.

The influence of this *wizard behind the curtain* manifests as a person's point of view. Because our communication with others is really addressing this *wizard* in their unconscious mind, it is necessary for us to carefully select our tone.

The model I use is discussed in the Point of View Commentary under Temperament. As illustrated in the Personality Styles Diagram [120], it uses a simple four-sided model: Driver, Analytical, Expressive and Amiable. The model helps me to be mindful of reader diversity. Much of my writing is technical—not just technology technical, but also conceptually complex. As I write, I try to imagine some of my less analytical friends reading my material. Will it make sense to them? Perhaps not, no matter how many ways I make the point, but at least more will understand than if I only wrote for my way of learning.

Learning styles hints at a person's point of view. That is the Idea I tried to convey in the Science Commentary as I described the various communities of interest and their focus. You may remember in that commentary that I had three ways in which the community is categorized:

- The Dominant Views of Reality in the paranormalist community are psi field, Trans-Survival and transcommunication.
- These are considered from the human psychology, emergent science, human potential, social paranormal and metaphysical points of view.
- The communities of interest that consider them are parapsychology, citizen scientists, mediumistic practitioners, hobbyists and seekers.

It is the point of view that must be given the most consideration, but each community of interest presumably attracts people with a specific point of view.

Key 9: The Hermit: From the Major Arcana of the Tarot in the Waite deck. (Colors are not true)

Realistically Speaking

A key concept from ancient wisdom schools is that our lot is to learn, and having learned, our lot is to teach. In the Progression, Teaching and the Community Discourse, I begin with:

The Way of Progression

Through community comes knowledge

Through teaching comes understanding

It takes a collective

From the perspective of the Implicit Cosmology, we are born with an imperative to gain understanding which is expressed as progression. It is through the act of teaching that we compile the disparate bits of information we collect in our lifetime into a kind of gestalt understanding. It is this understanding that we take with us to the other side beyond transition. Teaching is one of the best ways to integrate that understanding because it requires us to organize our thoughts via conscious consideration of worldview.

As an idealist, it is natural for me to occasionally dramatize a concept with imagery intended to appeal to our inner wizard. You may recognize the Hermit, key 9 of the Tarot's Major Arcane (59) from the ethericstudies.org banner. (If you are studying the Tarot, you will know the robe is supposed to be gray and the background is supposed to be black or deep blue.)

No, it does not symbolize my ideal me. It is intended to represent *The Way of Progression* noted above. In the Tarot, the Hermit represents the consciousness of a person who has attained some notable amount of understanding. Having been a seeker, the Hermit is now the wayshower or teacher. He holds the light high above the path so that others may know the way. In effect, *"As I am, so may you be."*

An example of a wayshower is found in John 14-5 and 14-6 of the Christian Bible. Jesus is preparing his disciples for his arrest and certain death. Here, he is explaining that they too would be able to go where he was going (to heaven) and that they would know the way.

> John 14-5: *Thomas saith unto him, Lord, we know not whither thou goest; and how can we know the way?*
>
> John 14-6: *Jesus saith unto him, I am the way, the truth and the life: no man cometh unto the Father, but by me.*

Jesus is saying that he represents the three aspects of a teacher. The Hermit also represents those three aspects: a seeker who has achieved a degree of understanding, has integrated that understanding into his or her life and has paused to show the way to those who would come after.

So the Hermit is symbolic of what I believe all of us who count ourselves as seekers will become. He is also symbolic of the path we will eventually find as we become the Hermit.

The Village Model

In the Point of View Commentary, I described the Sheep-Goat Model in terms of Villager-Explorer [199]. You will remember that the sheep in Gertrude Schmeidler's model were the believers in paranormal phenomena and the goats were the nonbelievers. I prefer a more flattering model of explorers for people who believe in paranormal phenomena and villagers for those who do not.

A second concept to consider here is the blind men and an elephant parable. In that, a number of blind men examined an elephant by touch with the expectation of describing what it looked like. As the story goes, each individual perspective produced a different description of the elephant. I used this parable in the Community Commentary to make the point that our community's acting as many isolated communities of interest cannot possibly see enough of the paranormal phenomena to properly understand the whole.

If the villager-explorer model is expanded to also consider their perspectives, then it might be useful as an updated version of the blind men and elephant parable. Think in terms of the village center as mainstream society, the village outskirts as our paranormalist community and the wilderness as reality. The idea is that the village is surrounded by and profoundly influenced by the wilderness. And in actuality, the village is a human construct, a human expression of reality.

The people living on the outskirts of the village are more directly influenced by the wilderness, and as explorers come and go, they hear tales of wonders that the mainstream citizens cannot comprehend. Here, the explorers are the conscious seekers who hold the light for others to see.

Legacy

While I do not see myself as any kind of wayshower, I do understand that some of what I have learned may be of value to others. The tools I came into this lifetime with are well suited for comprehending complex concepts. If you are familiar with the story of the Tent Maker, then you will understand when I say that the life path I stumbled upon has given me objective tools, such as technical writing and training in technology, which serve to facilitate the preservation of what I have learned for others to study.

Personal ability and life experiences matter little if they are not directed into useful fields of study. Lisa has been an important navigator for me by involving me in interests that I would not have considered on my own. Spiritualism, Electronic Voice Phenomena and mediumship are good examples. More likely, I would have focused my life on technology. While I like ancient wisdom and complex theories, it is the hands-on application of technology that called me.

So if you find value in this book, thank Lisa, as it would not exist without her. If I did not wish to put some of my more controversial words in her mouth, her name would be on this book as co-author.

You are the rest of the story. It is for you to step onto the path. Perhaps the Implicit Cosmology is not your way, but surely you will recognize the power it suggests you have in shaping your destiny. You came into this lifetime with an urge to gain understanding. In my mind, it is impossible to fulfill that goal without eventually teaching others. Your personality understands this, as it is expressed in everything you do.

These twin urges do not dominate you, perhaps they only make you restless. Probably, they are lost in the noise of your body's instinctual urges and its natural inclination to play. Thus, your first initiation is to recognize that there is a path and that consciously deciding that entering into the mindful way is the only way to satisfy those urges.

While we say "Do unto others as you would have them do unto you," perhaps we should be saying, "Teach me as I teach you."

Section III
Transcommunication

Introduction

The *Trans-* Prefix

In the Implicit Cosmology, the *trans-* prefix signifies *from the etheric aspect of reality to the physical aspect.* This can be considered in the same sense as *across the veil of forgetfulness.* Transcommunication then, is two-way communication between the etheric and physical. This might also be phrased as a *trans-etheric influence.*

Technically, all trans-etheric influences are considered intentional expressions of a personality; however, *transcommunication* is a term generally reserved for influences that result in some form of objective information. This includes mediumship, After-Death Communication (ADC) and Instrumental TransCommunication (ITC). Psychokinetic activity such as levitation and apports are considered indirect form of transcommunication because they are assumed to occur under intelligent control for a specific purpose.

Out-of-body experiences, near-death experiences and dreams are not usually included; however, there might be circumstances in which such experiences result in objective information, and as such, those instances would be considered transcommunication.

As a historical note, Ernst Senkowski (129) coined the term, Instrumental TransCommunication (ITC) to describe *"meaningful anomalies including extraordinary voices, computer text, recorded images and images found on video displays suggestive of a continuance of life after death."* Note that after-death communication and mediumship should be considered forms of transcommunication but are not necessarily ITC. (Because of common usage in this field, terms with a *trans-* prefix can be used without the hyphen.)

Transcommunication

Study 1:
Foundation Principles

Introduction

Transcommunication is explained here as it relates to the Implicit Cosmology. To understand the nature of transcommunication then, it is necessary to be familiar with a few important concepts. These are explained in detail in the body of this work and are briefly covered here. Please refer to the referenced articles for more complete explanations and please do use the Glossary of Terms.

Once again, you are invited to register on the ATransC Idea Exchange discussion board at ATransC.org/forum/ to ask questions, discuss concepts and share results.

Organizing Principles

All of the principles listed in the Organizing Principles Discourse apply to transcommunication. The principles listed here are the most influential.

Expression: *Reality is expressed via personality's attention on an imagined outcome with the intention to make it so.* Intended order is the influence expression has on the environment. It may be in the form of conscious awareness of an output of the perceptual process, speech, order impressed on a chaotic process or an ideoplastic form. Expression is initiated by conscious intention but the formation of expression is a mostly unconscious process of the Attention Complex. See the Perception and Expression Discourse.

In transcommunication, order in the form of recognizable images, objects or sounds is known to *emerge* out of chaotic signals. In Instrumental TransCommunication, the physical process that enables this emergence of intended order appears to be stochastic amplification acting on chaotic, broad-spectrum energy. See the Perception and Expression Discourse.

Perceptual Agreement: *Personality must be in perceptual agreement with the aspect of reality with which it will associate.* This organizing principle provides a mechanism by which order can be imposed on a system without many complex rules limiting the behavior of individual life fields (personalities). It results in the existence of the Veil of Forgetfulness (130), which is the practical consequence of the etheric-physical interface. The link between conscious self (*I think I am this*) and personality (*I am this*), as it is associated with a human avatar, is filtered by worldview in the Attention Complex. See the Perception and Expression Discourse.

The practical consequence of this principle is that the etheric personality is not able to directly influence the physical, but must rely on its conscious self, which is entangled with a human in an avatar relationship.

Entanglement: *Personality remains associated with its expressions via attention.* The Principle of Entanglement is seen as a property inherent in the process of expression. In this model, expression is in the form of an aspect of a personality (*Little Me*) or an ideoplastic structure. These remain associated with the creating personality by way of attention and the extent to which personality intends that the expressions continue to exist.

Personal Reality: *Perception of reality defines personal reality.* The *database* defining perception is contained in worldview as part of the Attention Complex. At birth, worldview is populated with instincts from the human body, with some level of understanding about reality and an urge to gain further understanding inherited from etheric personality. The body's instincts are dominant in the person's early years, but personality, by way of conscious self, is expected to eventually gain control of worldview as the person matures. Even so, the instincts and beliefs taught to the person by family, school and media tend to dominate the person's sense of reality, right and wrong (cultural norms).

Important Concepts

The relationship between our etheric personality and our human body is modeled as an avatar relationship. Understanding the following concepts will help you understand this relationship:

Avatar: In Hinduism, an avatar (from Sanskrit for *descent*) is a deliberate descent of a deity to Earth. The term can be translated into English as *incarnation*. A person is an etheric personality entangled with a human body in an avatar relationship. See the Life Field Complex Diagram [xxv].

Life Field: The field concept dominates this cosmology. Fields are generally referred to as etheric fields of which there are many forms. In this concept, reality consists of life fields and expressions of life fields. See the Life Fields Discourse.

The Life Field Complex Diagram [xxv] shows the entangled relationship between personality, body mind and physical body. All of the functional areas are in the etheric except for the physical body itself. This includes mind, memory, intention, instincts and body image. Practical consequences of this avatar relationship include:

- **Unconscious expression:** As described in the Perception and Expression Discourse, expression is the output of the Attention Complex. It is informed by worldview, so that expressions are colored by cultural beliefs and human instincts. Expression can be *outward* from the life field into the environment, but it is also *inward* as thought, memory, dreams and imaginings. Expression is bounded by perceptual agreement.
- **Unconscious worldview defines personal reality:** How a person consciously experiences the world is based on worldview. In childhood, human instincts tend to dominate worldview, and as such, a young person must be taught to be a citizen. Maturing is the process of learning how to manage instincts with understanding gained through experience.

  Right and *wrong* are cultural artifacts which the person must be taught by the community. Instincts of the body must also be managed by the entangled personality. From a cosmological perspective, the Organizing Principles represent enduring understanding leading to perceptual agreement.
- **Body Mind is not the same as worldview:** The body is governed by body image which is managed by body mind. Via the Perceptual Loop, gradual changes in the body are capable of modifying the body image. Aging is a gradual change. Also, there are culturally developed expectations that age equals old. In fact, this gradual decline may not be necessary but is driven by that culturally developed expectation.
- **All perception is psi; objectivity is assigned:** All of the functional areas are in etheric space except the physical structure of the human body and the ideoplastic structures [104] that result in physical objects as the person's expression. That which we consider physically objective is a convention used to color perception. In the Implicit Cosmology, the only difference between perception of a distant object sensed via clairvoyance and perception of a nearby object sensed via the body's sight and touch is how we have been taught to assign meaning.

Perception and Expression: The Perception and Expression Diagram [xxiv] illustrates the information flow for how we think based on the most current science I can find. (See the Perception and Expression Discourse for more details.) All of the functional areas are in the etheric and are better thought of as nonphysical fields which are bound into a system by the influence of attention. As such the Attention Complex represents our unconscious mind. Important features to note in the diagram include:

- **Transcommunication:** Personality interfaces with other personalities via the Attention Complex. Interaction is limited by perceptual agreement as defined in worldview.
- **Mind is like an *Internet IP Address*:** Only external signals (psi influences, physical signals) of interest to the person are passed by the Attention Limiter [111].
- **Worldview limits perception:** In effect, the Perceptual Loop asks the question, *"Is this real?"* of the information passed by the Attention Limiter. The answer is based on worldview which functions as the moderator for what is *real*.
- **Worldview can be incrementally changed:** Worldview represents habit, and as such, experiencing what is familiar requires much less effort than that required to experience something new. A slight or ambiguous difference between worldview and sensed information is most apt to change worldview while a great difference will likely be rejected (ignored by the conscious self).
- **Perception develops moment-by-moment:** The Perceptual Loop is a streaming process so that environmental information is continuously sampled. The effect is an evolving perception as more information becomes available. This may explain stream-of-thought dreams in which one moment suggests the next but not always to produce a logical experience.

Stochastic Amplification

The most common form of EVP is known as transform EVP because they are known to depend on the presence of broad-spectrum audio energy for formation. Also, most visual forms of ITC are induced by causing the presence of chaotic visible light which is transformed into features, apparently, according to intended order. Precipitation phenomena are sometimes associated with the availability of raw material in the physical which is thought to be apported to and then deposited in its new location according intended order expressed by the communicating personality. An apport is modeled as decomposition of objects at their original physical location and precipitation of them in a new location.

The common factors appear to be chaotic energy and intended order. As such, our working hypothesis is that the creative process involves attention on an imagined outcome to produce an intended order in a chaotic system. Here, a *chaotic system* is any physical system, field or material that is not confined by design or function to a particular order. Audio-frequency noise with many irregularities, including short spikes of noise, is useful for transform EVP formation. Medium-density light is considered an ideal chaotic source for feature formation in visual ITC. An electronic device that is engineered so as not to allow processing of unexpected signals is not a chaotic system. An example is a shortwave radio with squelch control. Within the constraints of the energy barriers of the material, semiconductor junctions may produce chaotic signals.

Stochastic resonance is an effect that results in a weak signal being amplified in a nonlinear system when a large noise background is applied. (131) It is speculated that a weak psychokinetic psi influence (signal) is made stronger via stochastic amplification. In electronic circuits, the active regions of tubes and transistors provide the necessary nonlinear condition for amplification. It is the combination of the electronic environment and the presence of a chaotic signal that we think makes psi influence of electronic equipment so common. By comparison, a physical object such as a cupboard door does not provide the necessary leverage to amplify psi influences.

One Thought

In the front material of this book, I said that the theory, community and transcommunication sections were three parts of one thought. I understand this is a long book. As my proofreader pointed out, the EVP section alone seems like a book. But this is the time for patience, as many of my comments about these phenomena make most sense if you understand the theory and the environment in which it has developed.

The Discourses are the teaching, the Commentaries are the living experience and the Studies are the application. These are the three parts of a whole that can lead to understanding.

Transcommunication

Study 2:
The Nature of Transcommunication

Introduction

Because there are so many pertinent concepts, the following comments are offered in a shorthand version. Please take advantage of the links, other studies and Glossary of Terms for supporting material.

Mechanics of Transcommunication

From the physical perspective and with the exception of our human organism, everything about us is nonphysical, made *real* by way of the Attention Complex and by convention taught to us since physical birth. Even mainstream scientists are beginning to think the physical world we live in might be a product of our imagination, not in a sense that it is not objective, but in the sense that everything objective must first be visualized in our unconscious mind before some version of the world is expressed to our conscious self.

Quantum theory is beyond the scope of this cosmology, but you can find ample support for this statement via an Internet search for *Collapse of the quantum wave function* and *Wave function represents reality*.

Avatar as Trans-channel

In this model, a personality in the etheric is only able to express intended order in the physical by way of an entangled avatar. In simpler terms, this means that long dead Uncle John must speak through a physical person, which is someone who is an etheric personality entangled with a human body (avatar).

In practical terms, this means that the message in EVP, mediumistic messages, automatic writing, the images of visual ITC, apported objects and precipitated images must be formed by an entangled avatar. That is, by way of a physical person.

It apparently does not matter who is providing the channel. For instance, in EVP, the channel could be an interested observer. The observer could be in New York City talking to the practitioner in San Francisco. In the same way, a mental medium could be accessing information via any etheric personality, entangled with an avatar or not.

Apparently, the only limiting factor for this information access is the Principle of Perceptual Agreement. In that, if the person's Attention Complex cannot develop a visualization of the information based on worldview, it cannot be expressed as awareness to the conscious self.

In a practical sense, this means that if the channel (person providing the entangled avatar) believes all flowers are red, then the message from a discarnate loved one that said *"These blue flowers are for you"* will likely be expressed as *"These red flowers are for you."* In other words, for the channel, communication from a discarnate loved one is just another environmental input. It must pass through the Perceptual Loop, and therefore is subject to interpretation by the channel's worldview.

If the person providing the channel understands a second language, the message could be in that language or a mix of first and second languages. This is one of the reasons EVP are thought to be mostly in a language the practitioner or an interested observer understands.

A possible exception to the need for an avatar relationship for trans-etheric influences is the Contact Field. In that, links of rapport connecting a cooperating community are thought to produce an intentionality field. If the intention is to facilitate trans-etheric influences, then it may be possible that the intention field fills the same function as the channel provided by an etheric personality entangled avatar. This may be the underlying mechanism of séances and development circles. We are pretty sure we experienced this amongst ATransC members during the time of the Big Circle [427].

What We Sense

That which we sense begins as that which enters into our Attention Complex by way of our Attention Limiter. The information is first limited by the extent to which we are paying attention: are we interested, does it matter to us, does it involve survival, is it a threat? The Attention Limiter filters uninteresting information from the environment. Human instincts have a lot to do with what is deemed important and anything seeming to threaten or help survival (procreation, food, and protection of offspring) will readily pass the limiter. Personal interests such as considering the purchase of a new car will tend to alert the conscious self every time that model of car is near.

The two assumptions of First Sight Theory proposed by James Carpenter (13) can be paraphrased as: First, everything and everyone produces a psi signal that perturbs the psi field that connects all of us. People psychically sense these changes in psi space just as they physically sense changes in physical space. Second, every expression from the Attention Complex includes a related psychokinetic influence.

Carpenter proposes thirteen corollaries describing the implications of these assumptions for how people process information. The Perception and Expression Discourse includes a description of how they apply to the Perceptual Loop. Factors such as how well a person focuses on information, what a person is interested in and how well a person is able to sense *into* the unconscious activity of the Attention Complex determine the clarity with which a person is able to experience the world. The extent of this clarity is referred to as lucidity.

What We Express

What we express is based on our personal reality as it has developed in our worldview. This is an important point because *as you think, so do you experience and so do you express*.

An example is how we react to a stranger. First will come an alert of something unfamiliar, then an unconscious attempt to remember something familiar about the person (in worldview). If we are culturally taught to be suspicious of strangers, and if we tend to be fearful of the unknown, then it is more likely that we will describe the person in a prejudicial way. We will find terms like *outsider* or *intruder* in our memory to describe the person. If we are open to new experiences, the description might be more like *someone new*.

An interesting aspect of the Perceptual Loop is that we unconsciously attempt to visualize or create a mental model of incoming environmental information based on what is already in worldview. This is accomplished in the iterative process of the Perceptual Loop. Your mind almost instantaneously tries a series of possible solutions that either produces conscious perception or rejection of the information. You may have detected this process, most likely as you try to remember something. Be mindful that what you finally remember is determined by that modeling process and may not accurately represent the original input. The difference is message coloring.

Message coloring is easy to detect if actions and speech are considered in the context of the situation. Transcommunication begins as a concept, and it is *clothed* in familiar meaning as it is made physical. As such, it is natural for a communicated concept to be expressed in a way that is characteristic of the practitioner (or an interested observer).

As is discussed in the Personality-Centric Perspective Discourse, we share worldview with our human host. That means our responses to environmental influences are colored by our human instincts. When confronted with a threatening situation, our human will probably respond with something like a fight or flight reaction, but depending on our degree of lucidity, our conscious self will more likely respond with curiosity and desire to understand. One way for us to know why we respond as we do to a situation is to consider how much rational reasoning is involved. This is true of the way we relate to food and our relationship with others, especially people of the other sex.

Expectation Colors Perception

Amongst Spiritualists, it is called *coloring* when a medium's thoughts seem to change the message that is supposed to be from the sitter's loved one. Typical coloring might be as simple as changing the intended *"These flowers are for you"* message to *"This gift is for you."* Such a change preserves the intent but is probably expressed as well as the medium's lucidity will allow. It is when the medium delivers the intended *"I love you"* spirit greeting to *"I worry about you"* for a sitter the medium thinks looks unhealthy that coloring turns mediumship into probably unconscious social engineering.

This is especially evident in EVP. An observer of a group of people who are recording for EVP on a walkabout will likely notice how captured EVP tend to reflect the practitioner's worldview. For example, a person who thinks the world is more than a little fearful will more often record confirming utterances such as *"Help"* or *"I hate you."* A person who thinks discarnate people are earthbound and hang around cemeteries will often record *"Help me"* messages around cemeteries.

This point is one of the foundation concepts for the Implicit Cosmology because it clearly demonstrates that the messages in EVP are colored by beliefs as they are transformed in the Attention Complex. The remedy for the coloring of messages with belief is mindful living. This is more specifically discussed in the *Being a Good Witness* Discourse on EthericStudies.org. (68)

Coloring is also a problem for the witness (sitter). If the witness is not open to new concepts and attempts instead to explain everything in familiar terms, what the witness consciously experiences will often be different from the actual nature of the event. There are a number of terms for different aspects of this idiosyncrasy of human nature, such as incredulity blindness [442] and agnosia [424].

Hyperlucidity [152] is a human behavior that is just the opposite of incredulity blindness. Paranormal phenomena are poorly understood and explanations for them are usually not well considered. Add to that the tendency for people to reject the authority of more experienced people in favor of their own beliefs and the result is sometimes a tendency to see phenomena where there are none.

Misattribution of mundane phenomena as paranormal is usually just a matter of a witness not understanding the situation or what to expect; however, the reason for this behavior can be more complex. The person might genuinely believe phenomena are present despite explanations to the contrary. This is not just a matter of persistence. In Hyperlucidity, a person is described as being convinced to the point of irrational insistence.

Be careful not to think that the possibility of hyperlucidity supports accusations of fraud. The existence of paranormal phenomena is well established. It is just that poor discernment can lead a person to wrong conclusions. Objective phenomena can be witnessed by many people. In EVP, for instance, hyperlucidity shows up when a person reports hearing specific messages while no one else agrees. As one person complained, *"She hears the entire Gettysburg Address in a ten second sound file."*

Accounting for Human Nature

As we became more aware of the characteristics of ITC, the mostly unconscious influences on the perceptual process are increasingly recognized as the dominant factor for how phenomena are experienced and reported. The phenomena have been substantiated as being real. The problem is that witnesses and practitioners alike too often lack the necessary theoretical understanding to reasonably characterize their experiences.

This is addressed in the Commentaries of Section 2. Here, I will say that information received via transcommunication must be processed in the mostly unconscious Attention Complex before becoming available to the conscious self. What does become available is a version of the intended information as it is translated in the Perceptual Loop. That is, it is almost inevitably colored by worldview. As far as we can tell, this applies to information found in all forms of mediumship including ITC.

Evidence of Transcommunication

Of the dominant proofs of survival—reincarnation, near-death experience and mediumship—mediumship is considered the least convincing by mainstream science as well as many parapsychologists. The main concern is that psychic access to information as an explanation for the many verifiable messages in mediumship cannot be completely ruled out. Of the three, mediumship is the one counted as a form of transcommunication.

Instrumental TransCommunication is studied by citizen scientists with more of a decisive determinism [223] approach than by calculating deviations from chance so commonly preferred in parapsychology. By that standard, both audio and visible forms of ITC have been shown to be objectively real and reproducible by competent practitioners using accepted practices. In my opinion, the missing element required to substantiate these phenomena for the scientific community has been a workable theoretical model. Of course, that is the purpose of the Implicit Cosmology, so once it is published, what will remain in an ideal world is the need to educate the community. (And evolve the cosmology with thoughtful collaboration.)

So the perspective I take in my study is that the objectivity of these phenomena is based on shared experience as measured by witness panels [401]. And that the theory can be reasonably applied to enhance people's understanding about the nature of reality and their relation with it—not as a matter of faith, but as objective, actionable understanding.

Information from the Environment

All forms of trans-etheric influence are modeled in the Implicit Cosmology as expressions from personality *into* the physical via an avatar relationship with a human body. The entangled personality-human complex is referred to as a person, but as is shown in the Life Field Complex Diagram [xxv], the person's personality is held separate from the person's conscious self by the Attention Complex and worldview.

Information exchange amongst personalities is also moderated by the Principle of Perceptual Agreement. In that, the exchange of information is limited by how well one personality can perceive information from another as it passes through the Perceptual Loop in the Attention Complex.

If this model is reasonably correct, then information communicated to a physical person (the sitter) from a discarnate loved one would necessarily be by way of an exchange between the loved one's personality (in the etheric) and a personality (in the etheric) that is entangled in a physical avatar relationship (the medium). This means that a mental medium psychically brings a message from a loved one by way of his or her Attention Complex in the same way ordinary information is sensed.

In parapsychological terms, this communication looks like a psi influence. That is, the medium senses the psi field for information about the sitter in the same way that it is sensed for every other unconscious function. Presumably, psi signatures from other people's thoughts and actions are always presented to a person's Attention Complex. The difference between an average person and a medium is that a medium has learned to pay attention.

Sensed Information as Evidence of Survival

Evidence is not always evidential. That little gem of a saying represents the dividing line between the academically trained researcher and the lay practitioner. To be evidential, an example must have a very well controlled and documented *chain of custody*, and the protocol by which it was collected must be known, repeatable and observable. And of course, meaning of evidence must be based on understanding.

For instance, from an online listening study, the words in the EVP example *"It's Frank,"* recorded by Karen Mossey, were correctly reported an average of 48.6% of the time. (21) This establishes that the voice is objective in the sense that people can hear the words and that 48% of them correctly understand the words. However, the objectivity of the example only means it exists. Who spoke the words, and when, are entirely different questions. Ignoring the usual explanations offered by the skeptic community that the example is a fake or that people are delusional to think there are words in the recoding, reasonable arguments are that there were unnoticed speakers or that the practitioner is confused as to when and where the recording was made.

Here, the experienced practitioner has cross-corresponding evidence because Frank was a cat owned by the family for which Mossey's group was conducting an investigation concerning a reported *ghost child* in the house. A mental medium had just stated that there was a child in the closet about the time that Frank wandered into the closet and the recording was made. A mental medium can say anything. Even a well-meaning medium can mistakenly think there is an entity nearby. Who would know?

Having eyewitnesses hear the medium's statement, and then see the cat named Frank walk into a closet, is not very good evidence by itself. The fact that the group was looking for a *ghost child* may have simply predisposed the medium's interpretation of what she sensed. Also, the EVP was heard later during the analysis of the data. The EVP containing a child's voice commenting on the cat as it walked into the closet in which the medium had just said a child was present amounts to corresponding evidence that is, indeed, evidential.

It should be noted that the cross-correspondence of the medium's report and the child's voice in the EVP provides strong evidence for the existence of psi functioning, but it does not assure there was a survived child in the closet. Collectively, the investigators had full awareness of the story about the ghost. A contact field had likely formed, which theoretically would enable the common awareness to produce expected phenomena.

Nonlocal

The *everywhere is here* characteristic of the psi field is referred to as nonlocality. The argument made by parapsychologists that EVP is only proof of psi functioning is properly based on nonlocality. The importance of the *"Doja, no!"* (33) example is that, when the recording was made, there was no pre-existing memory of the event. Nonlocality is not an explanation for that example. See Nonlocal and Nontemporal [94] in the Etheric Fields Discourse.

Nontemporal

It is sometimes argued that information in ITC could have come from a person's future knowledge of an event. When Martha returned home and saw the mess, she could have thought, *Cathy would never have let this happen if she was still here.* And based on the theoretical nontemporal nature of reality, she could have psychokinetically injected the EVP onto the recorder as the recording was being made before she got home.

There is some indication that presentiment, the physiological reaction to an event before it actually occurs, is best explained with First Sight Theory [126]. Presentiment is a short-term phenomenon while most reports of precognition indicate a long-term effect. In that, the concept of emerging thoughtforms representing potential futures might better explain the effect. Rather than precognition as sensing the future, there is better support for the idea of sensing the potential of a particular future. As such, a nontemporal characteristic of etheric space is probably incorrect. See the Nonlocal and Nontemporal in the Etheric Fields Discourse.

Apports

EVP are sometimes formed by more than one voice, as if one person begins a phrase and a second finishes it. Especially in telephone recorder EVP, there is the occasional sound of what a telephone technician might describe as a *break-in*, as if the transmission signal carrying the message has to override the first signal. For instance, the recording might begin with a little noise abruptly interrupted by a burst of louder noise which trails off as the voice of the EVP begins. This is demonstrated in the *Transcommunication Via Telephone Calls* article on ATransC.org. (132)

In some form of EVP apparently produced by more than one person, the first speaker might begin with *"John ..."* followed by a second speaker saying *"...is okay."* The effect is very much as one would hear if a sound engineer had spliced two statements spoken by different people together to produce one phrase. This is the kind of sound track one might expect if parts of the EVP are apported from existent recordings in other parts of the world.

This possible *spirit plagiarism* is seen in visual forms of ITC as well. For instance, the precipitation art Hoyt Robinette produced for me has the likeness of a man dressed in Civil War era military clothing. (See the example on page 36.) The feature looks as if it was taken from existing media. The features produced by Robinette on squares of cloth are pretty clearly from existing media. (See an example on page 35.) Yet after much examination of the materials and physical environment before, during and after they were formed, there is no evidence indicating anything other than paranormally produced art.

Apport is a term used to describe the movement of a physical object from one location to another. Usually, it is experienced as an object suddenly appearing, perhaps in the air and then falling to the floor. Flower petals, stones, even figurines have been reported. Apported objects are described as apports.

The apport process is thought to be the dissolution of an object in one location, *movement* of the object's formative information through the etheric, and precipitation of the object back into its original form in the new location. Remembering the nonlocality of the etheric, *through the etheric* is conceptual. Yes, this sounds a lot like teleportation, and that is a good model for visualization of the process.

Physical objects are described in the Implicit Cosmology as ideoplastic structures. This means that they are formed as expressions from the Attention Complex. In this model, our sense of the physical world is expressed in our mind as being objectively real, when in fact, reality is a mental construct based on a collective intention to make an imagined object of reality *real*. See The Nature of Reality Discourse.

In physical terms, transcommunication is an energy-intensive process. The etheric equivalence of physical energy is intention, and so in etheric terms, the creative process is likely demanding and its expression is seen to be limited by a personality's ability to maintain the necessary focus of attention, clarity of imagination and intention to express a desired outcome. So in physical terms, trans-etheric influences in general are probably very energy intensive.

Using available ideoplastic structures already existing in the physical that satisfy an intended expression may well be more efficient than producing a new ideoplastic structure. While some EVP are clearly in the voice of the known speaker, some can be more sensibly explained as existent bits of speech opportunistically apported to the sound file.

A planned study is the analysis of a collection of synchronized sound file pairs: one containing an EVP and one recorded at the same time but without the EVP. The objective is to determine if there is a net increase in energy in the file containing EVP. See the call for examples of transform EVP on ATransC.org. (133)

Apportation may well be one of the most common forms of transcommunication. With this in mind, researchers and practitioners alike should be cautious about too quickly deciding the authenticity of examples just because they are based on existing material.

Precipitation

Another form of trans-etheric influence is known as precipitation. In that, physical material in a chaotic form is deposited into the physical under the influence of intended order. Probably the most common form of this is precipitated art. Precipitation is also seen as the end-phase of apportation.

Precipitation medium Hoyt Robinette produces very impressive examples of precipitated art. In the one shown on page 36, Robinett placed crayons, felt-tip pins and pencils in a basket with a package of newly opened three-by-five white paper cards. The basket was covered while he conducted billet readings for about an hour, and then the basket was opened and the cards removed. For each of the twenty-some sitters, there was a card with their name and one or more drawings on it. Robinett performed a similar feat using squares of cloth and many opened colored inkbottles as raw material. (See an example from a cloth on page 35.)

Intended Order

Intended order is one of the common denominators for trans-influence phenomena. If reality consists of life fields and the expressions of life fields, then reality is an example of intended order. It is a characteristic we should expect to see in all of our phenomena.

The Global Consciousness Project involves the apparent influence of some form of collective consciousness on random event generators. The influence is seen as increasing order in a process very carefully designed to be random.

Virtually all forms of EVP involve the impression of intended order on noise. The ideal noise [307] is currently understood to be broad-spectrum chaotic noise, heavily weighted in voice frequencies and frequently perturbed by noise spikes. It is thought that the spikes facilitate the impression of the intelligence signal on the stochastic process, as if a kernel around which the voice can form.

Most if not all visual forms of ITC are formed in optical noise. Video loop ITC, for instance, involves the production of very chaotic optical noise by pointing a video camera at a television screen on which the output of the camera is being displayed. This *camera recording what it has just recorded* produces wide-spectrum, chaotic optical noise in which recognizable features such as human faces and animals.

Some specialty devices and computer programs involve white noise or other chaotic sources as the core process. In some techniques, memory addresses containing bits of speech are selected, based on changes in this chaotic input. The streaming output of that process is expected to produce meaningful messages.

The common factor in all of these is the impression of intended order on a chaotic process. As a common factor in many forms of transcommunication, the evidence of intended order is both a way to evaluate techniques for producing the phenomena of transcommunication and a way to model the nature of those phenomena.

EVP as a *Lab Rat* for Psi Research

An important aspect of EVP is that an experienced, confident practitioner can be expected to produce examples on demand. In this way, recording for EVP provides a means by which other forms of phenomena can be studied.

It appears possible for an EVP to be produced by a physical person if the information is currently known by someone or if it exists in the etheric. As such, locality does seem to be a factor.

It does appear that time is a factor. The potential field for something to occur might produce an EVP as it is sensed by a person, but it is very difficult to argue that a specific EVP such as *"Doja, no!"* might be the product of presentiment. In fact, we are aware of no examples in which an EVP example foretold an event that could not have been explained as a practitioner or interested observer sensing the potential for that event to occur. (There is more on the *"Doja, no!"* example on page 38.)

There is currently no known way to shield from psi. This means that any psi experiment has the potential of contamination by the experimenter and interested observers. This is certainly true of EVP sessions. However, if a protocol for EVP can be designed in which no physical person is aware of an event, its form or when it will occur, then it may be possible to enlist discarnate helpers to observe and report via EVP. Such a protocol would be useful for studying psi shielding techniques and validation of survival evidence.[56]

Implications of the Implicit Cosmology

The phenomena of transcommunication are usually considered from the perspective of known physics, human nature and psi research. The Survival Hypothesis is typically dismissively considered as a preamble to the efforts to explain proposed evidence of survival as mental aberration. If that fails, then the Super-Psi Hypothesis is turned to as the last word.

While the evidence offered by mediumship is usually dismissed as insufficiently convincing or easily explained with Super-Psi, there is virtually no consideration within academic circles of evidence offered by ITC. However, when ITC is considered, it is seen to be consistent with the evidence of mediumship. (We consider ITC a form of physical mediumship.) From my perspective, it is easy to develop a consistent model that makes sense of both psi and transcommunication phenomena.

If the Implicit Cosmology is correct, then all forms of transcommunication should depend on essentially the same concepts and their behavior should

---

[56] It would be necessary to ignore other evidence, but for the sake of research, it is reasonable to argue that the practitioner is the source of the utterance if researchers insist on the Super-Psi Hypothesis.

be reasonably well predicted by including those concepts. In the same way, examples of phenomena that seem to contradict the cosmology should be considered with some hesitation. They are either not what they seem or they present new concepts that should be, but may not have been considered.

It is important that phenomena that agree with predictions of the Implicit Cosmology inherit credibility from one another. As proof of survival, mental mediumship should be seen as more plausible if EVP is seen as evidence of survival. EVP should be seen as more plausible if precipitation can be established as *real* process. In the same way, apportation as a means of producing some EVP should make sense if there is seen to be a net increase in audio energy in an audio signal containing an EVP. It all depends on understanding the model and understanding the phenomena.

Transcommunication

Study 3:
Audio ITC a.k.a. EVP

Introduction

Electronic Voice Phenomena (EVP) is the audio form of Instrumental TransCommunication (ITC). There are anomalous, intelligible speech produced in electronic devices. They may be heard as a real-time output, but because the voice is usually difficult to understand, they are more generally heard on review of a subsequent recording. No currently understood physical processes account for the existence of EVP.

Most practitioners use a three-tier scale for classifying how well an EVP example is expected to be heard and understood by a listener. These are Class A for easily heard, Class B for marginal and Class C for very difficult to hear and understand. See Classifying EVP in this Study.

There are two very different schools of thought about how these phenomena are experienced. One is that these phenomena just spontaneously happen and the receiver must be lucky enough to have someone on the other side make contact. The second is that many forms of these phenomena are able to be induced. Given the evidence, it is reasonable to think that the second point of view is held by the more recently informed.

History

Speculation about EVP can be traced back to the 1920s. In a Scientific American interview, Thomas Edison was quizzed on his views regarding contacting the dead. Edison said that it might be *"possible to construct an apparatus which will be so delicate that, if there are personalities in another existence or sphere who wish to get in touch with us in this existence or sphere, this apparatus will at least give them a better opportunity to express themselves than the tilting tables and raps and Ouija boards and mediums and the other crude methods now purported to be the only means of communication."* There is no indication that Edison designed or tried to construct such a device. (134)

The Edison National Historic Site at home.nps.gov/edis/faqs.htm. Had a Question and Answer item concerning a possible Edison's machine for transcommunication (the item has been removed):

> Did Edison make a machine that could talk to the dead?

> *This seems to be another tall tale that Edison pulled on a reporter. In 1920 Edison told the reporter, B.F. Forbes, that he was working on a machine that could make contact with the spirits of the dead. Newspapers all over the world picked up this story. After a few years, Edison admitted that he had made the whole thing up. Today at Edison National Historic Site, we take care of over five million pages of documents. None of them mention such an experiment.*

In 1936, Attila von Szalay began capturing paranormal voices on phonograph records and then in the mid-1950s he was joined by Raymond Bayless. Together they acquired many evidential EVP on their new tape recorders and they published their findings in the *Journal of the American Society for Psychical Research.* (135)

The person credited with bringing EVP to the public is Friedrich Jürgenson, a Russian-born Swedish film producer. In 1959, after recording birdsong on his tape recorder, Jürgenson heard on playback what appeared to be a human voice. Subsequent recordings contained a message which seemed to be coming from his dead mother. (136)

Konstantin Raudive is credited for learning about EVP as a student of Jürgenson and introducing EVP to the English-speaking world with the Colin Smythe publication of the English-language book, *Breakthrough.* (137)

Colin Smythe is credited with coining the term, *Electronic Voice Phenomena* (EVP) as a more inclusive alternative to *Raudive Voices*, as the voices recorded by Raudive were referred to at the time.

Sarah Estep founded the American Association of Electronic Voice Phenomena (AA-EVP) in 1982. In 2000, she sent us an email titled, *"Would you"* in which she asked us to assume leadership of the Association. We agreed and our lives became regulated with the quarterly cycle of research, writing and publication of the ATransC NewsJournal.

In 2003, we wrote and published *There Is No Death and There Are No Dead* (1) as a sort of textbook for members and people who wished to have an understanding of EVP. All of the proceeds of the book have gone to the Association for operations, outreach and research. To date, it has earned $37,100.

In 2014, after uninterrupted publication of 129 issues of the Association newsletter, 57 of which we published, we changed the model for the Association from membership supported nonprofit to a general public supported one.

That is the essential history, but there are numerous other historically important events, such as the development of video loop ITC and the many blind alleys such as Spiricom and radio-sweep.

We have never had much interest in history, preferring instead to focus on the present and potential futures. It is important in this study to remember that any information older than ten or twelve years is probably out of date. In fact, the history only contaminates present understanding, making it difficult for contemporary observers to see past the false notions it has brought.

Types of EVP

EVP are spoken of in the plural because they are formed in at least three very different ways, all of which are dependent on the availability of a relatively chaotic signal which has been shown in research to be more readily influenced with psychokinesis than more ordered signals.

Transform EVP

Traditionally EVP formation has involved the transformation of available audio-frequency energy into voice. This transformation is thought to occur in the electronic equipment, specifically in an analog stage before or after the signal is digitized for storage.

The resulting signal is seen as a simulation of human voice which may closely mimic the physical voice of the person thought to be speaking. This includes nuances of voice indicating accent, age, sex and attitude. Analysis of the resulting voice often shows novel arrangement of formants (clusters of frequency groups occurring characterized as octaves of the voice box frequency produced during passage through the mouth). Fragmented voice box frequencies (Formant 0) are also characteristic of EVP. (138)

Prior to wide availability of personal computers, the static between radio stations was often used as background sound for transform EVP since radios were a readily available source of sound. Current best practices involve the use of noise, such as that produced by a fan. The use of any form of *live voice*, such as a possible voice from a radio broadcast, is strongly discouraged.

Random Selection

This is also known as *opportunistic EVP* because it is thought to require the availability of already formed voice fragments. It depends on a random process which is thought to be influenced by the communicating entity. In the application known as EVPmaker (22), a pre-recorded sound file containing voice is stored in a buffer with each buffer address containing a small segment of the original sound file. A random process is then presumed to be intelligently influenced by the trans-communicator to select buffer memory locations presumably containing the necessary bit of sound for message formation. The resulting sound fragments are combined in a streaming output sound file. Meaningful messages are sometimes found in that output stream.

EVPmaker developer, Stefan Bion (22), has provided an input sound file on his website which contains speech fragments known as allophones which are generated by a speech synthesis program. In this application, it is argued that, if a word is present in the output, it must be formed by a chance arrangement of allophones or it must be the product of intended manipulation of the random process. The deciding factor is whether or not the utterance is meaningful for the circumstance.

Environmental Control of Speech Synthesis

Some applications employ environmental energy sensors to control the operation of a speech synthesis process or to select addresses in a word or voice fragment buffer. It is believed that the communicating personality's intention to communicate might cause meaningful perturbations in the environmental energy, thus selecting intended buffer addresses to form a voice message.

Depending on the design, the environmental energy sensing mechanism may also be directly influenced by the communicator. This goes back to the problem that the mechanism can only report changes as it is designed.

What is Probably Not EVP

Review the characteristics below for EVP discussed in this Study. Based on those common characteristics, and the results of other studies, it is becoming evident that certain technologies probably do not produce EVP.

The most commonly used of these questionable technologies is radio-sweep, which uses modified radios popularly known as *ghost boxes* or *spirit boxes*. The concept is that repeatedly tuning a radio by sweeping through the local radio stations will produce an output of meaningful sounds. It is thought that the communicating personality will arrange for the right sounds to be present in ambient radio transmissions to produce the desired message.

Radio-Sweep: A Case Study, which I conducted in 2009 (73), is typical of the kind of study indicating radio-sweep technology probably does not produce EVP. *A Research Study into the Interpretation of EVP* (74), which is a 2013 study conducted by parapsychologist Dr. Mark Leary, indicates that radio-sweep examples thought to be EVP are typically not correctly understood by a listening panel. See Errors in Perception [349] in the Artifacts and Perceptual Error Study.

Characteristics of EVP

These characteristics will provide a sense of how EVP sound and the nature of the phenomenal voices.

Transform EVP

The first group of characteristics is specifically for transform EVP.

1 **EVP are distinctive**: EVP have a distinctive character of cadence, pitch, frequency, volume and use of background sound. The voices have a distinctive sound to them that is difficult to describe. For instance, EVP messages often have an unusual speed of enunciation; the words seem to be spoken more quickly than normal human speech.

2 **A need for background sound sources:** Research has shown that the voice in EVP is formed out of ambient sound. (138) Because of these characteristics, it is standard practice to assure the availability of ambient sound for voice formation, even while isolating the recording device or process from uncontrolled ambient sounds, such as crowd noise, so as to avoid mistaking unnoticed normal voices as EVP.

3 **Frequency range**: EVP are formed in available background sound. As such, if there is a high frequency component in the background sound, say caused by whistling wind, it is possible that the EVP will be of similar frequency range. If there are both higher and lower frequency components in the background sound, it is possible to find EVP formed in both regions of the sound. In some instances, two different voices might overlap.

4 **Missing frequencies**: Spectral analysis of EVP samples has shown that the fundamental frequencies of voice associated with the human voice box are sometimes missing. One researcher describes the typical EVP as a *thickening* of the background noise to form the voice. (138)

5 **Precursor sounds**: Sounds are often heard prior to an occurrence of EVP. Although these vary in nature, they tend to be within tenths of a second of a phrase and are a *popping* or *clicking* noise reminiscent of the *squelch* sound caused when the automatic gain control engages as the push to talk button is depressed on a Citizens Band radio. (139)

6 **EVP show evidence of being limited by available energy**: Utterances tend to have about the same amount of audio power in their associated sound wave from one EVP sample to another. That is, a short EVP will tend to be louder than a long EVP. A very long phrase might be composed of two or more average-length phrases separated by brief pauses. Also, an utterance may trail off at the end, as if the energy is being depleted before the message is finished. It is as if the communicator is attempting to manage available power as *packets* of energy. (85)

7 **EVP are complete words or phrases**: Message are typically one to two seconds in duration and are not truncated at the beginning or end. If EVP were radio interference, they would often begin in the middle of a word. EVP messages are usually complete thoughts, as well. (85)

8 **The voices in EVP are often recognizable**: It is common for an EVP to contain the recognizable voice of the discarnate person thought to be speaking. It is also common for that entity to say something that was typical of what he or she would have said while in the physical. Personality clearly remains intact even though the person no longer has a physical body. (33)

9 **EVP is found wherever the practitioner listens**: This suggests that the source of audio noise is not a factor for EVP, so long as the audio energy is suitable for voice formation. In practice, the majority of techniques for recording EVP involve sound conditioning, rather than unique forms of psi detection. For instance, upscaling infrasound so that it can be heard by human ears or downscaling ultrasound really constitute techniques of sound conditioning. The resulting EVP is not evidence that the utterance was formed beyond human hearing, but that it was formed when the audio frequency energy was made available to the recording process.

All Forms of EVP

These characteristics are common to all forms of EVP:

1 **EVP are in a language the practitioner understands**: Alexander MacRae has conducted experiments in a place that has no English language radio or television stations, yet resulting EVP were in English, which is his primary language. It is typical for the EVP, no matter where they are recorded, to be in a language that the practitioner understands. There have been exceptions to this which were apparently intended as a demonstration, but as a rule, EVP will be spoken in a language understood by the practitioner or an interested observer. (140)

This brings up an interesting point of speculation about psi-based communication. Mental mediums often report that they receive communication from nonphysical entities as images which they must interpret. These images are not just mental pictures. They are packets of information that are sufficiently complete for the receiver to fully understand their meaning. Robert Monroe (53) referred to this form of information as thought balls [463].

2 **EVP are not ambient sound or broadcast programming**: Studies have been conducted to determine if EVP are stray radio signals, ambient but unnoticed voices or other sounds. EVP have been collected in an electrical, audio and radio frequency shielded room. (140) In one study, a radio and a recorder were placed in a padded chamber which was then buried under ground. The recorder did not record radio programming but did record EVP, which were transformed from the noise produced by the radio. (141)

3 **Party line**: Some EVP sound as if they are comments intended for someone other than the practitioner. This is much like momentarily listening in on a party line telephone call. It is not uncommon in both field and controlled recording situations to record comments that seem as if unseen people are discussing the practitioner's actions in much the same way that you might discuss the activity of someone that you were watching.

4 **EVP are appropriate to the circumstances**: There are numerous examples of EVP that are clearly direct responses to questions recorded just prior to the EVP phrase or to the circumstances. An example of an EVP being appropriate to a circumstance is an EVP

recorded by Lisa. We were asking a woman about the upstairs lighting and sound room for the Frank Sinatra Theater at the Cal-Neva Casino at Lake Tahoe, California. They had heard that the heavy door to the room often shut for no apparent reason, scaring the crews setting up lights and sound systems for shows. The woman told them that she never, ever went to the control room. Lisa's recorder was on while she thanked the woman for her assistance. On the recording, Lisa can be heard saying, *"Thank you very much."* Underneath her voice, is a clearly heard paranormal voice saying, *"Please don't come."* However politely said, it seems obvious someone did not want to be disturbed. (142)

5 **Precognitive responses**: Answers to questions may be recorded prior to a question being asked, so that the answer, as a phenomenal message, is on the sound track followed by the practitioner asking the question. More research is required before making informed speculation about this observed characteristic, but the indication is that the entity may be sensing what the practitioner is about to ask as a mind-to-mind exchange following the avatar model. See Nonlocal and Nontemporal in the Etheric Fields Discourse

6 **Vocalized questions elicit more EVP**: There is evidence that the communicating entities are able to read our thoughts, as illustrated by the occasional EVP which clearly responds to a comment just seconds before the comment is expressed. However, analysis by MacRae suggests vocalizing our comments and questions produces more frequent EVP. (85)

7 **The *newness* effect**: The practitioner's excitement in trying a new detection device or recording technique may be the cause of improved EVP collection. As the new approach becomes *normal operating procedure*, the improvements generally fade back to a more *normal* Quality and Quantity (QQ) of EVP collection. This suggests that it is important for the practitioner to maintain piqued interest during experiments. This is also one of the reasons it is speculated that the practitioner is an integral part of the recording circuit. The practitioner is apparently supplying the necessary psi influence to produce a useful contact field to enable a nonphysical to physical transfer of information.

8 **Effective devices unique to the practitioner**: Exceptionally effective EVP and ITC collecting systems have been developed; however, these

typically work well for the developer, but do not necessarily work as well for other practitioners. This paradox supports the belief that the practitioner is part of the recording circuit. It has also reinforced the concept that the communicating entity may be specific to the practitioner.

9 **EVP can be thoughts of living people**: Two experiments appear to show that at least some EVP might be initiated by living people who were sleeping or distracted at the time. In these prearranged experiments between a practitioner and a sleeping person, questions were clearly answered by a communicating entity, and the answers were appropriate for the sleeping person. This fact of EVP suggests the possibility that EVP can become an important tool for consciousness research. For instance, is it possible that a patient in a coma might initiate an EVP? (143)

10 **Understanding EVP may be like learning a new language**: As discussed in the EVP Online Listening trials report, people with little or no experience listening to EVP will typically correctly report words in Class A transform EVP on average of 20% to 25% of the time. In contrast, an experienced practitioner should correctly understand close to 100% of Class A utterances. (21) This number drops to 0% to 5% for studies of radio-sweep (Spirit Box, Frank Box, Ghost Box) and 0% if single-syllable utterances are omitted. (73), (74)

Theories Proposed to Explain Observed ITC Phenomena

The existence of EVP is not realistically contested. Any reasonably capable person with a device capable of recording audio can expect to eventually record an understandable utterance that cannot be explained with known science. The same can be said for some forms of visual ITC. A workable protocol for recording EVP [306] is provided in this Study. The real questions are what or who is causing them and how they are caused.

It will be helpful if you understand the three dominant models used to explain the nature of reality: Physical Universe Hypothesis, Super-Psi Hypothesis and Survival Hypothesis. There is a detailed discussion of these in the Trans-Survival Hypothesis Discourse.

Normal Explanations

Alternative explanations for EVP that do not require paranormal phenomena are typically *sounds good* ones that ignore current understanding about transcommunication. With that said, the following sounds can be mistaken as paranormal. It is for the practitioner to gain the necessary experience and understanding to tell the difference.

Radio Signals and Unnoticed Conversations

Probably the most obvious explanation is that the voices of EVP are recordings of stray radio signals. Inexpensive portable electronic equipment designed to process audio signals is, under certain conditions, able to detect and record stray radio signals. This is not normally a problem, but it is possible and should be considered when evaluating possible EVP.

The proliferation of digital technology is making it less likely that radio signals might be mistaken as EVP. Analog radio is about the last signal source that a voice recorder can detect so that the broadcast information can be understood. If the signal is in a digital format, it must be converted to analog to be understood and audio recorders are not designed to do this. If you wish to assure yourself that EVP are not radio signals, consider the article "Eliminating Radio Frequency Contamination for EVP." (141)

It is usually the near zero level of the recorded signal that becomes contaminated by stray radio signals and that can become audible with amplification. This is one of the reasons that very much signal processing to *bring out the voice* is discouraged, especially the use of more than ten or so decibels of amplification.

Simple content analysis of EVP is usually sufficient to establish that the utterance in question is unlikely to be crosstalk or unnoticed voices from people in the room. Assuming the practitioner is not attempting to provide scientific evidence, it is usually adequate to note whether or not the utterances are complete statements, if they are appropriate to the circumstances and if they are understandable. If they were stray radio or television signals, they would contain the telltale signs of commercial broadcast; they would often be partial statements and they would usually be nonsensical, considering the circumstance. Until you have gained confidence in the validity of EVP you record, it is important to ask friends to be your listening panel. The Best Practice: Witness Panel Study will help. (67)

It is also a best practice to use a second audio recording process as a control for situations that might have unnoticed voices in the room or stray radio. (144) The need for background sound in transform EVP makes it less likely that a higher quality recorder will pick up an EVP. This makes the use of a video recorder to make a visual record of fieldwork an excellent technique for making a control audio recording of the session.

Practitioner Imagination

This is an important possible explanation for any form of phenomena involving unexpected sounds, images or events. As is discussed in the Artifacts and Perceptual Error Study, modern technology is capable of producing a wide variety of unexpected results. A person can easily mistake an artifact for phenomenal unless the person is an expert in the use of that technology.

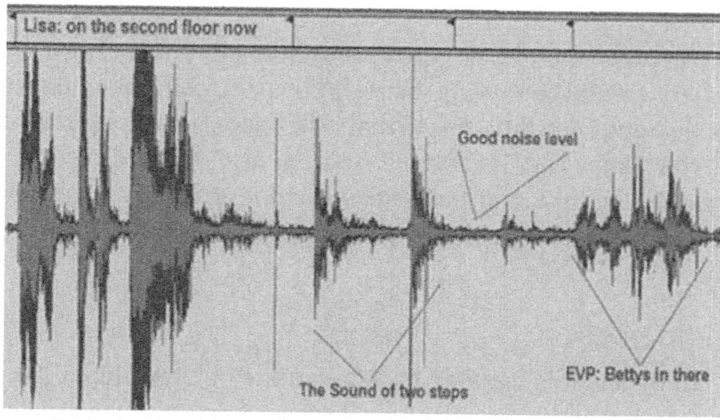

EVP Waveform: Figure illustrating a sound track with the utterance, *"Betty's in there."* A workable level for background sound used in voice formation is shown. The increase in amplitude of the voice is believed to be due to the accumulation of energy prior to formation of the utterance. This is transform EVP. (210)

Sounds in an audio recording can be made by very mundane influences yet sound very phenomenal. For instance, it is common for a person to take a breath of air just before speaking. The sound of an intake of breath can be mistaken for something phenomenal when a person is examining every little sound in the recording.

All aspects of Etheric Studies are plagued by a general lack of critical thinking amongst people seeking to experience these phenomena. In view of the complexity of the involved technologies, this is natural but there are precautions one can take to moderate problem. The recorded sound of a sigh can be eliminated as being paranormal by making it a practice to discard all single-syllable, standalone sounds. While most people resist discarding any possible EVP, it is better to err on the side of the mundane.

Recent study is indicating that the expectations of the practitioner or witnesses can have an influence on how sound is experienced. In the Phantom Voices study (66), nearly half of the participants reported hearing some form of voice in two audio files that were clearly marked as only containing noise and no voice. A clever practitioner might describe simple noise as voices and provide an explanation as to how those voices are meaningful. This is referred to as storytelling [350]. Distressingly, witnesses will too often agree with the story as evidence, especially if they trust the experience of the practitioner. Hyperlucidity can apparently be a shared experience.

This characteristic of EVP is still being studied, but is clear that all of us must be very careful to manage expectations and cultural influences. EVP is objective evidence and that means that other people can hear the same thing without coaching. The inverse of this is also true. Concepts such as inattentional blindness and incredulity blindness suggest that people who have a strong disbelief in things paranormal are less likely to be able to experience EVP.

Psi Explanations

EVP are not necessarily initiated by a discarnate personality. As shown in the Life Field Complex Diagram [xxv] discussed in the Personality-Centric Perspective Discourse, a person (etheric personality entangled with a human in an avatar relationship) is thought to be required for trans-etheric influences. As such, it is necessary to identify which personality expressed the initial intention to communicate.

Thoughts of the Practitioner

Once the *normal* explanations have been asked and answered, there remains the possibility that the voices are paranormal but that they are caused by a physical person. It is here that the argument changes from *"Is EVP real"* to *"Who is talking?"*

A number of experiments have been conducted that indicated it is possible to record the thoughts of people known to be still in the physical. (143) Current thinking is that the conscious aspect of a person is what survives physical death, and when a still living person is in some way disassociated from the physical body, his or her conscious self is much the same as the etheric communicator. In other words, it is predicted that the thoughts of a living person should be able to be recorded as EVP.

While many of the observed characteristics of EVP could be explained as thoughts of the living, there is evidence that some are not. On occasion, the voice of a person will be recorded, say *"John,"* who is later shown to be the discarnate son of a person who, at that time was unknown to the practitioner or interested observers. However, subsequently, the person joins the group and recognizes John's voice saying the sort of things his son would have said while in the flesh.

EVP are Echoes of the Past or Residual Energy

The Super-Psi Hypothesis (145) proposes that the mind is a product of the brain and the information being accessed via EVP the other forms of apparent etheric-to-physical influences are either caused by the practitioner, some other physical agent or a form of residual energy. Physicist David Bohm and neurophysiologist Karl Pribram described this residual energy as a field of energy that underlies all of reality. (146) This field of energy is thought to have three characteristics that might account for the observed etheric-to-physical phenomena:

1. It is thought to behave according to quantum principles, especially that of entanglement, which shows that once objects have become associated at a quantum level, influencing one object will have a similar influence on the other, even though it may be very far away in distance and time.
2. This field exhibits the character of nonlocality, meaning that information in it is equally available from any place in the field. This characteristic is required to explain such phenomena as a remote viewer accessing information that is miles away and a medium supposedly getting information from discarnate entities.
3. This field is formed of the residual energy of consciousness, and therefore contains all that was ever known, thought and experienced. It is this residual energy that provides the psychically gathered information. An alternative version of this is that, if the information is in someone's memory, then it is accessible to others, at least subconsciously.

Recording Techniques

The following information is intended to provide a technique which is most likely to produce results. This is for transform EVP based on the material provided by ATransC in the Techniques section of ATransC.org. Please also consider the best practices which are drafted in the ATransC Collective at atransc.org/bp/Collective.

Contact Field

A contact field is modeled as a special instance of a thoughtform which is entangled with a group of people or location. The concept is dependent on the concepts of rapport and cooperating communities. Contact fields are thought to develop from rapport amongst people sharing a common point of view, focused attention (usually long term) and the expression of relatively intense emotions such as pain, anger or love.

A person can intentionally develop a contact field by setting aside a quiet place for meditation and transcommunication. The idea of a dedicated place in which focused intention and attention are often expressed is thought to be the basis for Sarah Estep's instruction to find a quiet place and try to make contact via EVP every day for at least a week. (147)

In field recording, a contact field may already be present in known *haunted* locations. As a general rule, look for a place with longtime human occupancy, and/or places in which there has been the expression of intense emotions such as a church or hospital. Theaters are often associated with haunting phenomena, possibly because of the sometimes intense emotions expressed by the actors.

Types of Recording Sessions

We recommend people begin working with EVP in controlled conditions in which ambient sounds can be controlled. This also helps develop a personal contact field. Controlled conditions also make it easier to use a second recorder for experimentation and to help eliminate false positives.

Experience has shown that once a person becomes proficient recording for EVP under controlled conditions, it is probable that the person will be able to be successful wherever he or she records. The implication is that the person develops a personal contact field which is anchored with his or her attention, rather than a location.

Field conditions are especially challenging for EVP. For instance, when recording with a group in a haunted location, always use a second recorder to help eliminate false positives. Using a video camera to record the investigation is useful because the audio track acts as a second recorder. As a reliable rule, if the suspected EVP is on both recorders, it is more likely not EVP and should be discarded. An exception to this rule might be if the voice is thought to be direct voice [434].

When on a walkabout while recording for EVP, it is a good practice to announce for the recording where you are and to acknowledge ambient sounds so they will not be mistaken for EVP.

Recording Equipment

Any device capable of recording voice frequency sound will work for EVP. A computer is recommended for storing sound files and a way to transfer audio from the recording device to a computer will be needed. The computer should be equipped with an audio management program, such as the open source program, Audacity. (148)

Digital voice recorders are recommended for transform EVP. Less expensive models produce more internal noise which is useful for voice formation. High quality units will probably require added background noise. A computer can also be used but will probably require added noise. As has previously been explained, noise with many perturbations is more useful than steady-state noise. As such, an inexpensive recorder that produces a lot of internal noise that is steady-state may not be as useful for EVP.

Ideal Background Noise

Early attempts to produce EVP were based on recording in silence; however, it was quickly discovered that the voices tended to occur near or in noise. Radio static was used for background noise in the days when radios were about the only dependable source for noise.

Over the years, just about every practitioner joined the quest to find the ideal background noise. The first hint of what makes useful background sound came with the Panasonic RR-DR60 digital note taker. It was one of the first battery-powered, hand-held digital audio recorders on the market. It tended to have a lot of internal noise in the sound track, and one of the characteristics of that noise was what sounded like the noise one gets while adjusting the squelch level on a Citizen's Band radio. I characterized it as intermittent noise spikes.

The DR60 produced a lot of EVP, but because of the noise, the quality of EVP was usually very low. Even so, all of our examples on the ATransC.org were made with a DR60. The *"Doja, no!"* (34) EVP was also made with one.

Newer recorders tend to be a lot quieter internally, so a general rule is the quieter the recorder, the more noise needs to be supplied. Also, steady-state noise, such as white noise, or the rushing sound heard as background in recording made with inexpensive recorders, is not useful for EVP.

Radio-Sweep as a Noise Source

There is confusion about when an EVP is the direct result of radio-sweep [296] or transformation of the noise produced by radio-sweep. Most radio-sweep devices use a two- to three-second sweep. Too often, entire syllables can be heard in such a slow sweep; however, a sweep of less than a second, or fast enough so that no understandable voice can be heard, often produces an output sound that has all of the characteristics needed for transform EVP. The result is that some transform EVP are formed during radio-sweep sessions. It is for the practitioner to learn to tell the difference.

Background Sound

Most recording situations have some background sounds, but it may be necessary to add noise with something like a fan or running water. Always consider how sounds in the environment might seem like EVP in a recording.

Be inventive. In controlled conditions, it is also possible to use high-quality recorders, because you can supply sound. Experience is showing that a microphone is only important to introduce additional noise if the noise generated internally by the recorder is not useful for voice formation. Of course, microphones are also important for maintaining a record of the recording process.

Live Voice

Some people use foreign language radio, crowd babble or prerecorded allophones as background sound, but research has repeatedly shown these techniques also have far too many false positives to be useful.

In one informal study, we used the same foreign-language input file for Big Circle recording sessions over a period of six months. While the input file was played on one recorder, Lisa and I recorded using individual recorders. Nearly every utterance one of us identified as an EVP could be found on the other person's recording and the original sound file, clearly indicating that we had not recorded EVP

The current best practice for field recording is to use two recorders and discard anything found in both recording processes. (144) This is considered a best practice because it is well established that EVP occurs in one analog segment so that two recorders will not normally record the same EVP.

Using live voice, EVP are expected to be formed as a transformation of the live voice into a language understood by the practitioner or interested observer, but based on our study, naturally occurring sounds in a foreign language are too often mistaken as different words in other languages. This is enough of a problem to warrant recommending that, when using live voice, two recorders should be used and both output files examined to assure suspected EVP are only in one of the files. (Be sure to account for differences in atmospheric attenuation and recording process frequency response.)

Better yet, don't use live voice for EVP, including radio-sweep technology which access local broadcast signals. There are just too many false positives.

Preparation

Begin with meditation and a short prayer to ask for only those intending the highest good and an invitation to friends on the other side to participate. This is not a religious function. The conscious expression of intent helps to set your expectations and sets bounds for your communicators. Even if you do not accept the hypothesis that EVP are initiated by discarnate beings, whoever or whatever is initiating the messages, setting intention helps the process. Also, it is best to record when personal energy is the highest.

Scheduling

Entities will speak at any time, day or night. In the beginning, however, it is advisable to record at a regular time and place. By doing this, the entities learn when there will be an opportunity for contact and your expectation of the upcoming session helps focus your attention on the process. Try to find a place that will be quiet and free of interruptions. Background sounds are okay, but it is important to be aware of them so that they can be distinguished from EVP during analysis.

Recording

Begin by stating your intention for communication and then say that you are turning on the recorder. The communicators will often come through as soon as the recorder is turned on and these beginning messages may be the loudest, so it is a good idea to turn on the recorder and wait a few seconds before speaking.

Vocalize your comments during an EVP session. Questions should be recorded, and about ten seconds should be left between each comment for the communicators to respond. At the end, ask if they have something to say and leave time for them to respond.

Playback

Some audio recorders allow you to monitor what is being recorded, but in most situations, EVP are not sufficiently clear to be understood in real time. So as a general rule, EVP are not heard until playback. Practitioners report that the voices tend to become stronger and clearer as the communicators and the practitioner gains in experience, but at first the voices may be whispers. Voices may not be recorded in every session and it may take several sessions to discover the first voice.

Learning where to look for the voice amongst the noise and your voice takes a little practice. Many practitioners describe it as a learned ability because the utterances are often spoken with odd cadences and emphases on syllables.

It might take thirty minutes to examine a three- or four-minute recording, as each part is often listened to several times. Experienced practitioners learn to use the waveform as a clue as to where an utterance might be. The *Betty's in there* waveform [303] will give you an idea of what to look for.

Think of EVP as a form of communication. The attitude held by experienced practitioners is that the communicators will leave their message wherever they think someone will listen. It is a good practice to expect to find the voices where normal human voice would be found. It is not a good practice to amplify a nearly flat-line waveform to see if speech is present because enough amplification to make it clearly audible may amplify radio-frequency contamination.

Classifying EVP

Most EVP are difficult to understand. Using a classification system helps you have a sense of the quality of your examples and helps to set the expectations of your listeners when you share examples. The three-tier system is most commonly used; however, the ATransC is testing a slightly expanded system intended to also include visual forms of ITC. It is explained in the Best Practice: Classifying Phenomena Study.

Keeping a Log

Keeping a recording log is helpful for tracking your progress. If you do keep a written log, it is a good practice to include the date, time and circumstances of the recording. If there are multiple recordings in the recorder, then it is always important to speak your location and to note ambient voices and such at the beginning of each recording. When you find an EVP that is clear enough to keep, note the date and time and seconds into the recording in your log. Also note the words thought to be spoken and if you were doing something the entity commented on. Be sure to label and save the audio file so that it can be found at a later time.

Keeping a well-organized storage system and notes will be very helpful to you when you have many recordings, or want to revisit an old one and need to know when and where it was recorded. Experience has shown that keeping everything is just unreasonable. Unless the message has special significance, it is a good practice to discard all but the clearest examples. Depend on the help of friends acting as your listening panel. If they do not hear what you hear, discard the recording. It is just too easy to fool yourself.

Digital Voice Recorders

Digital voice recorders are recommended for EVP experiments. Today, all sound tracks—digital or analog—should be listened to in a computer and with a headset. Unlike tape recorders, the built-in microphone in digital recorders is usually satisfactory for EVP.

Analog Technology

The requirement of a nonlinear process in electronic circuity is an important characteristic. For instance, in digital voice recorders, EVP are thought to be formed in the analog first stage where the analog voice is transformed to an electrical signal ready for digitization. Once the signal is digitized, it is known to be very stable. However, it is possible that a new EVP might be formed in the output of the digital memory, since it is necessary to pass that through at least one analog stage to be heard. The difference would be that the secondary utterance would not be present when the file is played a second time.

Analyzing the Recording for EVP

Always use headphones when listening to the recording. The earmuff style that completely covers the ear is best, but also good are the soft rubber ear buds that are inserted into the channel of the ear.

It will be necessary to use an audio cable for computer analysis of a sound file from a recorder with only an earphone output. In that case, the signal must pass from digital to analog in the recorder and then analog to digital in the computer. There is little or no evidence that an EVP will be changed while in a digital format, but it is possible for change or even new utterance to be injected while transferring in the analog phase. There is no digital-analog conversion in the newer recorders using a USB interface

It is possible to reasonably reproduce the average human voice with a sample rate of 8,000 KHz, so it is recommended that the audio file be saved in the computer at typically the next higher sample rate of 11025 KHz, 16-bit text file. Mono or stereo is a personal choice. Once the audio file is in the computer, it should be saved as a *.wav file for storage and editing.

Current best practice for sharing an EVP example on the Internet is to convert the file to mono, *.mp3 format, but using editing tools on mp3 is discouraged if it is for research. A 200 kb audio file can be reduced to around 15 Kb when converted from a *.wav file to an *.mp3 file. This makes it easy for sharing files via the Internet.

Under no circumstances should an audio example be shared in video format. These days, I receive a lot of requests to examine examples on YouTube. In that format, it is very difficult for me to break down the sound file to examine the possible utterance. I just don't have time to work with such examples, and I know many of my more qualified friends feel the same.

It is also best practice to provide a raw clip of the EVP, along with a bit of your voice for reference, if possible, followed by a short silence and then the whole clip again with any editing you may have done. Explain what you have done so that your listener will know what to expect.

Finally, ten decibels or so of amplification, perhaps a little noise reduction and high-end filtering are about all that should be done to a soundtrack to make the utterance more easily understood. It is possible to change the meaning of an EVP with very much processing. The general rule is to discard the EVP if it cannot be understood with only slight enhancement. Certainly discard nonsensical and single syllable utterances!

A listening technique is to select a few syllables of a possible utterance and then play it over and over to allow your mind to look for familiar sounds.

People are trained to recognize common arrangements of sound as words, but EVP are often formed from odd arrangements of sound, depending on what is supplied. Also, the usual cues are often missing. A very loud, well-spoken phrase could still be difficult to make out for a person not accustomed to hearing EVP.

In every case, try to use a listening panel [400].

Storage and sharing

Be sure to set up a method of saving your recordings in your computer that will allow you to easily locate examples. A good practice is to save the raw recording session in a dated folder and then save clips containing the EVP in the same folder. We save field recordings under the name of the location and the date. It is helpful to keep a separate folder for your Class A examples for easy retrieval for demonstration to friends.

Unless you are conducting EVP sessions for research, consider it a kindness to only send the portion of the sound file containing the EVP. We too often receive examples that are embedded somewhere in a ten or fifteen-minute-long file. It is just too demanding for your witnesses to search for the intended utterance in a long file.

As a final note on this, after recording for over fifteen years, we finally learned that it is senseless to keep all of our recordings and logs. It is a good practice to cull the best from the files as you go and discard all but the most meaningful. EVP are actually very common amongst practitioners, and there is little special about Class C and B examples.

What You Need to Know

The following observations will help you gain a perspective on the scope of this study.

Lucidity

The degree of clarity between conscious self and etheric personality is referred to as lucidity. Thinking there is a clear connection, when in fact, there is hardly any at all is referred to as hyperlucidity. Hyperlucidity seems to most often occur when a person works alone or does not accept input from others. It is usually seen as a complex of behaviors including delusion, argumentative interaction with those who do not support the delusion and rejection of the opinion of subject-matter specialists. See the Mindfulness Discourse.

Hearing the Voices

A series of online listening studies showed that the average ATransC website visitor participating in the study could correctly identify only 25% of the words in the Class A examples. (21) Most experienced practitioners have learned not to expect a person who is unaccustomed to hearing EVP to correctly understand examples. This is especially true if the person is skeptical to begin with. The message to all of us is that, even though the example may be very clear and obvious to us, it is unrealistic to attempt to show an example to the police or a grieving person if there is not an opportunity to prepare the listener.

Before attempting to share an example with untrained witnesses, we recommend that the example be passed by a listening panel without prompting. Only if the majority of a listening panel is able to hear what is believed to be in the example is it ready for untrained ears. Doing this helps to protect the practitioner from accusations of taking advantage of people.[57]

Safety

A common concern is whether or not it is possible to attract a discarnate personality (ghost) and later be bothered by it. People are realistically concerned that they might *bring one home.* In fact, we have no verified record of anyone being harmed by their communicators. Any harm that has been reported has been due to the practitioner's overreaction to imagined danger.

There is a potential problem for people who are fearful of the unknown, as their imagination can create circumstances which might be dangerous for them. If you are very fearful of the unknown, if you are inclined to do what you imagine you have been instructed to do or if you have a history of mental illness, it is advisable that you avoid working with any form of transcommunication.

---

[57] An important note to remember is that, if you are attempting to use radio-sweep technology for evidential EVP, most experienced practitioners I know will refuse to confirm your work. We are publically on record stating doubts that such technology produces EVP.

Study 4:
Techniques: Visual ITC

Introduction

Just as voice phenomena are sometimes found in audio recording processes, visual phenomena are sometimes found in optical recording processes. When we think of visual Instrumental TransCommunication (ITC), we are thinking of the presence of anomalous features in the media, such as faces and recognizable shapes.

To be ITC, phenomena need to be objectively experienced, meaning that they are sufficiently substantial to be recorded in physical media that can be witnessed by others. Also, technology of some form must be employed to enable the phenomena.

Paranormal experiences that are witnessed by one or more people as mental perception, such as a shared deathbed vision, are not considered ITC because they cannot be recorded. While probably *real*, in a pragmatic sense, they are not considered objective. Poltergeist and other psychokinetic activity, direct voice and materializations that are able to be recorded are considered transcommunication, but unless they involve technology to manifest, they would not be considered ITC.

There are two primary categories of visual ITC. The most common form is often referred to as spirit or apparition photography. As a general statement, spirit photography involves the unexpected presence in recording media of paranormal features that are considered indicative of discarnate personalities or their influence.

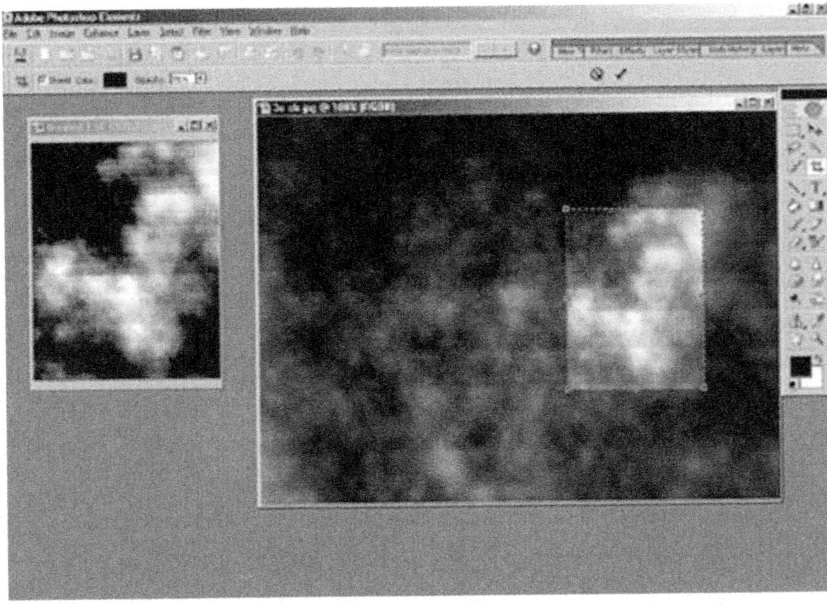

Chaotic Visual Noise in Video Frame: This is a video frame taken from a video-loop ITC session. Contrast in a region of the frame has been changed to make the feature more visible. The same region has been copied to a second file and contrast has been further enhanced. The feature appears to be a woman or girl wearing a bonnet and looking to your right shoulder.

The second primary category is induced visual ITC in which a technique is used with the expectation that the resulting media will contain paranormal features. Current speculation is that all induced phenomena are via the transformation of noise which is produced by the technique. A typical technique produces visual chaotic noise with many regions of different brightness. See the Chaotic Visual Noise example shown here.

Apparition Photography

Before beginning this subject, look ahead to the Artifacts and Perceptual Error Study. I will address the main forms of apparition photography here, but it is important to be aware of the many ways that ordinary phenomena can be mistaken as paranormal.

It is difficult to establish a confidence level for the authenticity of apparition photography because the examples are usually spontaneous, as opposed to induced ITC which is usually collected under controlled conditions.

A second reason for difficulties in authentication is that there is no well-established set of characteristics. Unlike the usual transform phenomena of induced ITC, spontaneous phenomena tend to be unique with few obvious similarities in how they are formed. With these considerations in mind, the small number of examples presented here does not represent all of the ways spontaneous phenomena may form.

Face on turned off TV 1: In this case, the mother's discarnate sister is watching baby's first step. ©

Faces on Turned Off Television Sets

A turned-off television screen provides an excellent surface for the production of neutral-density chaotic light as ambient light is reflected from the surface. However, in the typical face on a turned-off television screen example, reflected light does not appear to be a factor.

The first example was taken by a Las Vegas family when the baby boy had pulled himself up with a chair to stand for the first time. The woman, who can be seen on the television screen, was recognized as the photographer's sister-in-law, who had transitioned a year and a half earlier. The television was turned off. And by the way, this is not a picture of a well-known actor.

The second picture was sent to us in 2005 by Laurel Kenton. In it, you can see him sitting with his grandson. The turned-off television set behind him displays the partially formed face of his mother who transitioned in 1992.

Face on turned off TV 2: Photographer's transitioned mother on turned off TV screen, as if peering in to see him and his grandson. Insert is enlargement of TV screen. ©

We have collected a number of these *faces on turned off TVs* photographs. Some of the faces are full-color while some are black and white, and some even appear to be line drawings. A common factor is the presence of a child in the scene, as if the person on the screen is peering into the room to see the child. Like the two offered here, some faces are recognized, but many are not.

A characteristic of this phenomenon that agrees with other forms of visual ITC is the missing lower part of the face as if to emphasize the eyes. An artifact of taking still pictures of a turned-on TV is the presence of frame lines, and partially formed images. This has been considered and rejected as an explanation here.

Orbs

As discussed in the Artifacts and Perceptual Error Study, most orbs are mundane artifacts; however, the optical noise found in orbs is often useful for the formation of paranormal features. As such, a naturally occurring orb in a photograph might contain an instance of visual ITC as transform phenomena.

The Orb 1 example was taken by Peter Berg. It shows an orb artifact obscuring part of the Reverend Gene Pfortmiller's face. There is no reason to think the face-like feature in the orb has anything to do with the circumstances of the photograph. It may, but without other, more definitive reasons to think it does, the prudent thing to do is to set the example off as interesting, but indeterminate.

Orb 1: Probably mundane photographic artifact that has a possible paranormal face formed in the optical noise. The orb is enlarged on the left and background is erased around feature on right. It appears to be a woman facing to your left. ©

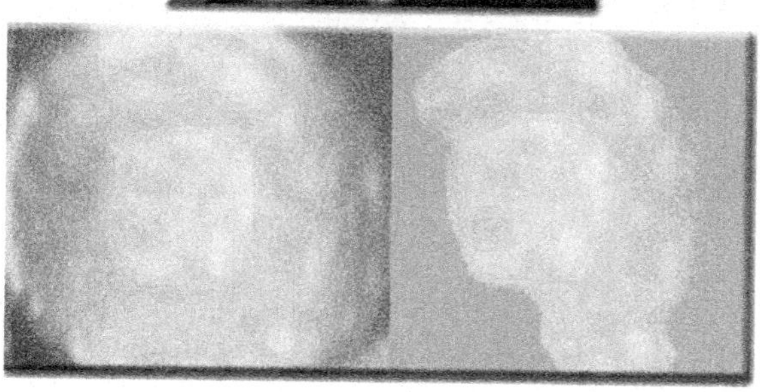

There possibly are paranormal orbs. Lisa and I helped Universal Studios market the movie *White Noise*. For the DVD bonus features, they had us doing walkabouts in reportedly haunted locations while being interviewed for the DVD. We were climbing a staircase in the Los Angeles Hollymont Castle when a fuzzy black ball a little larger than a softball moved past the top of the staircase. In the Orb 2 video frame from that scene, Lisa can be seen walking ahead of me.

We did not see the orb at the time and only learned of it from website visitors. The camera operator used available light video, which means the orb is probably not a camera artifact. It moved a little faster than a person walks, and seemed to be getting out of our way.

Orb 2: Orb (white arrow) moved from center frame to left exit in *White Noise* DVD, *Hearing is Believing.* ©Universal Films

Mr. D. contacted us concerning a series of experiences he had in his apartment. One aspect of these was the occasional presence of two teardrop-shaped orbs which could pass through walls and apparently delighted in tormenting his dog. In one video Mr. D. sent us, his dog can be seen suddenly jumping off of the couch, closely followed by a white orb.

As can be seen in Orb 3, the orb appears to be self-luminous. Mr. D. reports that the orb sometimes hovered in the air, which is why he was able to take this picture. He reported that it seemed to wiggle its tail before taking off.

As a best practice, orbs found in bright light situations, or when the camera flash is used, should probably be discarded as artifacts. Faces found in orbs are not necessarily caused by loved ones or the local *ghost*, and are virtually always so poorly formed that they should be discarded as Class C examples. Self-luminous orbs and black orbs may deserve more study.

Orb 3: This orb was photographed standing still (here) and recorded in motion. ©

Apparitions

Certainly this survey of visual phenomena would not be complete without an example of apparently human apparitions. Both examples presented here were recorded by Rashelle Eastman Morrison. In the first, a nighttime photograph of her grandmother's backyard produced this image of a woman apparently dressed in a 1800s dress. Shell thought she was hovering above the ground, but that is difficult to tell for sure. She also thought it might be her great-great grandmother.

Apparition 1: Taken in Rashelle's grandmother's yard. Woman apparition hovering a few feet off the ground is thought to be Rashelle's great-grandmother. ©

In Apparition 2, Shell was attempting to capture orbs outside of her window. The window is a tall, narrow shape and was opened a little. The flash washed out the window frame but chimes are visible. A little girl appears to be looking in from the dark. Shell said that no one was there when she took the picture.

Apparition 2: Rashelle was attempting to photograph orbs and found this little girl looking in when she examined the photographs. ©

Instrumental TransCommunication

The two main differences between Apparition photography (spontaneous phenomena) and Instrumental TransCommunication or ITC (induced phenomena) is that, in induced ITC, the communicating personality is thought to be attracted by the practitioner's intention to make contact via technology. The features appear to be made possible by the ability of the communicating personality to impress intended order on a chaotic process. As such, it is functionally the same as transform EVP.

The physical process known as stochastic resonance is a means by which a small signal is amplified when combined with a chaotic signal in a nonlinear electrical circuit. This process is hypothesized as the mechanism by which a weak psi influence is able to impress an intended order on the chaotic signal, thereby producing a visual ITC feature.

Video Loop ITC

In the video loop technique, a video camera is connected to the *Aux In* of a television set. The camera is pointed at the screen so that it is able to record the camera output as it is displayed on the screen.

The circuit shown in the Video Loop ITC Diagram is typical for producing chaotic optical noise suitable for visual ITC. We have been using a Cannon 8mm ES2000 analog camcorder placed about three feet from a Sony KV20TS32, 20 inch Trinitron color TV. The camera is used in *Portrait Mode* and is focused just beyond the screen surface.

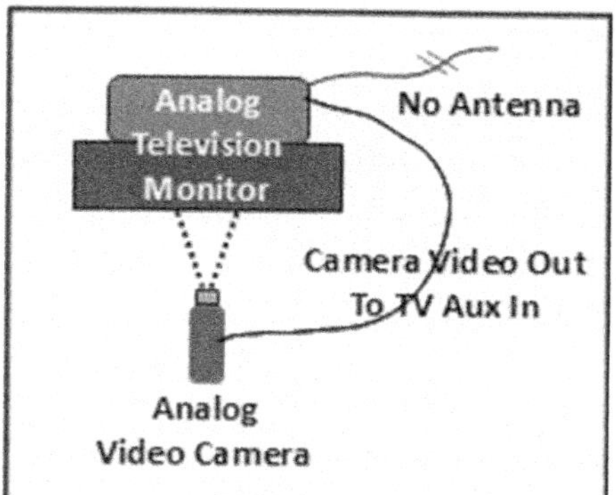

Video Loop ITC

The camera lens is zoomed so that only about 75% of the screen is visible to the camera. We usually have the camera at about a 45-degree angle from the screen and pointed just to the side of the center where the rolling image seems to begin.

It is routine for us to try different angles, and camera distances from the screen in an effort to find the best configuration. It is interesting that, some days, we are hardly able to develop the desired rolling scene. We have made no effort to track environmental conditions except that we noted a difference, apparently due to location. We were able to develop a good rolling scene in the corner bedroom but the best production of the phenomenal features was in the middle bedroom. The only difference between the two adjacent rooms is that, in the corner room, the equipment was about ten feet closer to an outside wall supporting the house power distribution box. The Chaotic Noise Example screen print [316] will give you an idea of what has been an optimum display to produce the features.

Clarity is a Function of Equipment

Each configuration of equipment used in video loop ITC will produce a different visual effect, probably even amongst units of the same brand. The rolling effect on the screen is a cycle that is about a second in duration. An average of thirty frames per second of full video quality only produces five or six frames per cycle that contain noise suitable for phenomena. A session is usually about fifteen seconds so that there may be ninety or so frames that may contain features. To save time, we only save the ones that appear interesting as we step, frame-by-frame through the saved video file. A typical fifteen second session will produce thirty or so frames interesting enough to carefully examine.

If we have the equipment set right, and if a number of as yet undetermined environmental factors are okay, we may save ten or so useful features from the session. There have been many sessions in which we saved none. Also, the texture of the noise may be more or less grainy, depending on where we had the focus plane.

Software for Frame Analysis

Finding suitable video capture and frame examining software has been a challenge. Adobe's Premier 10 was not very expensive and came bundled with Photoshop Elements 10. Currently, we are trying CyberLink Power Director with some success.

Each likely frame should be rotated in 90 degree increments while being examined. Frames are numbered something like 1-2-14-2007, 2-2-14-2007, 3-2-14-2007 and so on. Individual features are each copied into a new file, and those worth saving are labeled 3-2-14-2007a, 3-2-14-2007b 3-2-14-2007c and so on. The idea is to be able to associate a feature with the video frame from which it was captured.

Sharing Examples

The video loop examples shown here were made with an analog TV and a digital video camera that uses magnetic tape to store the video signal. The *grabbed* frames are saved as JPG files which is a file compression format that loses detail and introduces noise. Your computer monitor may have a number of different resolutions, but it is reasonable to assume there will be a resolution mismatch between the original TV display and a viewer's monitor.

If an example file is stored on the Internet for sharing, it will potentially be further compressed and then a person viewing the example has a monitor of unknown resolution, and it is unknown how well a viewer will be able to see/resolve the feature.

We conducted an online viewer test with a number of pretty good examples. (20) Each example was displayed in raw format as it was in the video frame stored in my computer before any processing. Also, as shades of gray, and a third, enhanced to increase contrast and brightness. 61% did not see the features we thought were in the examples.

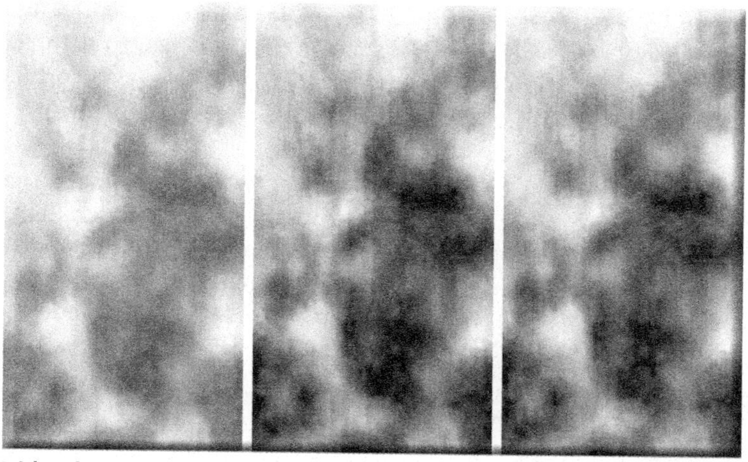

Video-loop ITC Example 2 of the Perception of Visual ITC Images study. Website visitors correctly described the feature 28% of the respondents. The dog shown in the Trans-Survival Hypothesis Discourse was correctly described by 81% of the respondents.

Detail of a saved video frame is a function of the resolution of each device beginning with the relatively low resolution of the television set. As such, you can add all of the pixels you want in your photo manager, but the actual information in the example will never be better than the lowest resolution component of the capture equipment.

As with EVP, sharing visual ITC examples with others can be a little tricky. The usual approach is to send an example via the Internet as a JPG file, but you should anticipate needing to help your viewer. A long as it is not for research, one of the best ways to help your viewer is to describe what you think they should see. For instance, in Video Loop ITC Example 2 above, I would include that *"I see a face looking to your left and slightly tilted as if the person is slouching. The person, probably a man, appears to be in oriental, ancient clothing with either a black hat or black hair done up like a crown. The face is skin tone and he appears to be wearing a light colored vest of some kind."*

The only enhancement is to change contrast. As you can see in Video Loop ITC Example 2 [325], the *master frame* (left) was left unchanged to provide a control. Your good judgment is required as to how much enhancement is done (middle). Sometimes changing to grayscale helps (right).

I will sometimes erase or smudge out the background of a feature to make it more evident to witnesses. As such, in the Troll example, the contrast was changed first, and then the background was smudged.

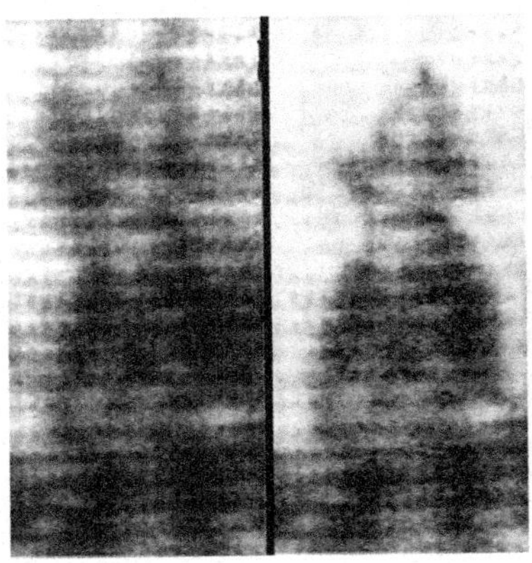

Example of *Enhanced* Example: The *troll* from the alien composite [333]: unaltered clip from the original video frame (left) and after the background was deemphasized (right).

Any enhancement should be noted if the original is not provided for comparison. I am guilty of routinely adjusting color density and contrast, because at times, that is the only way to make a feature stand out from the shadows for examination. Use discernment in such processing and always be prepared to provide the raw original.

Another way I sometimes enhance a picture to highlight a feature is by changing the contrast, and then using the smudge tool to further lighten the color density to highlight a feature. In the Cow example [330], the original (left) is not much different than the version in which the contrast has been increased and lightened (center). The real change is in the small areas indicated by *A* and *B*, which are shown heavily enhanced on the right. I discuss the small features in Face Fractals [330], below.

About the Technology

From our experience, every hardware setup is different. We use color analog equipment and many of our features seem to have true skin colors. When we began in 2000, it was generally understood that, if there is any color in the final example, it should be removed for proper display. It is difficult to know if we are simply doing it wrong or if our technique is an improvement. Most of our efforts to use all digital equipment lack color as well.

Analog Verses Digital

In EVP, it is pretty clear that the voice is formed in the electronics, probably in a single transistor junction. Stochastic amplification requires an analog, nonlinear process which is available in the active region of semiconductor devices. There is no apparent requirement for which transistor junction is used, except that it should be in a part of the circuit that permits recording of the results. Because of the nature of the video loop, it is not clear which device is actually used.

An all-digital circuit would not seem to provide the necessary environment for stochastic amplification. For instance, when a webcam is used that has a digital image sensor, and that connects to the computer via USB cable, the only analog stage before display on the computer monitor is the image sensor. While we are pretty sure transform occurs in a single active region, there is no reason to think that images can be impressed across an array of charge coupled devices as is usually used for image sensors.

So to make this simple, our speculation is that video loop ITC will be most successful if an analog stage is intentionally supplied. For instance, an analog camera might be used. We have not explored all of the alternative techniques. It is an important area for further study, because analog is less available to the average person. Some of the most astounding visual ITC has been captured by the early experimenters using older, modular, analog technology such as a video recorder, video camera frame synchronizer and video amplifier. A modern equivalent is needed.

Much more experience is needed to understand how the technology affects formation of the features. For instance, how does high definition video change the dynamics? One thing that needs to be understood by modern practitioners is that simply replicating the function with modern equipment (combined video camera, recorder and playback) will not replicate the early experiments. On the other hand, the only way to increase the quality of the examples is to use much higher resolution capture and reproduction.

Perhaps Not ITC

We have been seeing a lot of mirrored examples in which one half of the feature was captured, copied, flipped horizontally and joined with the original so that the original is mirrored by the copy. This is beginning to look like a visual equivalent of radio-sweep [296] because one of the characteristics our mind uses to identify a face is symmetry. As is shown in the Mirrored Image picture below, the typical mirrored shape looks like an alien—almost every time. While this is an interesting characteristic, I think further study of how we see things will show that the mirror-formed *faces* are mundane artifacts of perception.

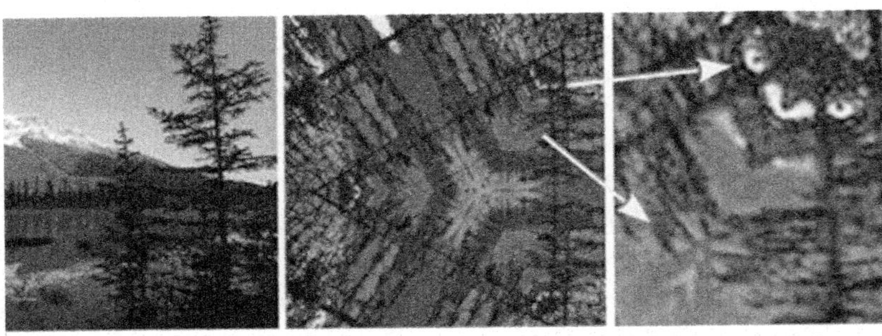

Mirrored Image: The photograph at the left is modified in the middle with a kaleidoscope tool using three mirrors. The arrows are pointing to two face-like features on the right. Any combination of eye-like shapes and symmetry will potentially produce a face-like feature. (215)

A second trend we have noticed is the technique of erasing all of the picture but the presumed paranormal feature. That is what I did with the Orb 1 Example [319]. I see the vague shape of a woman's head. To help you see it, I have erased all but the feature. Fortunately, I preserved the original so that you can see what I changed.

The Orb 1 Example is useful to illustrate a difficult to see concept, but it is a risky practice. In a way, selectively erasing parts of a picture to show a particular feature is a lot like storytelling [348] in EVP. If it is done, the original should be displayed alongside the enhanced version. Exceptions to this might be instances in which the feature is well-outlined as is the Troll [333] in the alien composite.

Witness Panels

Complying with the Witness Panel Best Practice is perhaps the most important tool available to the inexperienced practitioner. As has been stated many times in this book, we too easily fool ourselves. By definition *fool* means we don't know we are being fooled. The way to guard against being fooled is to ask others to review our examples and give us a candid description of what they see or hear.

In some of the examples I have seen on the Internet, I doubt an uncoached witness would see in the raw example what the practitioner highlighted. Even for a Class A example, probably not all of your friends will see what you see. To validate the feature, at least a few should independently describe what you think is in the example. The Perception of Visual ITC Images (20) study will give you a sense of how people see visual ITC examples. It is online in the Journal section of ATransC.org.

What are the Paranormal Features?

Just about any way you can imagine to produce chaotic optical noise is suitable for the spontaneous formation of anomalous features. That is why face-like features are often found in photographic orbs. I say *spontaneous* because they usually have none of the characteristics of communication. However, with that said, there is evidence that some such features are impressed into the noise as communication with the practitioner or an interested observer.

Our working hypothesis is that the patches of brightness in the looped video are areas of relative order that naturally form in chaotic noise because of stochastic amplification [276]. In that process, small irregularities in the uniformity of *baseline* energy are amplified, and in video loop ITC, those emergent areas of increased order appear to be intelligently controlled.

What Have We Learned?

We use color video equipment and we have found that resulting features are often displayed in true colors such as skin tones and color of lips or hair. We also note that, in all forms of visual ITC, it is common for the bottom portion of the face to be obscured.

In some examples, there is apparent interaction between features. More rarely, a feature may be in a scene. In a Class C example, I found a man apparently dressed as a medieval lord, his foot on a low wall as if he was posing for the camera. In another example, the man is clearly posing for the camera in the Man with a Dog example [332].

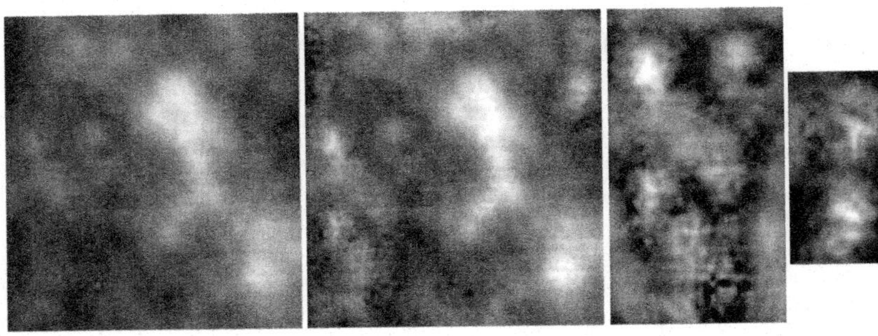

Cow: Video-loop ITC: Unenhanced on left, enhanced with Photoshop Smart Fix in middle and regions A and B isolated and enhanced with contrast tools. I see a number of face-like features.

As is found in many EVP, the features are approximations of what they appear to represent, as if the transform process was not completely controlled. EVP will often decrease in volume and/or become garbled near the end of a phrase. In a similar way, faces might be distorted, or as noted, not completely formed.

Face Fractal

In the Implicit Cosmology, life fields are described as the fundamental building block of reality. In effect, reality consists of life fields and their expressions as ideoplastic structures. Using fractals as the model, Source would be the top fractal and aspects of Source would be Source-like fractals. The Cosmology of Imaginary Space Discourse will give you a sense of how you can imagine this cosmological model.

Also in the Implicit Cosmology, the etheric personality of a person is able to interact with other personalities within the bounds of perceptual agreement. Conscious awareness of this interaction is filtered in the Attention Complex so that a person is normally unaware of the resulting information. That is, how much we become aware of things etheric depends on our lucidity. All of this information is provided here to help explain the apparent fractal effect of visual ITC.

Probably the most disconcerting aspect of visual ITC is that faces seem to compete for space in the display. Faces seem to be a most common fractal as if they are some etheric background noise. For instance, if the naturally occurring physical noise is suitably chaotic, there might be faces formed of faces. That is, a primary face will be present, but on closer inspection, usually after the viewer *resets* attention or steps back from the display for a fresh perspective, many less well formed faces will become evident.

The right frame of the Cow example above [330] shows the two darker regions around the cow as they have been enhanced to make the smaller features more obvious. Again, it is important when viewing these examples to remember that the actual nature of the anomalous features is obscured by the limitations of our equipment. With that said, it may be possible that the cow is also formed mosaic-like by many face-like features.

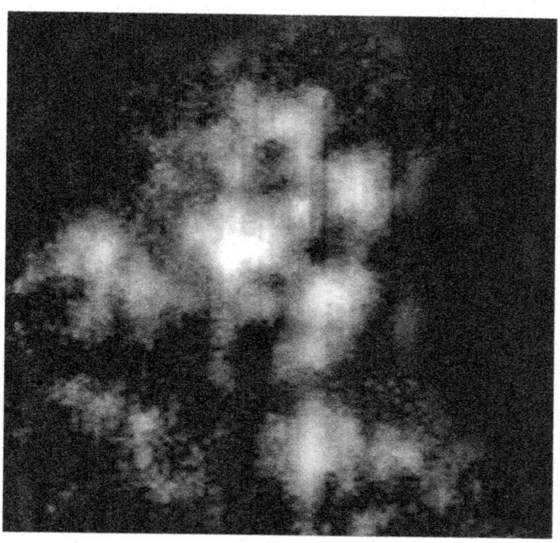

Two Lovers: The pale colored area in the middle-left of the frame appears to be a man's head facing your right. The yellow-orange area right of center appears to be a woman's face partially obscured by the man. Her left eye is showing. They are together as if caught in an intimate pose. Background has been obscured for clarity. See original on website.

The Lovers example shows the face-like secondary features and also two primary features apparently interacting. Focus on the white area at the middle of the frame. Let your eyes find a man's face looking slightly down and toward your right. His skin is white. Once you find him, look at the yellow-orange area behind his head. I see a woman's face as she is looking at him, maybe whispering in his left ear. He is obscuring the right side of her face. All of the yellow-orange area represents her face and neck. Close examination of the lighter colored areas in the background will show that there are numerous faces in the picture as well.

Animals and Aliens

It is common to find a person apparently proudly displaying a favored pet. In the Man with Dog example, the man's skin color appears to be correct. If by this, we can guess other colors are correct, he is wearing a white jacket that looks like a dress uniform. The dog appears to be white.

Man with Dog: This appears to be the portrait of a uniformed man proudly holding his little white dog. The head of the man is badly distorted but the skin colors appear to be true. He appears to be wearing a white jacket. The white dog is looking directly at you.

According to the Implicit Cosmology, and considering the nonlocal character of reality, it is logically consistent to expect people from other lines of evolution (extraterrestrials) to appear in visual ITC. In fact, a number of convincingly alien features have been collected. The Aliens example here is representative of this.

As I noted before, the colors often seem to be true, so that appropriate skin, lip and hair colors are sometimes seen. With this in mind, I refer to the first feature on the left in the Alien example as the *Blue-faced Man*. He appears to be wearing a chest-plate with a dark fur collar.

We have collected a little, spindly-armed creature with a pointed head and ears which point right and left directly away from the skull. This is a poor Class C example and cannot be displayed here. However, it is very much like the gremlin captured by Jose Garrido and Alfonso Galeano, and the one captured by Erland Babcock.

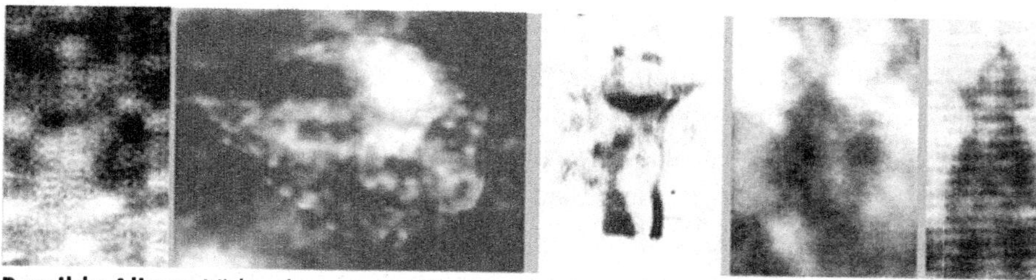

Possible Aliens: Video-loop examples that seem to be of alien people. From the left, blue-faced man (recorded by Butler), Gremlin (Jose Garrido and Alfonso Galeano) ©, possibly same *gremlin* species (Erland Babcock) ©, apparent grasshopper-like being (Butler), possible troll (Butler). The background has been suppressed to make the troll more apparent. Some contrast and color intensity adjustments have probably been made on these. Ask if you need to know for sure.

It is important to note that Babcock used much older technology than we have and produced what I think are considerably better examples. If you experiment with visual ITC, be aware that success is as much influenced by the technology you use as the human factors contributed by you, interested observers and the trans-communicating personality.

Induced Visual ITC

In one visual ITC example, we had asked my father to show himself in our visual ITC session. The example below is what we found. Again, it is common for the lower part of the face to be obscured. It is also interesting to note that I had the photograph of my father on my desk, as it was being moved from one storage area to another.

Possibly My Father, Clayton: We asked my father to show himself in a session. We later found this feature. It has a similar pose to a photograph of him on my desk.

Of course, I will not say the *Clayton* example is definitely the product of my father's intention to respond to our request. I feel that it is. Because after carefully examining hundreds of video frames without seeing anything even remotely like my father's face, it is unreasonable to expect his face to be found in a video frame from the session in which we asked him to appear.

Another reason I am inclined to say the feature is my father is that his pose in the video frame is the same as in the photograph I had on my desk. We have also noted that the appearance of an anomalous face does sometimes closely match available photographs. While this characteristic may be explained by the communicating personality's access to the practitioner's memory, it might also be explained as an apport.

Apported Material for Feature Formation

A characteristic of apports is that they are physical-world objects such as stones, trinkets and flowers. In field conditions, they either mysteriously appear, such as one that Lisa found in her shoe, or they fall from the air as if they have emerged from an alternative reality.

Apports come to us as physical objects because that is the way we are able to experience them; however, one of the characteristics of ITC is that the features sometimes seem to be at least partially formed with preexisting physical world objects. It may be more efficient to adapt something already embodied in the physical than to create it from pure thought. If so, then it makes sense that the pose of the Clayton feature is similar to what was on my desk at the time. Again, remember that these features are representative of the actual thing, and are often very distorted.

Light Reflected from Moving Water Technique

Short of simply taking a picture of medium-density texture and brightness surfaces, probably the simplest method to produce ITC features is photographing light reflected from moving water. All you need is a camera and a container for water as shown in the Water-ITC photograph below. Any camera will do but if you use a video camera, you will need a computer program for examining the individual frames. Otherwise, handling of the individual pictures is the same as for video loop ITC.

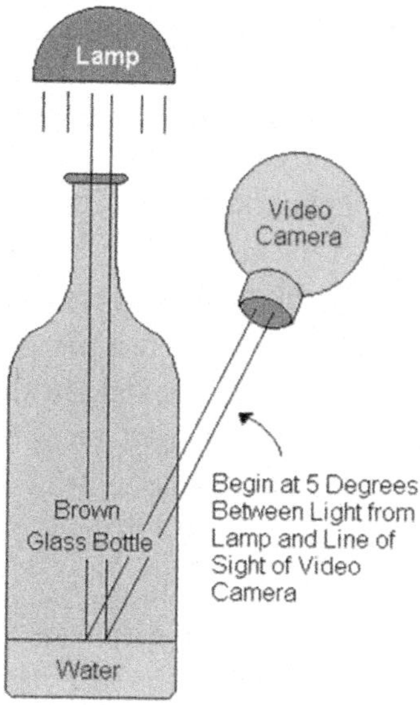

Water Technique: Arthur Soesman of the Netherlands introduced us to this technique in 2004 with this illustration. Most practitioners have substituted the jug with a sauce pan. ©

As shown in the Water Technique Diagram, this technique began for us with Arthur Soesman's report of using moving water in a bottle but quickly evolved into the use of any container. Since the features are found in color photographs or videos, a colored container or added colored lighting has been popular.

Sauce Pan for Water ITC: Glass sauce pan in a black pot with about three inches of water. Black pot is just for contrast. Any overhead light source will do to reflect from water. Take pictures of water as it is disturbed with finger or spoon. The objective is for the moving water to produce many areas of medium-intensity reflected light.

As previously noted, clarity of video loop ITC features is limited by the video equipment and method of their display. Since a very high resolution camera can be used, the actual limiting factor is the resolution of the display technology. However, the reflected light tends to be very high contrast, which limits the amount of usable neutral-density regions of reflected light necessary for feature formation.

Before his transition, association member Erland Babcok was very good with photographic phenomena. As shown in the From Erland Picture below, he found a large amber glass jug for his experiments and often added a violet light source.

It was common to add *high-frequency* light to experiments because of the persistent belief that *The Other Side* is only separate from the physical because it is of a higher frequency. The light was intended to help bridge the gap. While it is still uncertain if it helped, the colored light did produce interesting anomalous features with often beautiful accents.

From Erland: A one-gallon brown glass jug (left). Erland situated his camera at about 1/3 the way up the side of the jug with the camera very close to the jug. The jug was wrapped in black paper with an opening for the camera. Light was applied from above. ©

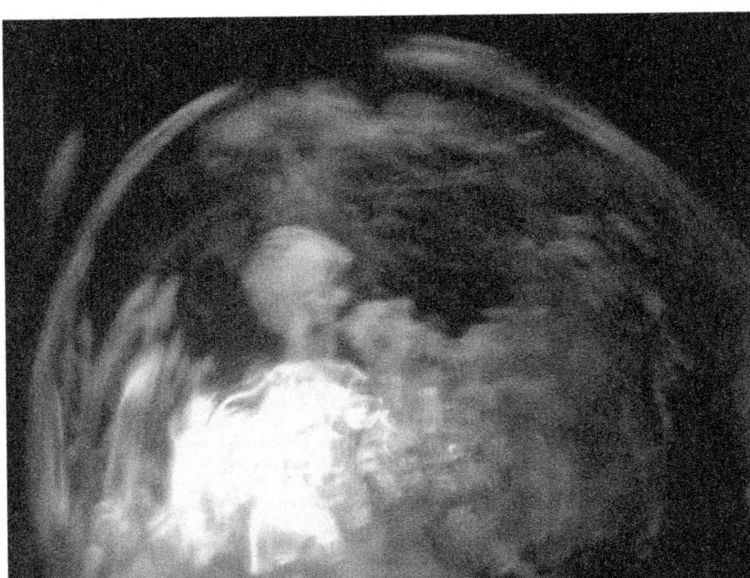

Elves: An example of Erland Babcock's moving water technique. ©

Erland produced many excellent ITC examples with the moving water technique, and some were of such beauty that he even had a showing in his local library. The one offered here in the Elves example above is especially interesting. You may need to look for a while to see what I describe, but please do. It seemingly provides a mysterious glimpse into an unknown world.

In the Elves example, it appears as if a strong wind blows from your left toward your right. Look for a tall man-like creature in the background with long hair blowing in the wind. His right ear sticks out from his hair more like a mule's ear. He is looking toward you but down at the girl.

Directly under his chin is a translucent dome hat (think mushroom cap), in which you can see the head and hair of a little girl facing your right. Hair on the left side of her head is blowing past her right cheek. It looks like her bare shoulders are pulled up, as if resisting the wind that is at her back.

Just in front of her is possibly a boy wearing the same type of hat. His chin is sloped into his neck and you can see his open mouth just under his nose that is facing up. There may be a dog in front of and facing him.

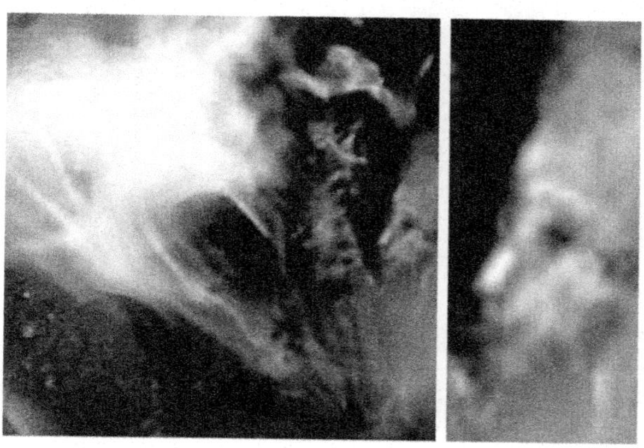

King: Except for the face, the left picture is typical for light reflected from moving water. The violet comes from the glass sauce pan. The paranormal face at the right of the left frame has been enlarged in the right picture.

Most of our light reflected from moving water features are what we would refer to as Class C, meaning that recognizable facial features are present but are so distorted that they probably would not be recognized by witnesses.

The King example is an exception. You should see a human face looking toward your left. The Class A face is very close to a second feature I would consider a Class C.

Margaret Downey is a very capable ITC practitioner. In the Grandfather example below, she found a face in her video that looked like her great-great-grandfather. Daniele Gullà, with the Italian Interdisciplinary Laboratory for Biopsychocybernetics Research (Il Laboratorio) (122), used forensic-quality face recognition software to compare Margaret's ITC image with a photograph of her great-great-grandfather while he was in the physical.

Gullà wrote that *"I have processed the human face as a 3D model and rotated it to superimpose over the ITC face. I have compared the repere [reference] points (only 5: eyes, lips, nasal and subnasal) in the human face and ITC face. The final result is that the difference in the two images is inferior to [less than] 5%. They are very similar!"*

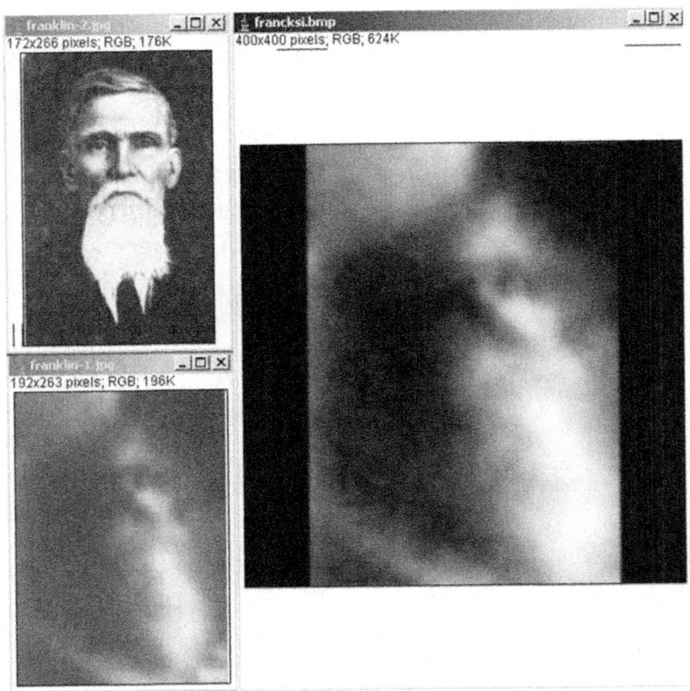

Grandfather: Margaret's great-great-grandfather Benjamin Franklin Main Sr. (upper-left), ITC image recorded by Margaret (lower-left) and the two images superimposed by Gullà (right). ©Composite prepared by Gullà.

Theory

Effort has been made to eliminate mundane explanations for the examples of visual ITC presented in this study. This is not to say that one will be found, but based on current understanding, it is reasonable to consider the possibility they represent an etheric-to-physical influence. Not all examples are communication in the sense of a two-way exchange of information. However, it might be argued that all are caused by an intelligence that has expressed an interest in the situation in which the phenomena were experienced.

Many examples indicate a person looking into the scene, perhaps out of curiosity but more likely with a purpose. For instance, the presence of a child in the room is a common characteristic of faces on turned off television sets [317]. This seems to indicate a degree of family attachment.

A common belief amongst paranormalists is that people have personalities in the etheric who attempt to provide guidance for them without interfering with self-determination. Perhaps the *people looking in* are guides observing the progress of a physical charge. Certainly, the presence of these people looking into our world serves as a catalyst, causing us to reconsider what we know about the world.

The Implicit Cosmology remains the overriding theory in this book for how these phenomena are formed. As such, the worldview of the practitioner or an interested observer likely plays a part in how the features are actually formed. For instance, in the Apparition 1 example from Rochelle [321], the great-grandmother apparition may have been presented to her unconscious awareness as an intention field which was translated by her into familiar terms.

So Many Faces

It appears likely that stochastic amplification is the physical process by which the influence of intention acts on optical noise to produce an intended feature. This follows the concept of intended order.

The real challenge is in modeling the presence of so many face features found in noise. How we ration our attention seems to offer a possible model. A person entering a crowded room is aware of most or all of the people in the room. Some will be more familiar than others, but the person will become more aware of one person at a time.

The presence of paranormal faces is modeled as having been formed via the personality-avatar entanglement of the practitioner or interested observer. A video frame that has a face feature in it typically has one well-formed, primary face feature and numerous more or less well-formed secondary face features.

Based on this, anomalous features found in a video frame are a reasonably fair expression of the personality's personal reality. That is, my personality is aware of many others, probably in my collective. All of those personalities are liable to be represented in the transformation of noise into features, but more likely, the one to show up the most is the one with which I am attempting to communicate or that is trying to communicate with me.

Inherited Credibility

Examples in this book which we think represent trans-etheric influences may not provide conclusive proof of our etheric nature; however, when considered as a set, each inherits validity from the others. For instance, physical mediumship brings credibility to the idea that ideoplastic structures are possible. If they are shown to be possible, then influencing an electronic device should be considered a logical extension of those phenomena.

The well-documented effect of intention on random event generators (51) establishes a sort of behavior that must be applied to ITC involving random audible or visible noise. I say this because both involve the apparent effect of intention on the behavior of noise in a nonlinear system. The common factor is stochastic amplification.

One of the themes of this book is that Etheric Studies inherits some validity from psi studies and also that psi studies inherits some validity from Etheric Studies. However, this inheritance should not be construed as proof of survival. Much more study is needed before such a conclusion can be made. Because in the foreseeable future it is unlikely psi researchers will take up Etheric Studies, the task for metaphysicians is to find a way to model trans-etheric phenomena so that psi researchers will see the compatibility of the two studies.

I See It, Why Don't You?

It is part of paranormalist lore that the ability to hear and understand the voices of EVP is an acquired skill. EVP usually only approximate live speech so that the usual structural cues found in live speech, and on which we depend to understand meaning, are often formed in novel ways, making it difficult to decode meaning. The context of what is said may also be obscure so that the important circumstantial hints which help us assign meaning to the words of EVP may be missing.

The anomalous features in visual ITC suffer from an equivalent problem. Consider this, if an inkblot used in psychological tests is consistently identified by testees as a butterfly landing on a flower, is it reasonable to argue that the inkblot appears to be a simulation of a butterfly landing on a flower? Just because the feature was formed by a randomly blotted ink spot does not mean it cannot reasonably represent something real.

This is a common problem in visual ITC. The context in which a feature is formed can be used to assign intended meaning. If an inkblot on white paper is intended to be random and without meaning, but is consistently reported as a butterfly, should it be required to contain no meaning? What if Vincent van Gogh had splashed black ink on a piece of paper that looked just like the ink blot and titled it *Butterfly on Flower*? Would that give us reason to accept that the ink blot was a reasonable simulation of a butterfly?

One of my jobs in the corporate world was long-range planning. You may know it as strategic planning. The core ability for long range-planning is the ability to conceptualize trends as a possible future. In many ways, artists are long-range planners because it is necessary to visualize the consequence of every step as it will relate to a conceptualized system called the work of art.

When I search video frames for anomalous features, I try to see each part as a component of a system of related shapes and colors. Very often, once I see a face-like shape, it becomes difficult not to see the face. I will sometimes *reset* my mind by walking away and later taking a second look. In poorly formed features, it takes me a while to find the face. The important thing is that I do find the face.

All of this is to say that I have a native aptitude to see the features in visual ITC that not all viewers share. It is a fact of corporate experience and it is apparently an explanation as to why not everyone viewing my examples sees what I expect them to see.

As with EVP, there is something of a learning curve one must traverse to work with visual ITC. The only way I know to develop the ability is to work with the examples with the intention to see meaningful relationships amongst the otherwise random shapes.

If someone tells you they see a butterfly, take a little time to see the same. Of course, we must be mindful of the hyperlucidity trap and not force meaning where there is none. There is a balance which can be found with experience and by cooperating with a witness panel.

A second difficulty people have in seeing the anomalous features is looking in the right place. Often, the video frame will not have a primary feature, so just looking at the frame as one image is not enough. Look at the little parts where there is a spot of brightness amongst the darker, random shapes. A good example of this is the Cow feature [330].

The Future

Visual ITC is potentially one of the most promising forms of paranormal phenomena. The anomalous features seem to represent a moment in the life of a probably long-transitioned person. Sometimes they represent meaning for us or an interested observer. Most likely, they review an otherwise invisible characteristic of reality.

Visual ITC is one of the least understood forms of paranormal phenomena. Virtually every parapsychologist with whom I have discussed the features have advised me to stick to EVP because the features are simply too unlikely. So if you decide to study them, be aware that seeing the faces is not enough. At some point in your study, it will become necessary to stop just collecting examples and begin developing a conceptual context for what they are and why they exist.

ITC is clearly a metaphysical study, but there is an important physiological link, as the appearance of the features is biased by practitioner and/or interested observer expectations.

Transcommunication

Study 5:
Artifacts and Perceptual Error

Introduction

Perhaps the single biggest problem affecting how people perceive their world is that they do not understand how it operates. Of course, this is the root of superstition in primitive tribes, but we moderns have our own superstitions. For instance, I grew up with the idea of *earthquake weather*. That came about because the adults in my life as a child had experienced a strong, mid-summer earthquake. It was hot and muggy with virtually no breeze when the earthquake occurred. Thereafter, every hot and muggy day had the older folk around me on edge, if not consciously, at least unconsciously fearful of another earthquake.

I am not a geologist, but as far as I know, the weather has little to do with earthquakes. But then, that is the point. I have only a tourist-level understanding of geology and do not know how to argue that hot days are not a sign of coming earthquakes, other than it makes no sense based on my worldview.

And that is the second point. Mindful living [145] can be described as habitually asking yourself if what you think and do makes sense. Considering your intention, are your thoughts and actions reasonable. If you do not know the consequences of your actions, then it is incumbent on you to move with caution. Instead of *"It is this,"* at least say, *"It might be this,"* or even better, *"Is it this?"*

As an electronic engineer, I have a relatively good sense of how technology might bias the expected output of devices. For instance, there is a design flaw in Panasonic RR-DR60 digital voice recorders that causes an occasional, loud noise that was at one time called *Angry Man EVP*. Of course, it is just an angry sounding artifact of the device, but it took some time before we listened carefully to the noise and found that the trans-communicators were using the noise, perhaps even initiating it, to produce the audio energy needed to say such things as *"I love you."*

Reality is complex. Chaos theory is probably the best way to study reality. To be realistic, every experience we have is apt to have unexpected elements.

If you consider this point, it is a profound statement. The statistical norm probably applies, meaning that the weather will be about as it was predicted and work will be more or less as expected. Yet meteors occasionally fall out of the sky, the ground under our feet sometimes moves and a storm occasionally brings catastrophic flooding.

Technology and the environment combine to influence our experiences. A picture taken of a bird in a tree is apt to be ruined by a ghostly light caused by over-exposure from the sun which was just out of camera range. A magnetometer might register a psi influence as a change in magnetism when there is no such change. Apparent paranormal words might be formed by recording foreign-language speech, when in fact, the sound was only muffled by attenuation in the air and poor quality of the player and recorder. The rest is all listener expectations.

Photographic Artifacts

The list of potential environmental and technology photographic artifacts is much longer than the ones offered here, but these are the ones I most commonly encounter. By becoming familiar with these, you should know to look for others. This too is part of mindful living.

Low Light Latency

If you are an experienced film photographer, you will know about image latency. It is a term used to describe the presence of an image in exposed film which has not been developed. That concept does not apply here.

The latency we are interested in here is the delay between the beginning and end of a process. In imaging devices such as motion and still cameras, the concept is usually applied to command latency. This is the delay between pressing the button to take a picture (command) and the time it takes the camera to finish taking the picture (response).

When the process to take a picture begins, the light coming into the detector increases the electrical potential of each pixel in the detector, and it is the size of that charge that is entered into the electronic process to create image data. A brighter object in the scene will produce more charge in a cell than a dim one.

Latency: Example of latency in low light photography.

The effect of latency is easily seen in many of the nighttime photographs collected by ghost hunters. One example I spent a long time trying to explain to a website visitor was a twilight picture of a supposedly haunted house. A nearby house had a porch light on, and while the haunted house was dimly but well imaged in the scene, there was a bright yellow trace that wiggled around to its left. Even though the camera was on a tripod, when the person pressed the button to take the picture, he slightly moved the camera. The camera returned to a stable position for most of the exposure and produced a good image of the house. However, the porch light was much brighter than the haunted house and it saturated affected pixels for every slight movement of the camera to produce a wiggly yellow trace.

The picture here is of a large ferry at sea. I took the picture with the camera set to automatic. You can see that the dimmer parts of the ship are well imaged (see the name), but all of the bright lights produced the same trace pattern as I moved the camera to take the picture.

Daytime Latency

A common example of daytime latency is a photograph that has something in the scene missing a part. For instance, people walking in a museum without legs or a dog that is missing its head because it moved during the exposure.

Again, the camera is usually automatically setting the exposure for the average light it detects. A walking person's legs move faster than the rest of the body. If automatic camera exposure is perfect for the relatively large area of the torso, it may be too short to image rapidly moving legs. The result is a ruined photograph which is too easily mistaken as paranormal by a person with a predisposition to seek evidence of the paranormal.

Sun Flares

If the sun or a bright light is just out of range of the camera lens, it is apt to register in the resulting photograph as an often very pretty rainbow or series of colorful orbs. Most camera lenses are compound, meaning they are formed of more than one lens shaped element. The front surface of each element may image the bright light as an orb. It is usually possible to imagine a single straight line passing through the orbs and the sun.

Sun Fade

Bright light will sometimes cause a photograph to have what appears to be a fog in the scene. This is usually just saturated image detector or film, but sometimes the ruined photograph will have a ghostly quality.

Orbs

Photographic orbs are usually round, often translucent objects found in optical media. Because they generally appear as chaotic optical energy, they are prone to have recognizable facial features in the manner of visual ITC. Being mostly white with little texture, features they may contain are seldom more than outlines suggestive of recognizable forms.

Popular wisdom amongst some paranormalist holds that orbs are evidence of discarnate personalities or that they contain the face of a local ghost. It is clear, however, that most orbs are artifacts caused by light reflected from shiny surfaces or particulates in the air. Others are easily shown to be caused by bright light sources such as the sun. With that said, there are circumstances in which some orb-like photographic objects may be paranormal. These are discussed in the Visual ITC Study.

Audio Artifacts

Audio technology can produce artifacts as well. The technology introduces the artifacts, but it tends to be the way practitioners process sound files that makes the artifacts audible.

Over Amplification

Inexpensive audio editing programs make very sophisticated capabilities available to EVP practitioners. Of course, all of us have tried everything available to make that hint of a voice clear enough to share with others. After working with thousands of audio files since the mid-1980s, Lisa and I have come to the realization that the only safe processing is around ten or fifteen decibels of amplification and maybe a little noise reduction. Anything more tends to produce processing artifacts that can make the possible utterance say what was not intended.

Electricity at the power outlet is very noisy. This becomes obvious when it is examined with a device that will show small signal noise. The problem is that the many miles of wire tends to act as an antenna, allowing energy from radio broadcasts to produce a small signal in the wire. This can sometimes be amplified so as to be heard with the unaided ear.

Hand-held devices are also excellent antennas for radio-frequency contamination. The result is that amplification of any recorded sound track of more than twenty or so decibels will begin to make the contamination audible. We have had many sound files brought to us for examination that had been amplified and the resulting audible radio contamination mistaken as EVP.

Single Syllable Error

One of the more common mistakes EVP practitioners make is thinking naturally occurring single syllable-like sounds are paranormal voices. This is especially a problem for people using *silver bullet* technologies such as radio-sweep [296]. For instance, if the question is *"Who lived here?"* a *"Bo"* sound that was a fragment of a phrase spoken by the radio announcer as *"Boy did we win!"* might be identified as a paranormal voice saying *"Bob."* Since the practitioner does not know the name of the person who lived there, *Bob* is as good a name as any.

The majority of EVP examples we have seen posted around the Internet are mundane, single syllable-like sounds attributed as EVP. In some cases, a single syllable sound might have other, probably mundane sounds nearby in the sound stream, enabling the practitioner to claim they are part of a multi-syllable utterance.

As representatives of the ATransC, our recommendation is that possible single syllable utterances be either disregarded or reserved for personal use. Certainly, they should not be used for research.

Selective Reporting

The technology used in radio-sweep and EVPmaker [295] is designed to produce a stream of random sounds. Presumably, the output sound stream is influenced by a communicating personality to produce meaningful messages. One problem with these technologies is that they always produce a streaming output of often voice or voice-like bits of sound.

If a meaningful word or phrase is detected in the output sound stream, the sounds before and after in the sound stream may be included in the interpretation of the possible paranormal utterance. This is a reasonable approach, but it is usually a judgment call as to how much of the sound stream applies to the suspected EVP. As such, the practitioner must be mindful not to inappropriately include trailing or leading sounds in the reported EVP. This flexibility too often leads to storytelling or selective reporting as the practitioner attempts to show meaning where none is intended.

In transform EVP, the utterance may come after it has been unconsciously composed by the practitioner but before it is consciously spoken. With this in mind, an acceptable interval between question and answer in transform EVP is just a second or two, but certainly after the preceding question and before the next question.

Selective Reporting by Deletion

A second and more damning form of selective reporting is modification of a sound file to eliminate irrelevant sounds and only present the question and answer as if one was directly associated with the other. While it may be a technique to show the listener what the practitioner believes is present in the recording, selective reporting of this kind is more an ethical violation than a mistake.

If such editing practices are to be used, unedited sound file should be presented with the display version for comparison. What a person is guided to believe as real has the potential of changing that person's worldview. The practitioner presents examples under the cloak of assumed expertise, and therefore, has an obligation to be clear about what is thought to be paranormal and what is known to be mundane.

Storytelling

A common practice amongst practitioners is to record a likely sound and then concoct a story designed to make it sound like a meaningful EVP. Without noticing, a practitioner might have recorded the intake of breath from one of the ghost hunting team. Upon review of the recording, the breath sound is in the recording just before that person said, *"Is there anyone here?"* The practitioner then explains that the one syllable breath sound is really an EVP answering *"Yes"* just before the question.

An important practice in field recording is to use a second recorder, and then to compare recordings to see if the suspected EVP is on both. As a general and very reliable rule, EVP only occur in one recording process, so if the sound is on two, simultaneous recordings, then it is either direct voice (paranormal voice in the air) or it is an artifact. (144)

Mistaken Low Volume Sounds

Especially in field recording, there may be people speaking in other parts of the building. Such normal sounds often do not register in the practitioner's mind. The voices might be picked up on one recorder, but depending on the quality of microphone, a second recorder might not register them. While such distant sounds are often reported as EVP, the best practice is to discard them as, at best, poor Class C examples.

Errors in Perception

Here, I am not talking about the need for a person to learn how to hear EVP. Not being a psychologist, the best way I can explain this concept is to say that people tend to mentally fill in what their unconscious mind assumes is missing. This *closure*, as I have heard it described, is mostly a problem with live voice techniques, such as radio-sweep and EVPmaker, in which a fragment of speech is heard as a whole word.

A good listener is one who has learned to resist closure and to hear sounds more as they are rather than as they seem. However, it is common for a person to say *"I don't hear anything but noise."* But when told what is thought to be said in the EVP, the listener will be better able to make out the words—even if the example is only noise.

Research by Dr. Mark Leary has clearly shown this listener suggestion bias. (74) The following is from Item 2 in Part 3 of the cited report:

> *All EVP enthusiasts know that people's interpretations of EVP are sometimes affected by what they think other people hear. In fact, it is often difficult not to hear what someone else said they heard. In our second study, we found that agreement with the individual words in an EVP jumped from 6% to 23% when listeners were told what the recording investigator thought the EVP said.*

The following is from Leary's Part 2 Conclusions and Recommendations:

> *The results of this study suggest that investigators should be less confident in their interpretations of EVP than they typically are. On*

average, the most common interpretation of each EVP was shared by only 22% of other people. And, of course, all interpretations other than the most common, consensual one had even less agreement. In fact, most of the raters' interpretations were not given by any other listener! Furthermore, raters were not particularly good at judging the correctness of their interpretations. Thus, having the sense that 'I'm sure this is what it says' does not indicate that other people will agree with one's interpretation (or that it is actually correct).

These results lead me to offer four recommendations for the responsible interpretation of EVP:

1. *In light of the fact that any particular investigator's interpretation of an EVP is not likely to be shared by other people and that people's interpretations are biased by what they expect to hear, investigators should never interpret an EVP for other people without playing it for them several times and soliciting their independent interpretations.*
2. *If the interpretation of a specific EVP is particularly important (such as when it is being interpreted for grieving family members), investigators should use a scaled-down version of the procedure used in this study. Have at least 10 people independently listen to the EVP and determine the consensus interpretation, if any. Then report an interpretation of the EVP to others only if a majority of listeners agrees on that interpretation. In some cases, it may be helpful to report more than one potential interpretation, along with the percentage of people who agreed with each one. Providing listeners with such data is a more honest and responsible way to share EVP than to offer a particular interpretation that might, in fact, be idiosyncratic.*
3. *Investigators should be willing to refrain from interpreting ambiguous EVP. Providing a questionable interpretation as if it is certain is misleading, if not sometimes dishonest. Just because an EVP cannot be interpreted does not mean it is not a useful piece of evidence, so investigators should not interpret EVP that are unclear.*
4. *Paranormal investigation groups and EVP practitioners should have formal guidelines for the interpretation of EVP that minimize the likelihood that they will offer interpretations of EVP—whether*

to other group members, clients, or outsiders—that are expressed with greater confidence than the objective evidence warrants. Investigators should exercise greater care in sharing their interpretations of EVP, and procedures should be in place to ensure that clients, other investigators and the public are not inadvertently misled regarding interpretations of an EVP.

Faux Transcommunication

After years of abuse from the skeptics of Wikipedia, I have come to understand that some people do not understand the logical distinction between a phenomenon and something that only looks like that phenomenon. EVP, for instance, is defined as a form of transcommunication. It is experienced as instrumentally aided anomalous access to information. By that, it is intended that no mundane explanation has been found for the presence of apparently paranormal voices in recording media.

If an example can be shown to be fraudulently produced, that it is actually produced by an unnoticed person speaking in the room, or it can be demonstrated as an artifact of the technology, it is not EVP.

Two things must be true to properly debunk EVP. First, the practitioner must have followed best practices to record an example, including use of two recorders and a listening panel. Then the debunker must show a physical explanation for the presence of the reported EVP.

The idea is that there first must be an actual utterance honestly thought to be EVP. There are hundreds of examples on the Internet that are reported as EVP but for which there is no chain of custody, so to speak. Also, practitioner experience is generally not known, nor is the recording technique and whether or not the example has been altered or if time has been cut out. It is illogical to think a proper evaluation of EVP can be made with such examples.

Unexpected Technique Induced Artifact

Practitioners try just about everything they can imagine to improve transcommunication. Often, the technology they us is new and the community has little experience as to what causes what problem, what works and how best to apply the technology.

Of course, radio-sweep is a good example of this, but in visual ITC, technology is making it possible to mirror pictures, causing some very strange, face-like features to emerge out of an otherwise mundane frame.

Both mirrored pictures and radio-sweep involve effects which are not well understood by practitioners and witnesses.

The naturally occurring angry sounding noise artifacts originally identified as *Angry Man EVP* mentioned at the beginning of this study is also a good example. When a technique is used to produce phenomenal results, practitioners and witnesses alike are expecting the next output of the device to be transcommunication. This predisposes them to mistake mundane output as paranormal.

Retro Familiar Storytelling

In mental mediumship, information tends to come to the medium as a vague impression which must be clarified before it can be expressed as a message to the sitter. To be clear, it has been experimentally shown that the original information is often correct and verifiable. It is the human factor which causes confusion.

The more lucid the medium, the clearer perception of the intended message. The task of the medium is to express that information to the sitter as it was received. It is called coloring when the medium changes the message to agree with his or her worldview.

A good medium will resist translating the initial sense of the information and express the message to the sitter in a *give what you get* format. An example might be the original sense of a puppy and the accompanying understanding that it relates to the sitter. As such, the medium might say, *"I see a puppy near you."*[58]

The sitter's response to such a general statement determines the veracity of the message. If the sitter replies that he or she just brought a puppy home, then the message is complete and the medium should move on. However, this is when a form of storytelling often begins.

---

[58] From the Spiritualist perspective, the message in this example would be considered a psychic message because it relates to something that is probably in the mind of the sitter, rather than a mediumistic message initiated by a discarnate personality. From the perspective of the Implicit Cosmology, both sources come to the medium in the same way, and as such, there is no functional difference. One must consider the message, rather than the source.

It is also possible that a discarnate personality might initiate the message because the puppy is important to the sitter. It is very difficult to know what has occurred upstream of the message before it emerges into the medium's conscious awareness.

If the medium elaborates by saying something like "Yes, I see it exploring your home," it is likely that the medium is mentally filling in detail of the original impression based on what the sitter is saying. This feedback can produce changes in perception that can seem to be part of the original impression. But, in fact, feedback is from the sitter rather than the original source and is not paranormal.

Allowing the sense of the initial information to be modified by feedback from the sitter is what I refer to as *retro familiar storytelling*. In effect, it is the development of a false memory based on guidance from a witness, and then accepting that false memory as if it is the initial memory. This is probably a natural problem and a medium who allows this to occur should not be considered fraudulent. More properly, it should be considered a bad habit which needs to be managed.

Education

People seldom have the necessary technical or cultural references to be reliable witnesses; however, perceptual errors are avoidable with a little discernment and willingness to learn from subject-matter specialists. Perhaps the most effective way to avoid perceptual errors is the practice of suspended judgment [73]. (68)

As a person who is dedicated to the study of survival and trans-etheric influences, it is important to me that people make an effort to learn about the nature of their possibly paranormal experiences. The alternative is fueling the public perception that we are delusional. Perhaps more important is the negative influence such faith-based comments have on our efforts to understand these phenomena.

Transcommunication

Study 6:
Mediumship

a.k.a. Psi Sensing

Mediumship

A useful definition of mediumship as the term is intended here comes from the National Spiritualist Association of Churches:

> *A medium is one whose organism is sensitive to vibrations from the spirit world and through whose instrumentality intelligences in that world are able to convey messages and produce the phenomena of Spiritualism.* (2)

To paraphrase, a medium is a person who is able to sense the expression of etheric personalities (discarnate people) and convey the intention of that expression to people still in the physical.

The *phenomena of Spiritualism* are the forms of trans-etheric influences thought to be initiated by etheric communicators. Usually, these are various forms of mental and physical mediumship and healing intention.

It is important to note the emphasis on who is supplying the information in mediumship. In the traditional view, mediumship is characterized as an exchange of information from a friend or loved one in the etheric to a person in the physical (the sitter) by way of a medium.

The traditional Spiritualist point of view is body-centric. From the personality-centric perspective, which is the perspective of the Implicit

Cosmology, a person is a personality entangled with a human body in an avatar relationship during a lifetime. A person who has translated out of a lifetime (died) is usually referred to here as an etheric or discarnate personality. Both living and dead people have an immortal personality as their core intelligence. See the Life Fields Discourse and Life Field Complex Diagram [xxv]. There is also more on Spiritualism in the Contemporary Spiritualism subsection of this Study [366].

Psi Sensing or Being Psychic Verses Mediumship

A person is said to be psychic if he or she is able to access information by sensing the psi field. In the past, this ability has been referred to as Extra Sensory Perception or ESP, but more recently it is known in parapsychology as psi functioning. It is correct to say that a person is psychic or to refer to a psychic ability.

In the traditional point of view, a psychic is not necessarily a medium but a medium is always a psychic. This can be understood to say that a psychic only access information from a person's mind, presumed residual information in psi space (Super-Psi Hypothesis) or from objects. A medium is able to access that same information, but primarily accesses information by way of a discarnate helper or a sitter's loved one. In the traditional view, the difference is profound.

From the personality-centric point of view, a person is always psychic in that the Attention Complex interfaces with reality via the Attention Limiter. This interface includes psi signals from other personalities (loved ones) and a signature signal from objects in the environment (Ideoplastic structures). See the Perception and Expression Discourse.

In the sense of psi functioning, *psychic* and *medium* are functionally the same and any distinction between the two is in the origin of the information that is being accessed. That is, if the message is thought to come from a person in the etheric it is referred to as mediumship. If it is thought to come from some physical source or residual information in psi space, it is considered psychic.

This is a major deviation from traditional Spiritualism. It is also based on current science. All I have done is translate that new science as part of this cosmology. If you have been trained to distinguish between psychic and medium, it is good to recognize that science is an ever-evolving understanding of Nature. Being open to new ideas also means allowing old ideas to be replaced with new ones if appropriate.

Psi Sensing

Psi sensing is a more contemporary term for psychic or being psychic, and as noted above, it may as well be a new term for mediumship. The term accommodates the parapsychological term of *psi* and *sensing* describes the process by which psychic and mediumistic communication is accomplished.

Lucidity

As discussed in the Mindfulness Discourse, lucidity is used to describe the clarity of awareness between a person's conscious self and mostly unconscious personality. Take a moment to examine the Life Field Complex Diagram [xxv]. Note in the Life Fields Discourse that personality represents the core intelligence of the life field. It is also important to note that the conscious self (*I think I am this*) is isolated from the (mostly) unconscious personality (*I am this*) aspect of our life field by the Attention Complex.

In the traditional view, one of the objectives of training in mediumship is to teach the student to better sense *inner guides* and to *get out of the way* by not coloring the message. In more contemporary terms, this objective can be paraphrased as making the personality-to-conscious self connection more lucid (clearer).

In the Attention Complex, the Attention Limiter filters out all uninteresting information from the environment and the Perceptual Loop makes a decision about what is passed by that filter. Since it is the outcome of these process that emerges into conscious awareness, the first step in gaining lucidity is to gain control by expressing intention to do so.

The lack of attention about environmental information can be overcome by learning to be open to new ideas. It is very hard to undo a decision, but the conscious decision to suspend judgement provides a means of telling your unconscious processes to more often arrive at a *Maybe* outcome of the Perceptual Loop. Remembering that worldview can be changed in small increment, it is the *Maybe* outcome of perception that provides that incremental influence.

The Collective

An important concept for the study of transcommunication is the idea of a group consciousness or collective of personalities sharing the same local source. For instance, it is common for a communicating personality to indicate that it is part of a group or that it is only the speaker for a group. Jane Roberts' Seth (63) and the SORRAT's (149) Imperator group are examples.

As it is modeled in the Implicit Cosmology, a personality expresses aspects of itself to gain understanding through experiences in an imagined venue for learning. Once expressed, the life field representing an aspect is persistent and remains entangled with its source personality as part of a collective of similarly expressed aspects related to the same source personality.

The Principle of Perceptual Agreement is defined as *Personality must be in perceptual agreement with the aspect of reality with which it will associate.* Based on this, a person will be most in agreement with other members of its collective and in agreement to a lesser degree with people from other collectives. In practical terms, in a room full of people, a person might feel that one or two people are *soul mates* (collective mates) and while there will probably be a few that *rub the person the wrong way,* the person's response to most in the room will be neutral.

Collectives can be described as cooperative communities. That is, as it is modeled, members of a collective tend to help one another gain understanding. Each member of a collective contributes acquired understanding to the source personality, and therefore to each other. The objective is to bring their source personality's perception into agreement with its source personality. As it is modeled, this agreement includes acquired understanding which is sought by the source personality.

You can see that cooperation amongst members of a collective furthers their progression, and thereby, their source personality's progression. The same sort of mutual support can be seen in cooperative communities within our physical, social networks. Two understandings illustrate this:

The Golden Rule Redo: ***Teach me as I teach you.***

and

The Way of Progression

Through community comes knowledge
Through teaching comes understanding
It takes a collective

Sensing Reality

Sensing reality is a way of describing psi sensing and mediumship. While some people naturally come by mediumistic/psychic ability, most of us must develop the ability. In most cases, that means unlearning bad perceptual habits.

As explained above, a person senses reality by way of the Attention Complex. That is a mostly unconscious part of our mind which translates information from our psi and physical environment, according to our worldview, into conscious awareness. The expression of intention is how we indicate to our unconscious mind our interests and preferences.

Managing the perceptual process is an acquired discipline that can begin with understanding the Implicit Cosmology. Most of the thirteen Discourses describing the cosmology are focused on theory; however, the last four discourses, beginning with Discourse 9: Perception and Expression, include explanations for how the cosmology might be integrated into daily living.

Test Beliefs

From my experience, I have learned that if you have studied psychic or mediumistic phenomena, it is likely you accept a particular person's teaching. For instance, you may consider information from across the veil to be from a particular level of reality or you may assume that a particular group entity is the only source of truth.

The Implicit Cosmology is just a blueprint for our spiritual anatomy and our relationship with others. It is not a thing of belief, as most of it is based on current science and lessons learned from transcommunication. Anyone should be able to arrive at essentially the same cosmology after examining the evidence. If it is reasonably correct, then it is important that you use it as a test for what you believe to be true. That is, learning to define what you believe to be truth, based the structure of the Implicit Cosmology, will help you align your worldview with what I will call contemporary spiritualism.

Sensing Potential Fields

Every event has a psi signature and every event is preceded by a field representing the nature of the event and the gathering potential for it to occur. As it is modeled, a potential field is the output expression of the Attention Complex and is created when a person first imagines the possibility of doing something. This expression might be a thought, an action such as speaking or walking, an aspect of the personality or an ideoplastic structure.

As a product of the creative process, potential fields representing events are shaped by the creating person's quality of attention, clarity of visualized outcome and extent of intention to make it so. When first conceived, the event might have optional outcomes. In this model, each potential outcome would be represented by a potential field. Each of these fields would gain or decrease in potential as the expression of the event approaches, with the one representing the actual outcome becoming dominant as the others dissipate.

Following the concepts associated with perception, and as shown in the Perception and Expression Diagram [xxiv], a person unconsciously senses psi signals from the environment. If a potential event is of interest to a personality and passes the Attention Limiter, then it will be introduced to the Perceptual Loop. Working with worldview, that process will attempt to characterize the potential in familiar terms, and if that is possible, present it to conscious perception.

This process occurs for everyone but most of the usually unrelated events remain in the unconscious. They become more evident for those who are trained to *listen* to subtle cues from their unconscious.

Potential fields may precede their related event by considerable time. If for instance the event is an act of war or a crippling fault in a commercial aircraft's battery, the potential field might be sensed very early. Even a minor potential might be sensed if it involved safety of the person or a loved one.[59]

A person might sense all potential fields, or perhaps only those that make sense according to worldview. Since the moment for expression of an event might grow near only to be aborted by the creator personality, a medium might sense a future that had been possible, but never occurred.

Sensing the Physical

The Life Field Complex Diagram [xxv] will help clarify the way information is thought to be processed via the etheric personality-avatar relationship. Early versions of the Implicit Cosmology included the idea that the human senses made physical information directly available to the conscious self, but it appears more correct with current understanding to say that all environmental information is filtered before coming to the conscious self.

As modeled, if the information does not make it into the worldview, it cannot be directly sensed by personality. In effect, a person is blind to

---

[59] Remember the idea of the Butterfly Effect in Chaos Theory: it is said that the flapping of a butterfly's wings in one part of the world might produce a storm in another part. Thus is the nature of a chaotic system. The potential of that storm would begin with the butterfly. The potential would have many possible forms, depending on the climatic conditions. Only one potential would be realized. (150)

anything that falls too far outside of personal reality as it is held in worldview. Ambiguous information, the *Maybe* solution to the Perceptual Loop, will likely be sensed, but its true nature may be masked by a substitute version as mind seeks to find a familiar interpretation of unexpected information.

Once again, the importance of purging worldview of beliefs in favor of more objective understanding cannot be stressed enough. So very much of who you are is determined by your worldview and that is entirely under your control. See the Mindfulness Discourse.

Mind-to-Mind Sensing

An important factor in psi research is the current inability to shield from psi influence. For instance, if an experiment calls for a sitter to identify what is in an envelope, it is possible that the sitter may determine the contents by psychically sensing the information from the memory of the person who put the object in the envelope.

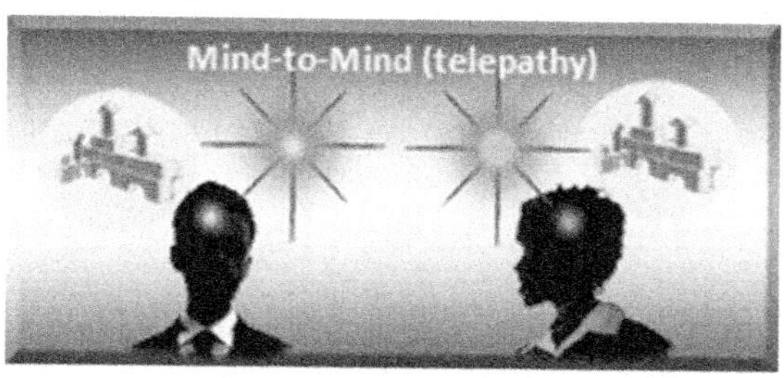

In a telepathic mind-to-mind exchange of information, the receiver becomes aware of information via the contact its etheric personality makes with the personality of the sender. Awareness of that exchange must be translated into physical consciousness via the person's perceptual process and is limited by perceptual agreement.

In remote viewing, the viewer likely views the target location by way of a mind-to-mind exchange of information between the viewer and someone who is either physically viewing the target or has memory of the target. See the Personality-Centric Perspective and Perception and Expression Discourses for clarification of how a person might be aware of telepathic exchanges.

Channel

A channel is said to be a person who is able to mentally *step aside* while in deep trance to allow an etheric communicator to speak through its physical body. The distinction between a medium and a channel is usually based on who is communicating and for what purpose. Generally, a medium is said to be a person who senses messages from discarnate people. It is usually thought that a channel brings through more profound information from *higher beings*. This may include entities who have not experienced a physical lifetime. A personality representing a group of personalities is frequently a contact.

Physical Mediumship

> Physical mediumship is discussed in a number of other locations in this book, especially in the Trans-Survival Hypothesis Discourse.

A mental medium typically functions in a state of awareness ranging from fully conscious to medium trance. A physical medium is expected to enter into a very deep trance.

As a deep-trance medium, a physical medium is thought to be able to allow the conscious mind to *step aside* so that a second personality may assume control of the body. In the Implicit Cosmology, the actual process is seen as a combination of increasing lucidity and a conscious decision to accept the direct influence of the communicating personality.

In most examples I have witnessed, the physical medium has been as if in a stupor, able to walk and talk, but when doing so, acting as if trying to hold on to a tenuous sense of lucidity. In all of the cases I have witnessed, this has been a trained mental state, making it difficult to distinguish between affectation and actual disassociation. Perhaps it is this *behaving as if* that is necessary to maintain an open channel.

There is little doubt that the resulting phenomena are real. Spirit lights, levitation, ectoplasm, reunions and apports during darkroom séances argue the presence of very sophisticated control one would not expect from a conscious person.

Another factor that is clear is the ability of the communicating personality to orchestrate the intentionality of the sitters. Virtually everything I have learned indicates that participation of a second personality is essential for the production of phenomena. (28)

Why the Old Beliefs Are Suspect

Let me first explain why the usual source for information about the other side is not widely accepted by researchers who seek an objective view. Most of what we are told about the other side is conveyed to us via a medium. That is, a physical person has made contact with an etheric personality, and in some way, has conveyed information to us from that personality.

Having a physical person in the circuit is seen as a problem because research is showing that the person acting as the medium is more involved in the message than popular wisdom would have us believe. In mediumship, what they call *"coloring"* is seen when the medium's own worldview contaminates the information coming from the etheric communicator. This is a known problem and mediumistic or trance messages are almost always taken with considerable reservation by researchers seeking objective evidence. (26)

A test of how clear the channel is would be if channeled messages have a cultural bias. In fact, they do. For instance, how heaven was described to us in the late 1800s and how the *Other Side* is described to us today has changed in much the same way that our societies have changed. The same can be said about the difference in information channeled by a USA medium, and, say, an Australian aborigine.

Transcommunication involves the movement of information across the veil from the very conceptual environment of the etheric to the mostly objective environment of the physical. A message exchanged between a medium and an etheric communicator appears to occur as the mind-to-mind exchange of a gestalt thoughtform containing everything one would want to know about the subject. It is then translated by the medium via familiar patterns and embodied into the physical as a spoken message.

Misunderstood Messages

Considering how trans-etheric influences are thought to be propagated, it is important to pay attention to how the Principle of Perceptual Agreement applies to cultural differences. For instance, a holy man of ancient times would have had a relatively simple worldview. Many of our strongly held beliefs of today were revealed in the early days of civilization and passed forward to us as systems of belief.

Belief such as the existence of karma and a silver chord connecting soul with our body are frequently accepted today in a literal sense when current understanding suggests a more objective alternative view. If old beliefs are understood in the light of the Organizing Principles, it should be clear that self-determination and the Prime Imperative to gain understanding make the karma concept obsolete. In the same way, the concept of entanglement helps us understand how the link of influence between personality and conscious self could easily have been understood as a silver cord by an ancient holy man.

The same kind of misunderstanding can be seen today. For instance, if the medium is very religious, then something like *"I am part of your collective"* might translate as *"I am God."* In this cosmology, each of us is a point of view of a group consciousness modeled as a collective.

We have found in EVP that a person will tend to record messages confirming expectations. We are reasonably sure the communication is real. It is just that the message must be processed in the practitioner's or an interested observer's Attention Complex. In fact, the colored messages represent an important part of our foundation evidence for the Life Field model we use here.

You can see that we think of information given to us via mediums and holy books as historically interesting but not very useful. Of course, it does have value as a reference point, and we have spent a lot of time normalizing such information in an effort to find the root message. (47)

The messages of ITC often have an objective component which gives us reason to have more trust in the information. This has given us a unique perspective not shared by those who study survival but who do not understand or accept the evidence of ITC. The Trans-Survival Hypothesis is the result of this normalization and consensus building based on psi research, mediumistic and ITC sources.

A Meditation to Train Psi Sensing

One of the better meditations I have found for developing your psi functioning, and by extension, mediumship, is to visualize your unconscious mind as a vast, deep cave. You as conscious self are standing at the entrance of the cave, your back to it, as you look at the bright, sunlit world before you. The sunlit world represents your conscious self.

Normally, people learning to meditate are taught to suppress their mental chatter by *clearing* their mind. So let us call this a contemplation because, while I want you to enter into the *body asleep-mind awake* meditative state, I also want you to think of that chatter as bubbles floating toward you from the back of the cave. Imagine the bubbles of colored oil that rise up toward the light in a lava lamp.

Each thought emerges from the darkness of your unconscious, and you only become aware of them as they enter into the dim light of the cave entrance. Each bubble is a fragment of thought beckoning you to pursue it to examine the story it can tell.

Consider each thought as it comes into your awareness. If it is familiar, if you know what it is about, then mentally send it back into the cave or ignore it. Familiar thoughts need no further action, other than to acknowledge them and then let them go. You are looking for the unexpected thoughts, the ones that are a little surprising.

Allow yourself time to get comfortable with this visualization. The cave is a safe place, but it is very large and very dark. While it represents your unconscious mind, it also represents a doorway to the rest of reality. Hidden at the back of the cave is a threshold through which you are connected to everything.

The daylight at your face represents your waking reality. It is what the world looks like from your body's eyes. When you turn to look into the cave, you are trying to see reality as that etheric part of you which is the intelligent core of who you are, your *I am this*. To do so, you must extend your senses deep into the cave and the way you do that is to distinguish the occasional, unexpected scrap of memory as it floats into your awareness. Assume that, if you do not recognize the thought bubble or if it surprises you, then it does not come from you; it is from beyond your *I am this*.

It will take time to learn how to get to the point in your contemplation to consciously examine the scraps of thought to know if they are from your etheric personality or from *beyond* your life field. When you do find a new thought, consciously see your mind's eye follow the thought bubble as it drifts back into the cave. Do not see your body follow the thought, only follow with your attention. You are now a free-floating observer who is able to enter the deepest reaches of your subconscious.

Work with this exercise over many sessions. The back of the cave is your Perceptual Loop, and it is possible to teach your mind to reach far into that loop to sense the information as it comes from the etheric environment. As you learn to do so, you will be able to direct your attention beyond the back of the cave, through that threshold and on to sense other personalities that are perceptually in agreement with yours. In effect, you can develop your natural psi sensing ability.

Contemporary Spiritualism

There are six foundation concepts I will refer to as Six Pillars of the Implicit Cosmology. Understanding them will help you understand what I intend by *Contemporary Spiritualism*:

> **Pillar 1, Reality:** Reality consists of life fields and expressions of life fields. (Reality Discourse)
>
> **Pillar 2, Person:** A person is an immortal etheric personality entangled with a human in an avatar relationship. (Life Fields Discourse)
>
> **Pillar 3, Purpose:** We inherited the urge to gain understanding about the nature of reality from our source personality. (Prime Imperative in the Organizing Principle Discourse)
>
> **Pillar 4, Mind:** Unconscious psi sensing precedes conscious perception with unconscious worldview as the arbiter and conscious intention as the motivator. (Perception and Expression Discourse)
>
> **Pillar 5, Creation:** The creative process is attention on an imagined outcome to produce an intended order. (Perception and Expression Discourse)
>
> **Pillar 6, Progression:** Integrating acquired understanding of the nature of reality into worldview constitutes progression and expands opportunities for greater understanding. (Perceptual Agreement Discourse)

What is usually thought of as *Modern Spiritualism* is based on 19th century understanding. Many of the pioneers are of that time as well. It is from that understanding that many of us have learned about mediumship. In fact, many of those early concepts have been shown to be correct within the context of a body-centric cosmology: however, much has been learned since the beginning of modern spiritualism that suggests a more contemporary view is needed.

I make a point of this *modern* verses *contemporary* nature of belief because the difference also applies to a large part of New Age thought. When compared to the implications of current understanding—pragmatic, objective understanding rather than magical thinking—concepts such as karma, vibration, ascended masters and planes of existence begin to seem more fairytale than beliefs worth shaping worldview.

Discussion

There are three common ways of considering the phenomena of mediumship:

> **Organized Spiritualism** sees mediumship as the exchange of information from a discarnate personality to a person in the physical (the sitter). A medium is one who provides the conduit for this exchange. Psychic functioning (information from the physical) is discouraged if it is disguised as mediumship. As such, in Spiritualism it is poor form to call oneself a psychic-medium.
>
> **Paranormalist culture** sees mediumship as psychic-medium, since the medium may access information from the environment, other people or the sitter and possibly discarnate personalities.
>
> In the **Implicit Cosmology**, all anomalous access to information is via the unconscious perceptual process. This means that mediumship and telepathy are seen as a mind-to-mind exchange of information between etheric personalities. This information must pass through the perceptual process, as does psi information from the environment and mental translations from physical senses. As such, there is really no difference between psychic and mediumistic functioning. To tell the difference, one must consider the message content.

This is also why I prefer to think of transcommunication and psychic functioning as psi sensing.

Ethics

A point that needs to be emphasized here is that, to people who are unfamiliar with the actual nature of transcommunication, a medium is seen as a spokesperson for discarnate loved ones. The ethical considerations of this are profound and potentially have legal implications.

As a general statement, mediums are unaware of current thought on the relationship between worldview and perception. In actuality, the average medium is not prepared to manage sitter expectations in an ethical or legal manner. In my mind, this is a measure of how important general awareness of a good cosmology is to our community.

A Personal Code of Ethics for Developing Mindfulness [150] in the Mindfulness Discourse addresses ethics and offers a means by which developing a personal code of ethics can be beneficial as a tool for personal development. Ethics is not just about right living. Awareness of right living advances progression.

Transcommunication

Study 7:
Healing Intention

Introduction

You may know this subject as spiritual healing. Perhaps a more progressive term is energy healing, but if you really want to be technically correct, I believe biofield therapy is the term becoming popular in parapsychology.

In their essence, the various modalities of healing involve the expression of a practitioner's (the healer's) intention to help a sitter (the person being healed). The healing influence is typically expressed with the intention to directly improving the sitter's health, or at least to improve the sitter's general sense of wellbeing.

This study focuses on the nature of healing intention in the context of the Implicit Cosmology. However, as is discussed in the Etheric Fields Discourse, it is the combination of what is intended, how it is visualized and the focus of attention that determines whether or not the expression is sensed as healing, or perhaps, only as friendship.

Spiritualism

According to the National Spiritualist Association of Churches (NSAC), a *"Spiritualist Healer is one who, either through their own inherent powers or through his or her mediumship is able to impart vital, curative force to pathologic conditions."* The healer's objective is to access help from a discarnate loved one to heal the sitter. *"Imparting vital, curative force"* should be understood as the medium's personal energy being used for healing the sitter. This appears to happen, but Spiritualism is all about putting the discarnate helper or loved one to work.

Implicit Cosmology

As it is modeled in the Implicit Cosmology, the healing expression may be given purpose from ether the practitioner's etheric personality or another personality, perhaps the sitter's personality. It is the conscious self of the practitioner that initiates the intention to help signal to his or her Attention Complex.

My Qualifications to Speak About Healing

Especially when it comes to discussing healing, it is good that you understand my qualifications. Developing a model such as the Implicit Cosmology is a logical process which requires me to be knowledgeable about a wide range of subjects. To speak about these subjects with a degree of authority also requires the kind of understanding that usually comes from personal experience and considerable contemplation. I feel qualified to speak on many metaphysical subjects, but there are areas in which I have not gained sufficient knowledge to speak with any authority. Healing intention is one of those subjects.

Much of my personal experience in metaphysics is with ITC, but over the years, I have been trained in and have practiced a wide range of healing modalities. I am a certified Reiki Master and an NSAC Commissioned Spiritualist Healer (2), and I am trained in a number of other healing modalities from various metaphysical schools. Even with this experience, I do not consider myself qualified to make blanket statements about healing.

With that in mind, what I say here about healing is in the context of the Implicit Cosmology. In other words, I am stating what makes sense if all or most of that model is correct. It is for you to decide whether or not this makes sense to you. Please take advantage of the references I provide along the way, especially the bibliography provided by Dean Radin on his personal website. (151)

Foundation Concepts

As is my custom, we will begin with some of the foundation concepts you will need to understand.

Intention

Intentionality is the motive influence which is necessary to realize a visualized outcome. The creative process entails attention on an imagined outcome to produce an intended order. In a practical sense, that process without intention is just fantasy. See the Perception and Expression Discourse.

Perception and Expression

The output of the Attention Complex is an expression of awareness to the conscious self (perception) or expression of intention to the environment (intended order). The creative process itself is a mostly unconscious one of which the conscious self only becomes aware when the process results in some form of *outward* expression. This may be movement of a hand, speaking, walking, a conscious thought or the awareness of something in the environment. Expression can also be unconscious, for instance, as the unconscious expression of doubt when consciously trying to achieve some goal.

Rapport

The outcome of the creative process which includes the intention to heal might be imagined as a thread of attention between the practitioner and the sitter. This thread is characterized as rapport [68], the nature of which is a factor of attention, visualization and intention.

Intention Field

Expression of intention manifests as an intention field [100] which represents all of the imagery, thoughts, understanding and beliefs associated with the expression, as well as the organizing rules necessary to establish intended order. Intention fields persist as long as the personality maintains attention on the intended outcome. This is a form of etheric field.

Biofield

As a biological organism, the human body has an electromagnetic field produced as a byproduct of biological processes. This field is detectible with instruments, but some people also maintain that they are able to see, or at least sense this field. This field is physical and is studied in the mainstream sciences.

The body also has an etheric field (psi field or morphic field) which is associated with the conscious mind, body consciousness, and if applicable, the entangled etheric personality. These fields are not currently recognized by mainstream science, but are studied in parapsychology and are an integral part of many systems of belief.

Mapping a Person

In the context of the Implicit Cosmology, and as shown in the Life Field Complex Diagram [xxv], there are three fields associated with a human body which must be considered to understand healing intention.

1. The life field is the primary field from the personality-centric perspective [115]. The immortal intelligent core of a life field is an etheric personality which is our *I am this*. As an entangled aspect of its source, it represents inherited purpose and prior understanding. It is

also part of a collective of personalities so that, although it functions as an individual, it is able to solicit cooperation from fellow personalities in its collective. See the Perceptual Agreement Discourse.

2. The life field represented by the human organism is described as a morphic field [103] in the Hypothesis of Formative Causation. (9) It holds the body's formative body image and includes the body consciousness. The formative influence for the body image is derived from a prior form of the organism (Nature's habit). The organism's autonomic controls are represented in the morphic field. The body consciousness brings instincts which are based on the intended survival of the species.
3. The functional areas representing perception, known here as the Attention Complex [110], are shared by both physical organism and etheric personality fields. This complex includes a worldview memory *database* which represents the person's worldview and personal reality (a person is a personality entangled with a human body). Worldview is similar to the body image memory field associated with the morphic field, but is more concerned with expression and understanding. This complex includes an Attention Limiter which filters out uninteresting external stimuli and a Perceptual Loop which, working with worldview, determines what a person becomes conscious of and how it is perceived and/or expressed.

The personality is entangled with the human body in an avatar relationship; however, in practical terms, the human body consciousness is dominant until personality has become sufficiently self-aware to moderate the otherwise instinctual behavior of the organism.

Worldview

Worldview represents the person's personal reality, and the education of the person effectively *fills the database* with beliefs that shapes the point of view and behavior of the person as a citizen of its community.

Worldview also supports inherited and acquired understanding. It is important to note that understanding is relative and gaining understanding is an iterative process. Inherited understanding predisposes a person to styles of learning [119], but during a person's youth, the avatar's influence should be expected to bias acquired understanding toward human instincts.

Attention Complex and Attention Limiter

Everything in the environment produces a psi signal which, if it passes the Attention Limiter, is sensed in the Attention Complex. Likewise, the physical body senses numerous physical signals. The Attention Limiter serves as a filter to ignore or process these environmental signals, depending on the person's interests and whether or not there are potential survival implications.

Perceptual Loop

The Perceptual Loop attempts to image each external influence based on worldview. The resulting image is then offered to worldview, and if it agrees with current personal reality, it will be perceived (consciously sensed), but as it is colored by worldview.

For instance, a farmer might ignore a plane flying overhead but quickly notice a hawk. The mayor of a town might see a new subdivision as a good thing, while a farmer might see it as more competition for water. A person not governing its avatar's instincts might overeat, while one who has learned to moderate such instincts might know to limit portions. A person taught to believe in evil is more likely to translate experiences as evil influences than a person who sees such behavior as just errant human nature.

All of these functional areas are etheric (nonphysical) in the sense that mind is separate from the brain. All but some aspects of perception are unconscious, so that what is consciously sensed is a representation of what was unconsciously sensed based on worldview (personal reality).

The Aging Paradox

As proposed in the Hypothesis of Formative Causation (9), the component parts of a biological organism are organized by morphic fields based on body image informed by Nature's habit. Body image is for the physical body as worldview is for personality.

The human body is formed as humans have always been formed, but evolved by the creative solutions to environmental problems inherited from previous generations via morphic resonance. Also, cells in the body have a much shorter life span than the body itself, and new ones are frequently formed according to the memory of morphic fields. See Organizing Fields [102] in the Etheric Fields Discourse.

There is clearly a mechanism built into the human body morphic field that causes aging. At first, it is the process of maturing to adulthood with ever-increasing vitality, but at some point, the growth stops and aging sets in, leading to eventual death of the body.

Things do go wrong during the growing up and aging process. Beyond mechanical problems, there are a host of potential malfunctions that plague the body. Aging, though, appears to occur independent of the normal wear and tear on the body.

One of the aging mechanisms built into the body is the telomere. (152) It is a component of the cell that, in effect, seals the end of a DNA string and keeps it from binding with other strings. Biologists say a cell starts out with some 15,000 base pairs but loses 25-200 base pairs each time it divides. When the cell runs out of telomeres, it dies. In effect, the cell is scheduled to die, and that seems to be echoed in the entire organism.

Based on depletion of telomeres, the human body is apparently not intended to be immortal. At the same time, a cancer cell is a problem because it is immortal. Somehow, the telomeres are not depleted in cancer cells and the cell can divide indefinitely.

So to me, the paradox is how the body image, which determines the formation of the body, fails to form the cells according to Nature's habit—unless it is Nature's habit for the organism to get old and die.[60]

No Magic

One of the assumptions of the Implicit Cosmology is that there is no magic in the sense that something might *magically* violate Nature's principles. Processes always proceed according to knowable influences and outcomes are always natural. There are no magical shortcuts, just an ordered reality.[61]

The assumption is that everything in reality is ordered by knowable processes that naturally follow from intention. The formation of our human body is also predictable, as are its internal processes. No magic! For this to be true, then the body image must have a mechanism as part of Nature's habit that is causing the organism to age.

---

[60] Of course, I am not a biologist. What I have described here is based on how I understand relevant articles. So please don't go up to a biologist and tell him *"Tom said ...."*

[61] Be mindful that *Nature's principles* and *knowable* are relative terms. In reductionist physics there is no such thing as the etheric and related principles. Everything paranormal is magical thinking to a reductionist. Magical to one may be science to another.

Scientists are working on techniques to stop aging. For instance, telomerase is an enzyme that restores telomeres. It is active in young bodies, but not very active in aging bodies. The idea is to find a way to increase the production of telomerase to keep cells young without promoting growth of cancer cells. It appears to be at least feasible that cells can be enabled to divide indefinitely, presumably bringing longer life for the body.

Given that body aging can be at least postponed, then the body image (Nature's habit) should be equally pliable. As such, a reasonable objective of healing intention is to influence the body image to *remember* its youth.

Incremental Change in Worldview

Studies indicate that worldview resists dramatic change, but is responsive to small incremental changes. This is the *Maybe* outcome of the Perceptual Loop. The human body image is the morphogenetic equivalent of worldview, and according to the models, changes in the body are preceded by changes in the body image.

The incremental change in worldview comes from changes in behavior in response to external influences. As the model goes, these adaptations do not normally occur unless they are small changes while proposals for dramatic changes in behavior are simply ignored, rejected or negotiated into a series of increments within the Attention Complex.

Rupert Sheldrake described this evolution of the body image in terms of creative responses to environmental influences. In some ways, this hypothesis is a modern variation on Lamarckism's 1809 Inherited Adaptation Hypothesis. (104) Lamarckism held that changes in a species were caused by adaptation to environmental influences and that these were inherited. That theory was set aside in favor of Darwin's 1859 Natural Selection Hypothesis (153), which proposed that random mutations were selected based on their contribution to survival and that stronger individuals were able to perpetuate beneficial mutations.

Darwinism does not allow for the existence of an etheric mind, and without the etheric body image concept, there is no apparent mechanism for creative responses to change the body. Sheldrake's Hypothesis of Formative Causation does include the possibility of an etheric body image (Nature's habit), and thus, a way for environmental influences to cause a change in the body via the body image. Conversely, the relationship of the morphic field with a collective provides a mechanism for propagation of a change throughout fields sharing the same characteristics.

Based on the Implicit Cosmology, it is possible that aging came about as an accidental response to environmental influences. There is nothing in the model that restricts evolutionary changes to be only beneficial ones. A series of incremental responses to environmental influences such as skin damaged by the sun or too much mineral in available water might have produced creative survival response changes to the body image as the species attempted to adapt.

Aging may be nothing more than an evolutionary response to environmental influences. Assuming aging is not predestined, then it should be possible to address the effects of aging by addressing the creative process associated with the body image. I leave it to people more qualified to design self-help regimes to explore this possibility.

Mind is Impressionable

An alternative explanation of aging is that body image is easily influenced by expectation. Looking again at hyperlucidity, it is clear that rational thinking can be overridden by imagination. This impressionable nature of morphic memory is possibly a factor in aging, as in modern times, aging is something we are all expected to do.

There is clearly a purpose for the aging process associated with the growth of a newborn baby into an adult. That aging process basically completes the development of the organism, as can be seen with the many provisions in the formation of bones and glandular secretions for growth. As is too often demonstrated in modern culture, a young person typically views a physically prime person as old. Is it possible that this concept of age differential has become so much a part of our culture that aging is considered an inevitable fate?

The Expression of Healing Intent

When a person thinks of another person, a link of influence is established between the two. In the context of the Implicit Cosmology, this is referred to as rapport. The nature of this rapport (the link) is determined by the person's attention, visualization and intention. For instance:

- A thought may be in the form of a casual glance devoid of intention. The object of the glance will become unconsciously aware of the attention, depending on how open the person is to uninvited attention and how receptive the person is to the intention signal accompanying the attention.

- The thought might be a stare inadvertently expressing admiration. In this case, the other might sense the admiration, but since it is not specific, it might never emerge into the person's conscious mind.
- The thought might be an expression of love, complete with a recalled mental image of the loved one and desire that the loved one know. In this case, attention on the image of the loved one, with the intention that the loved one is aware of the love, will likely be sensed by the other. However, not being accustomed to translating psi-received information, the other person might never become consciously aware of the thought. Even so, it would likely have an influence on the person's sense of wellbeing.
- The thought might be in the form of the intention to send healing to another. As with the thought of the loved one, if the thought is accompanied by focused attention, a clear image and specific intention, then the receiver can be expected to at least unconsciously sense the influence. It will likely not be consciously sensed unless that was the intention. Once again, if the person is not accustomed to consciously experiencing such psi influences, the person's worldview might not have the necessary data to translate the influence into recognizable form.

There is little systematic difference between expression in the form of telling a story, seeking information from the Internet and expression in the form of a desire to benefit the health and wellbeing of another. The creative process applies the same.

A difference in quality of rapport based on clarity of expressed intention, visualization and focused attention is predicted by the Implicit Cosmology. However, there *seems* to be a not so evident difference in quality between the expression of healing intention and a mundane act. The expression of reverence, love and intention to help another seems to have, as yet unidentified qualities.

An interesting exploration of this *other quality* is Robert Monroe's story of loosh he told in *Far Journeys*. (154) (155) The short version is that earth was developed to cultivate loosh, which is the thoughtform [99] associated with an expression of emotion that (in the story) has great value in the greater reality. A mother expresses a form of low-quality loosh when she cares for her child, but she expresses very high quality loosh when she risks all to save her child.

We instinctively respond to strongly felt/expressed feelings (their thoughtform). It appears that we are more moved by love than by anger, probably because our human responds to anger in a primal way while love likely appeals to our etheric instincts. Perhaps this is a subject better addressed by philosophers than by metaphysicians.

Expressing Healing Intention

There are many different healing modalities, and each is supported by a system of thought that is often based on centuries of tradition. Most are based on assumptions that include considerable articles of belief, but because healing intention is so influential, probably all of these modalities are effective. What might be questioned is whether or not they are effective because of the associated beliefs or because simply wishing someone good health helps.

The process of expressing healing intention involves important ethical considerations that are shared by all modalities I am aware of. As a general practice, a healer-practitioner expresses the intent to improve the sitter's wellbeing no matter how that might look. It is left to the sitter's etheric personality and body image to respond in a way that is best for the sitter. While healing is the intention, it is up to the sitter to accept healing, or how to apply the good intention. (This is generally an unconscious process for the sitter.)

Examine the Life Field Complex Diagram [xxv]. The physical body is really the outward expression of a very complex unconscious process which includes both the human body consciousness and the personality. In this model, all of the external influences must pass through the Attention Complex, and with healing intention, the expression of that complex is the sitter's response. Basically, the healing intention is just another external influence.

If the sitter's worldview includes acceptance of the concept of a healer-practitioner/sitter relationship, influence of the healing intention will be more readily accepted by the sitter. The same is true if the sitter thinks something might happen. If the sitter believes in healing but doubts it will ever personally happen (*I am not good enough, I am too sick, the healer is just an amateur*), then that aspect of the sitter's worldview will tend to defeat the process.

One of the techniques in healing is to find a way to make the sitter believe something might happen. A healer-practitioner's good reputation helps. A dramatic environment can evoke a strong *this might happen* response in the sitter, which in turn, opens the Attention Matrix to express beneficial information to the body.

The idea of fooling the sitter's Perceptual Loop into accepting the healing is just acknowledgment of human nature. For the practitioner, the sitter is a human, and when it comes to health, the human instincts are usually dominant. This means the body image is the target of the practitioner's intention. If the sitter accepts poor health (*I am just getting old, I am always sickly*), worldview will tend to defeat positive input.

Fear is another powerful block to well-being. From my studies, counseling related to alternative health practices involves a great deal of fooling the fear response. The very idea of cancer is so loaded with fear that it is difficult for the sitter to keep from incorporating acceptance of eminent death into worldview. Such practices as meditating to imagine tiny creatures running around in the body to eat cancer cells is really a way of cajoling worldview into accepting hope, and thereby offering the body image a more positive perspective.

Hyperlucidity

Remembering the hyperlucidity concept, inspirational writer William Arthur Ward told us that, *"If you can imagine it, you can achieve it. If you can dream it, you can become it."* Imagining something is true gives the mind a kind of permission to look beyond old beliefs ingrained in worldview. In hyperlucidity, this can become a problem if the person is unable to make a distinction between thinking one *can do* what is visualized and believing one *has done* what is visualized, even though that is not the case.

Hyperlucidity can be a delusional state, but this same mechanism is potentially a powerful tool for changing body image. The healer-practitioner is addressing the often irrational hindbrain and not the more rational etheric personality. As such, the practitioner can be instrumental in fooling the sitter's avatar's hindbrain into accepting the possibility healing can happen.

Who is Doing What?

In Spiritualist traditions, the healer is assumed to be a conduit for healing *energy* provided by discarnate loved ones and helpers. However, in most healing modalities the healer-practitioner is thought to provide the healing influence. In some, it is understood that the practitioner is more of a counselor and that the sitter is self-healing.

Considering the Implicit Cosmology, the practitioner is always a source of the expression of healing intention. Other personalities are probably involved in any such exchange between practitioner and sitter, and the more appropriate model is that the intention of the practitioner, the acceptance of the sitter and the circumstances of the session all come together to facilitate modification of the sitter's self/body image.

All of us are part of a personal and universal community integrated as a whole by attention, intention and the entanglement of rapport. The practitioner usually consciously expresses the intention to access assistance from this community, thus *opening* a channel for the expression of healing intention from the community. For an experienced practitioner, the act of approaching a sitter likely triggers an unconscious healing response which also opens this channel. In the same way, the sitter opens a similar channel by the act of being a sitter and the expression of the desire to gain help.

Given the fact of self-determination, it is this *opening of the channel* that is necessary for others to help. In the end, though, the sitter is the source and the means by which healing occurs.

Modes of Healing

It is possible to learn how to sense the biofield. Healer-practitioners are often taught to move a hand over the sitter's body, usually not touching but inches above the body, and imagine feeling a difference in the auric field. Depending on the modality, the practitioner might be taught to feel *stiffening* in the field near a problem area. Alternatively, the area might be expected to feel warmer or colder. It depends on the system of thought.

From my experience, it is possible to learn to feel a change, and this can be verified by blind tests. Of course, problem areas in the body might actually be warmer because of increased biological activity, but the stiffening is unexpected. In one modality, in which the thickening is related to an item in the sitter's memory, having the practitioner physically *break up* the thickening to make it smoother, might evoke related memories in the sitter's mind. This is related to a counselor-sitter role more than a practitioner-sitter one.

When dealing with people and human nature, it is difficult to argue that the effect is real and not just psychosomatic, but the results can be therapeutic. The point is that the interaction between practitioner and sitter can be tangible.

In some systems, a distinction is made between spiritual healing and magnetic healing. Here, *magnetic* is an old term that might be better thought of as vitality. The idea is that some of the practitioner's own vitality goes to the wellbeing of the sitter. In this case, it is common for the practitioner to report being depleted after a session.

Probably the more desirable approach to healing is what is commonly referred to as spiritual healing: the practitioner imagines him or herself as a channel for healing energy from loved ones on the other side.

In the Implicit Cosmology, vitality from the practitioner or spiritual healing form the other side are modeled in the same way. In fact, I do not know if there is an actual physiological difference, as I treat both the same. For me, it is all intention and clarity of imagined outcome. I do focus more on mentally addressing the sitter's body image than I do the personality.

As a historical note, in the 1700s, Franz Anton Mesmer (156) introduced the world to animal magnetism and mesmerism. Like a shaman, he used suggestion to entrain people's attention. Mesmerism was probably a precursor to psychotherapy and hypnotism, and the concept of animal magnetism certainly relates to the concept of biofield healing as opposed to body image healing. What I refer to here as vitality is probably just a more acceptable name for what Mesmer referred to as magnetic fluids in the body. Again, different systems have different ways of looking at essentially the same concepts.

Prayer

Prayer as a means of communicating intention with an unseen god has been fundamental to religions throughout the world and throughout history. Other than giving thanks, probably the most common purpose of prayer is to ask for divine help. Prayer in behalf of others has also been a common practice.

In the Implicit Cosmology, prayer is seen as the expression of the intention to realize an intended outcome. When thought of as prayer, the person's worldview likely includes the assumption that some intermediary personality will sense this expression and cause the intended result. For instance, asking a saint for protection or God for healing.

This desire to help another by way of what parapsychologists refer to as intercessory prayer is translated as spiritual healing when a person's health is concerned. Intercessory prayer is functionally the same as noncontact healing or distant healing, both of which have been shown to be effective.

In the Implicit Cosmology, the act of praying expresses an intention field which includes the purpose and visualized outcome. In principle, the quality of focused attention, how clearly the desired outcome is visualized and the degree to which the desire is intended to be true are important factors determining if the intended outcome will be realized. These factors apply to both the practitioner and sitter; however, research shows that benefit can come even if the sitter is unaware of the practitioner's effort. This is likely because, in the etheric, the sitter and practitioner can be expected to be mutually aware of the effort and intention.

The etheric personality of the person who is praying (healer-practitioner) is modeled as always being in contact with its collective of related personalities. By extension, this contact includes the collective of all other personalities, so it is correct to say in this model that the practitioner is not alone in this expression.

The ability of one person to influence another person's wellbeing appears to be limited to the practitioner's ability to manage the creative process and the extent to which the sitter is able to accept the influence. Given self-determination (free choice), there is no reason in this model to expect anything like divine intervention. But then this is just a model, the validity of which you must decide.

Body Image

In Rupert Sheldrake's Hypothesis of Formative Causation (9), physical organisms are modeled as a hierarchy of nested organizing fields which he describes as morphic fields. Morphic fields determine how the body will be formed. I am not sure if this is in Sheldrake's model, but as I translate his work for the Implicit Cosmology, the etheric body mind is necessarily the regulating influence that determines how the body will change over time. As shown in the Life Field Complex Diagram [xxv], the body mind can be seen as the physical organism's equivalent of the Attention Complex. Morphic memory translates as worldview, and body image translates as personal reality.

Cells in the body are frequently replaced and the new ones must be organized according to the body image. Yet, as the cells are replaced, they reflect recent changes that may not be represented in the body image. For instance, the aging process may reflect environmental influences such as cell damage from the sun and physical injuries, but it affects protected parts of the body as well. These aging and environmentally induced changes are passed on to the new cells, which means they must somehow change the body image despite morphic memory.

Also, based on First Sight Theory (13), as it is described in the Perception and Expression Discourse, worldview resists change, but is more responsive to incremental changes produced by ambiguous results of the Perceptual Loop. This momentum is also present in morphic fields. So, in a similar way, it is reasonable to expect that the body image, as it is informed by morphic memory, should dominate the continuous regeneration of body cells.

So the question to consider is how the aging process seemingly drives changes in the body image independent of the influence of morphic memory. I do not have an answer to this question, but it is clear that the mechanism is in some way part of the etheric body mind. The body is an ideoplastic structure expressed by the body mind, and as such, gradual aging of the body and perpetuation of injuries must be in agreement with body image.

The Way Back from Aging

To make a bold statement, aging is a response to expectation and not a response to time. Like worldview, morphic memory evolves incrementally, and as such, that evolution should be able to be reversed incrementally. I would not expect to see an old person suddenly become young in body, say under the direction of hypnosis. However, I would expect a bodily response to a life-long rejection of traditional expectations.

As the practitioner addresses the sitter for a session, the visualized outcome probably should be directed toward the body and not the personality. This may sound like nitpicking, but there is a difference in expressed intention and visualized outcome.

The practitioner can be expected to be *stuck* with the expectation that the person seems sick, old or in some way malfunctioning. That accepted impression in the practitioner's worldview limits the practitioner's ability to visualize the sitter in good health. The routine practice of visualizing the body in its prime might side step this presumption. Such mental devices are important when dealing with human nature.

As many systems of thought tell us, sickness comes first in the emotional body. In the same way, illnesses should also be considered an expectation artifact. It is that physiological response to expectation (in this case, fear) that makes a cancer diagnoses so dangerous.

A Well-Documented Example

There is a lot of material on the Internet concerning healing, and it is not my intention here to say one system is better than another. There are also a lot of questionable claims which confuse any casual research on the subject. If the basic assumptions of the Implicit Cosmology can be accepted, then it can provide guidance as to what to believe and what to ignore. In any case, I recommend that you err on the side of caution, as some simply silly claims can be very convincing.

One example I am willing to recommend for study is the work of William Bengston. (157) As he explained in his book, *The Energy Cure,* Bengston met and became a student of a man who was an able healer-practitioner. As an academic, he was naturally oriented toward research and clinically verified his healing ability via numerous studies in which he reportedly (literally) cured lab mice that had been injected with virtually always fatal cancer.

An interesting aspect of Bengston's technique is what he refers to as cycling. In that, he teaches the practitioner to visualize a list of life goals or things the practitioner wants to achieve. At the beginning of a session, the practitioner should set the intention to heal the sitter, but rather than trying to *push* or will the healing *energy* into the sitter, the practitioner should begin remembering the list of goals, visualizing each one in succession, rapidly cycling through them for the remainder of the session. Bengston emphasizes the idea that this should be very rapid cycling from visualized want to visualized want, over and over again.

Following the Implicit Cosmology, cycling would be effective in helping the practitioner sidestep self-doubt. In effect, it takes the Perceptual Loop out of the equation and allows the etheric personality to more directly influence the expression output of the Attention Complex.

Proxy Sitters

An interesting approach to healing intention, and one that helps sidestep practitioner's preconceived notion about the sitter's wellbeing, is the use of a proxy sitter. In this, the intended receiver of healing intention is not present. Instead, a healthy person who knows the intended receiver acts as the sitter, and while the healer-practitioner focuses attention on the sitter, the sitter focuses attention on the intended receiver.

A second version of this is use of a cuddly stuffed animal as a carrier of healing intention. In this, the healer-practitioner treats the stuffed animal as the sitter. This approach is useful for a development circle to send healing to a friend.

Tradition holds that physical objects are able to absorb and retain the characteristic signature of thought energy. Psychometry is based on this understanding, as is the concept of an intention field [100]. In healing intention, the intention of the healer-practitioner establishes a thoughtform which represents that intention. The intention field-thoughtform is entangled with the physical object and remains with it as long as the intention remains.

As I understand this concept, when the healing intention is expressed for the sitter, the entanglement links the practitioner, the stuffed animal and the intended receiver. In this way, the receiver is thought to benefit from the healing intention when in contact with the stuffed animal. A benefit of this technique is the sense of objective possibility, as opposed to the usual need to visualize help.

William Bengston talked about using natural cotton balls as a carrier for healing intention in *The Energy Cure*. (157)

Mindfulness Tool for Healing Intention

As I noted in the Mediumship Study, there are ethical issues that must be considered for practitioner-sitter relationships. That is one of the reason why I prefer *healing intention* instead of *healer*. The term better describes the healer-practitioner's probable ability and does not remove personal responsibility for self-healing by the sitter. Here are a few issues to remember:

- Because of perceived authority, comments by the practitioner about the sitters wellbeing may develop into a self-fulfilling prophesy.
- A promise of healing could cause the sitter to suspend medical care.
- Always, there is a concern about inappropriate touching.
- There may be legal issues about practicing without a license. Practitioners have some protection if working in religious service, but not so much if the sitter brings legal action.

See A Personal Code of Ethics for Developing Mindfulness [150].

Distant Healing as a Tool for Psi Research

The psi field and psi functioning are difficult to study. First, no physical substance is known to shield a person from psi. As such, a person functioning as a mental medium is always suspected of *reading someone's mind* to access desired information. Putting them in a different room or even in a different country is not believed to stop this possibility.

Most psi functioning effects are small, so that the best way to study, say, the psychokinetic influence on coin tosses, is to look for deviation from chance over many trials. That is one of the reasons statistical analysis has become so prevalent in parapsychological research reporting.

Biofield therapy typically involves very small effects, as well. Then there is the problem of subjective responses to suggestion. Research is usually complicated when human nature is involved. This is especially true when the subject is supposed to be impossible in the first place.

As it turns out, distant healing research provides a way to study the effectiveness of the healing concept as well as the nature of psi. If a benefit from distant healing intention can be shown to be real, then the psi influence for delivering that means should also be real.

There is a substantial body of research concerning healing intention, nonlocal healing and psychokinetic influence. I see that parapsychologist Dean Radin is maintaining *"Selected Peer Reviewed Publications on Psi Research,"* (151) so rather than attempting to develop a list of references here, I will refer you to the list on his personal website.

From Radin's comments on that page:

> *Commonly repeated critiques about psi, such as "these phenomena are impossible,"' or "there's no valid scientific evidence," or "the results are all due to fraud," have been soundly rejected for many decades. Such critiques persist due to ignorance of the relevant literature and to entrenched, incorrect beliefs. Legitimate debates today no longer focus on existential questions but on development of adequate theoretical explanations, advancements in methodology, the 'source' of psi, and issues about effect size heterogeneity and robustness of replication.*

Study 8:
Introduction to Best Practices

Foundation of Civilization

Of the thousands of questions from website visitors I have answered about ITC, most could have been answered with a few form letters. That is because there really are not that many variables. For instance, the main themes for EVP are how to record, does it attract evil spirits, who is communicating and how to share examples. If you consider these phenomena in general term, all of the phenomena are like that. That is why best practices are such a great idea.

A second benefit of best practices is that, properly executed, they are composed as a community project and represent a consensus from experienced paranormalists. They are also seen as living documents which evolve as new understanding emerges. Contrast that with *The World According to Tom*. While I accept the responsibility to teach, I am probably the first person to complain that everyone is an expert. By that I mean that a quick survey of the Internet will show that much of what I have taught has been ignored or superseded with new, usually baseless ideas. Happiness for me is pointing to a best practice, rather than one of my FAQs.

A community can be described as a group of people cooperating to achieve a common goal. One of the most important ways in which this occurs is when the elders take time to teach what they have learned during their lifetime to younger members of the community.

The ability to convey lessons learned to the next generation permits the community to move forward with new discoveries without the burden of needing to rediscover foundation concepts.

A business accomplishes the same perpetuation of experience by maintaining best practices. These may be in many different forms, but the basic idea is to preserve what has been found to work best for the company, instructions for operational standards and maintenance procedures, in a form that is easily accessible to new employees.

Living Document

Best practices are important contributors to the community's cultural norms. In the short term, this is a good thing, as the uniformity in practices across the community allows for a kind of *plug and play* environment in which members may reasonably expect certain behaviors everywhere in the community. In the long term, however, practices tend to discourage innovation if they do not evolve with changing circumstances. For instance, the education system represents most of our civilization's best practices. However, as we can see with the cultural tension between mainstream science and the paranormalist community, currently accepted practices are effectively stifling possible progress offered by the paranormalist community.

To be effective, best practices require a few important characteristics. First, there is a need for a mechanism to evolve them as circumstances and understanding evolve. In a business, this is usually accomplished by assignment of a dedicated practices manager charged with periodic review. Of course, the practices manager also needs to maintain a revision management system so that employees may know that they have the latest version.

The second need is a way of vetting practices. The people who do the work, who are usually subject-matter specialists, may not be directly involved in the mechanics of practice development and maintenance. A best practices manager will periodically convene a subject-matter specialist review of the applicable practices.

An important version of this in which I participated for a few years is sometimes known as strategic planning. In that, a few subject-matter specialists are given secondary assignments to consider current trends with the objective of understanding how practices might need to evolve to intersect with future requirements. An example of that today is how I have attempted to evolve the Survival Hypothesis to accommodate First Sight Theory.

Enforcement

At least in the paranormalist community, use of best practices is voluntary. They only become important when members of our community are willing to ask *"Did you follow best practices?"* A *no* answer should detract from the credibility of the work. In this way, practices are like a cultural norm in which there is a reasonable expectation of behavior.

The Collective

The Collective at *ATransC.org/bp/Collective* is a WikiMedia wiki sponsored by the ATransC. Its purpose is to enable members of the paranormalist community to develop best practices and related standards to further understanding within the larger community. As of this writing, it is very much in a development stage and has not been supported by others. I maintain it with the understanding that it will eventually gain support.

The Collective wiki is essentially the same document development and management program used by Wikipedia. Editing is easy but does require learning a few simple editing rules. Editing an article is intended to be a community effort with subject-matter specialist review. Editing rules are designed to avoid the kind of *tyranny of the majority* seen in Wikipedia.

Wiki articles written as a community effort can converge on useful standards as long as basic rules of collaboration are followed. They are excellent for living documents because editors are encouraged to update them as understanding changes. Being on the Internet, anyone can access them as a free and open library.

Another benefit of an online, publically viewable wiki is that version control is always built in. The history of changes can be easily reviewed, as well as possible discussions amongst the contributing editors.

A major failure of Wikipedia is that a few editors working together can dominate the process to foist their view of truth on the world. As such, it is important for readers to review the History and Talk pages, especially the archived Talk pages, to see how the editors arrived at the current version of the article.

While it is intellectually lazy to use Wikipedia articles as references, the Collective is intended to be cited as a potential community standard. If readers see something they disagree with, it is expected that they will either make a change to the article or propose a change on the Talk page. If anyone feels their input is being ignored or ridiculed, contacting the staff editors should bring a fair referee to help. Such offered changes may not be accepted, but they must be fairly considered.

Please consider becoming an editor. You need not be an expert, as many of the more important aspects of the articles involve grammar, research and common sense. For instance, the one Wikipedia editor I trusted, and who provided the most constructive guidance for me, seemed not to accept many of the paranormal concepts. His approach to editing was that all articles should be balanced to give the reader a candid but impartial understanding of the subject. As I remember, he was banned when multiple skeptical editors ganged up on him.

Please use the contact form at ethericstudies.org/contact.htm to request access, make suggestions or ask questions.

A Pragmatic View of Best Practices

Many of the concepts in Etheric Studies [205] resemble religious tradition. There are also many aspects of the etheric that are held to be true in paranormalist tradition but that have not been experimentally demonstrated. To add to the confusion, the metaphysical meanings of many paranormalist terms are not well known. This poses the difficulty of knowing what is based on religious belief and/or tradition and what is supported by experimental evidence or well-established, effective methodology. The use of best practices is an attempt to identify what works and what is problematic, what is substantiated by evidence and what is merely supposed.

If a person studying trans-etheric influences follows applicable best practices, the results can be described as being *standardized* by those practices. This means that anyone studying the results of such work will know what measures have been taken to avoid known pitfalls, and therefore will have a better understanding of the results. This also means that the practitioner will more likely avoid operational errors that might not be obvious without extensive trial and error.

As an example of how best practices can help, a person might report the recording of an EVP that, when measured against applicable practices, would indicate the possibility of unnoticed voices in other parts of the recording environment. The simple recommendation that a person recording for EVP in the field should identify in the recording each change in location is considered a best practice.

An ATransC website visitor used the contact form to inform us about the many EVP he had recorded. Because what he described was very far outside of the norm for EVP, we asked him if he had used a listening panel [400] to see if others agreed with what he heard. He ignored the question the first time we asked, but days later, when he described in a second contact that many of the messages called him by name and others were very evil, we asked again if he used a listening panel. He responded, *"Oh yes, my wife heard them and so did a few friends."*

The protocol for a listening panel is very specific about uncoached and independent listeners. Knowing the tendency for people to feel like they need to please the practitioner, a wife and friends are probably not the best test for the validity of phenomena. If he had indicated that he actually read the practice I referred him to, and then told me that, using that protocol, independent listeners agreed with his assessment of what was said in his examples, then I would have reason to think he had new and novel examples. As it stands now, I am inclined to think the explanation of his unusual examples is hyperlucidity.

Standard Practices Already in Use

Some organizations have included in their charter the requirement to teach members the techniques for working in that field of study. For instance, the ATransC attempts to teach about EVP, its nature and the best ways to record for voices. Other organizations include dependable methodologies for organizing a field investigation, as well as how to use the associated investigative tools and how to report the results. Also, experienced circle leaders have documented techniques for sitting for phenomena.

A well written best practices should not presuppose that a particular method of teaching is wrong. Instead, editors should first solicit input from the community and incorporate existing practices as part of a consensus of what is known to work.

Ideally, instructors within the community would base their lesson plans on the applicable best practice. In this way, it is felt that instructors can move the burden of *being right* to the community. One might say, *"Do it this way because it is a well-documented technique not because it is the way I do it."*

Scope

Best practices are concerned with methods and procedures only. They are not rules, but may be referred to as community norms or any similar phrase indicating common practice; they should never be required, as *informed* innovation is always encouraged.

A person reporting the results of an activity concerned with the study of trans-etheric phenomena can be expected to identify whether or not best practices have been followed. A person claiming to be qualified in some aspect of Etheric Studies, such as recording for EVP, should be aware of the applicable best practices. Anyone conducting research into some aspect of Etheric Studies and intending to submit a report for publication in the online ATransC Journal should be aware of the applicable best practices and should apply them except in the area of the research objective. For instance, for a study of the effectiveness of an audio filtering technique, best practices should be followed for the recording sessions and the determination of what was said, but may not be appropriate for the analysis phase.

One Must Do the Work

Probably the single most important factor separating hobbyists from serious researchers and practitioners is a willingness to do the work necessary to become informed about paranormal concepts. It is not necessary to become a paranormalist scholar, but the unfounded assumption of knowledge many members of our community have stifles progress toward a mature community. It also gives the general public reason to think all of us are just hobbyists and tends to confirm the skeptic's idea that we are deluded.

Many best practices will also serve as educational text. This is especially true of an article development and maintenance process that is well supported by the community.

Examples

Two draft best practices are offered here as examples. I say they are draft because I am the only author. In best practices, a one-author document has limited authority.

You are invited to register as an editor (contact me) and make changes to these as you see fit. Of course, you will need to participate in a consensus building process for the changes, but your work will become a valuable contribution to the community.

Transcommunication

Study 9:
Best Practice: Classifying Phenomena

Draft

Abstract

The classification of paranormal phenomena is important for setting the expectation of an observer and to help the practitioner gauge how well he or she is doing based on community norms. A three-class system has previously been used, and only for audio ITC (EVP). This practice is designed to accommodate static and transient forms of audio and visual ITC.

The two-type, three-tier classification includes Type 1 for noise transform phenomena (transient) and Type 2 for more persistent (static) features opportunistically formed from mundane sources. The Class A (easy to experience), B (more difficult to experience) and C (mostly experienced by the practitioner) remains so that there are Type 1, A, B or C and Type 2 A, B or C.

Justification/Introduction

A common dilemma for practitioners is how to classify examples. The phenomena of interest to the paranormalist community may take many different forms. Two primary categories are spontaneous and induced. For instance, near-death experiences are usually considered spontaneous while EVP are usually considered induced. After Death Communication (ADC) was considered a spontaneous phenomenon until EVP came along, so now we have spontaneous and induced ADC.

In most instances, classification is embedded in the introductory dialogue of an example. The practitioner or witness is expected to realize that exceptions to the general rule should be explained. For instance, an EVP that was stumbled upon during a video-take for a television commercial is an unusual spontaneous EVP and should be explained as such.

Psi sensing as human potential is considered very different from psi field phenomena related to survival. This is a differentiation seldom made by paranormalists who assume there is no such thing as survival.

There is a need to distinguish research intended to prove phenomena is mundane and research intended to understand phenomena. Anomalistic psychology [218] verses psi sensing [356] studies is a good example. How phenomena are described in the two different contexts determines how examples are introduced. In this case, it is important to make sure the audience knows which view the practitioner or witness is taking.

Trans-etheric Influences

This is a young field of study, and there are many forms of phenomena about which too little is known to classify. This leaves the practitioner or witness little alternative but to be conservative by describing such phenomena as *probably not paranormal*. For instance, the paranormal faces sometimes found in video loop ITC are typically present for just a moment (transient). That is why they are usually only seen in photographs or video frames. There is little doubt amongst people who have studied the technique that video loop ITC produces paranormal phenomena. Classification for the examples are discussed below.

As a comparison, a face seen in calcium buildup on a subway wall (persistent) may be paranormal, but it is tempting to ignore it as happenstance. Ignoring such examples may be intellectually lazy in that we really do not know the bounds of our etheric friends' capabilities or their motivation. Our reaction might be very different if we knew that a person looking a lot like the calcium pattern had been killed on that very spot.

Current Classification System

EVP can be classified as transform and opportunistic [295]. A three-tier system for classifying EVP has been successfully used for many years for transform EVP but has been proven inadequate for other forms. For the purpose of classification, the main difference is that transform phenomena is transient and opportunistic EVP are persistent. A similar problem has been encountered with visual forms of ITC, for which there has been no classification system. The three-class system for rating EVP is:

Class A: Can be heard and understood over a speaker by most people.

Class B: Can be heard over a speaker, but not everyone will agree as to what is said.

Class C: Can only be heard with headphones and is difficult to understand.

Class B or C voices may have one or two clearly understood words but it is how well the overall meaning can be understood that determines the class. Also, loud does not equal Class A.

Practice

As illustrated in Figure 1, in the existing classification system, the majority of examples are rated as Class C while a small percentage of examples are rated Class B and even fewer are rated Class A. It is estimated that there are a thousand or more Class C examples for every Class A.

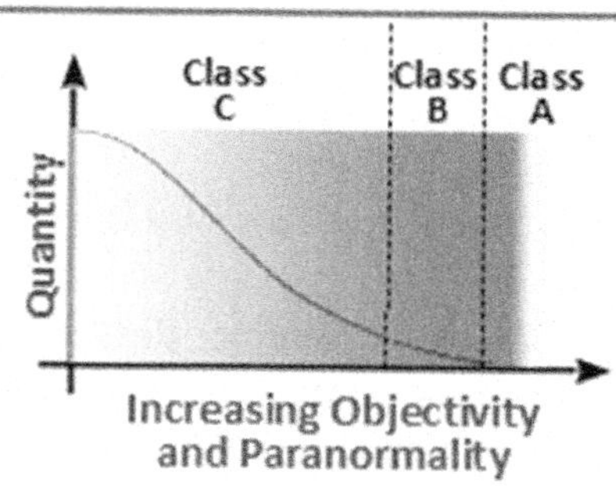

Figure 1
Expected Distribution by Class

Distinguishing Characteristics

The proposed system is based on two major types of phenomena. Phenomena that are typically only momentarily present in the physical are classified in this practice as Type 1 or transient. Phenomena that typically exist in the physical for a relatively long time are classified as Type 2 or persistent.

In Type 2, *persistence* can be quantified as sufficient time for a person to observe the feature. As a general rule, persistent phenomena appear to be formed by opportunistically adapting naturally occurring processes to express the message, assuming one is intended—a cloud formation that resembles a face, for instance.

Each type is divided into three-subclasses as Class A, B and C. As before, the classes are defined so that there are many Class C examples for a few Class B and even fewer Class A examples.

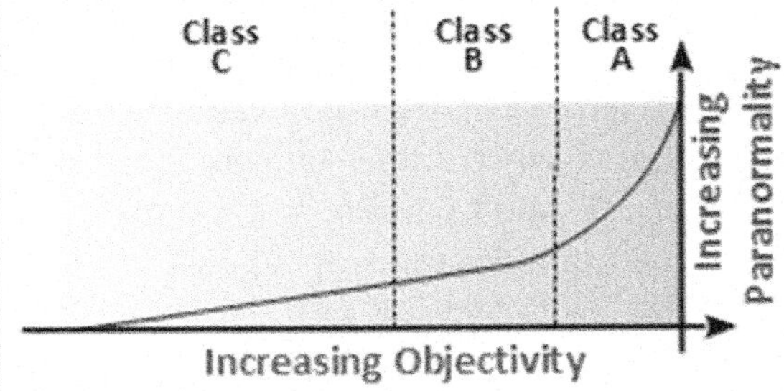

Figure 2
Perceived Objectivity by Class

As indicated in Figure 2, Class C examples (less objective) are considered much less evidential than Class B or Class A examples (more objective) because they are not easily shared. For examples that are perceived as less objective it is much more difficult to establish their paranormality.

Both audible and visible phenomena are classified as:

Type 1: Transformed physical media; not always present, a transient feature.

Type 2: Long-lasting presence; often as a persistent artifact.

Both Type 1 and Type 2 are divided into three classes. They are described as before, but in more generic terms to accommodate different types of phenomena.

Class A: Evident without explanation.

Class B: May require directions to witness.

Class C: May be vaguely experienced; mostly obscured.

Audio ITC

The input sound used in EVP helps determine the type. There will be exceptions, but as a general rule, the following holds.

Type 1 Audible: Input is audio-frequency noise, either ambient room noise or supplied, perhaps from a fan or a noise generator. The formation of voice is thought to be via transformation as the communicator imposes intended order on the otherwise chaotic noise. This appears to involve a stochastic amplification-like process.

Type 2 Audible: Input sound is typically live voice. This includes someone talking on the radio, in the room or pre-recorded, perhaps in a foreign language. The easily heard voice is supplied, but formation of the message is seen as the opportunistic selection of parts of the existing voice. Often, paranormal status comes from the context of the voice-like sound.

It is important to note that a Type 1 EVP can be formed in any sound, including noise or voice. As such, foreign-language voice can be transformed into new words. With that said, the practitioner can be expected to provide both input and output files for comparison. Also, since it is known that EVP occur in one process, two recorders recording the same input file should not produce the same EVP. (144)

Visual ITC

Features found in photographs and video frames of medium-density optical noise are considered transform features. They are transient, in that an observer does not see them at the time of recording, only upon review of the media.

By comparison, a pattern on a piece of toast that resembles a face is long lasting and visible without the need to examine a photograph. With these considerations in mind, the following holds.

Type 1 Visual: Input is optical noise, usually medium density which is not very light or very dark. Textured surfaces facilitate image formation, as do image compression techniques. Often, visibility of the resulting paranormal feature is limited by the resolution of the media.

Type 2 Visual: Naturally occurring surface characteristics which are more or less static can sometimes be arranged to form faces. Whether or not the faces are intentionally formed is not clear.

The feature may be mundane background shapes seen in a distorted reflection, unfamiliar combinations of shapes or naturally occurring processes such as the buildup of calcium on a wet wall that seems to resemble something familiar. In many cases, Type 2 visual examples may be mundane, but become paranormal based on when or where they are detected.

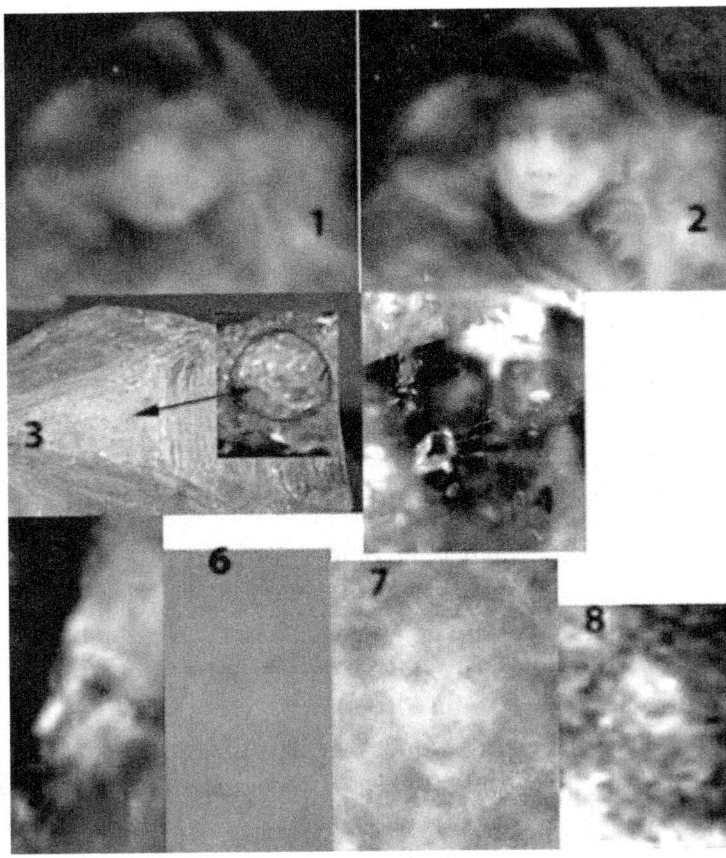

Example Application

The ITC images shown in this montage would be classified as:

1. **Type 1, Class B:** Original breath-vapor photograph. (158) ©
2. Not classed; Image 1 enhanced by the artist.
3. **Type 2, Class B:** The crystal Chris used to make his point. The arrow points to an enlarged, enhanced version of a face more or less on the surface. We argued that it is a fortuitously arranged artifact. (159) ©
4. **Type 1, Class A:** A face *in* a crystal photographed by the Scole Mediums. ©

5. **Type 1, Class A:** A face found in light reflected from moving water.
6. **Type 2, Class C:** Original photograph of a face on a wall of a Spiritualist church. You will barely see it. The congregation swears that it was not there when they painted the wall white.
7. Not classed; The same face as in 6 only enhanced. The camera used to take the picture did not have a good UV filter and detected the mostly blue image better than the naked eye. It could be argued that the feature in example 7 is now a Type 2-A, but since it is enhanced, the example probably should only be used to help witnesses make out the feature in the original, unenhanced version.
8. **Type 1, Class A:** The video loop image I use for my avatar.

Further Reading

1 Locating EVP Formation and Detecting False Positives, ATransC.org/journal/false_positives.htm (160)

2 The Formation of EVP, ATransC.org/journal/evp_formation.htm (161)

3 A Research Study into the Interpretation of EVP, ATransC.org/journal/radio-sweep_study2.htm (74)

Transcommunication

Study 10:
Best Practice: Witness Panel

Draft

Abstract

One of the biggest problems in ITC is the difficulty practitioners have in assessing the quality of examples. Even examples that are considered Class A, which should be correctly seen or heard without prompting, are shown to be correctly understood only on average 25% of the time by online listeners (21) and 61% for video loop ITC. (20)

An important aid for practitioners is the use of witness panels, people with at least average hearing, vision and comprehension who will examine examples, and without prompting, tell the practitioner what is seen or heard. The rest of the task is for the practitioner to accept the results.

This practice details a methodology for establishing and using a witness panel to assess the quality of examples.

Justification/Introduction

People asked to examine examples of reportedly paranormal phenomena often complain that the examples are not convincing. For EVP, the first response from witnesses tends to be that the examples are just noise. If voices are heard, it is difficult for witnesses to mentally orient themselves so as to place the voices in a context that will give the words meaning if all that is offered is just the presumed phenomenal voice.

Examples of visual ITC can be even more confusing. For instance, reflected light phenomena are too easily discounted as mundane if the part of the scene that is being reflected is not known. In another example, it is difficult to convince the witness that a face-like feature in any medium is paranormal if the witness can associate the offered example with the faces-in-clouds effect.

A little time spent on the Internet examining examples posted on various websites will show that this complaint is often deserved. Each time a website visitor responds to an example with *"That is just noise"* or *"All I see is odd patterns that look more like a Rorschach test than something paranormal,"* the credibility of anyone in the paranormalist community becomes easier to discount. The result is increased credibility for the sceptics and progressively less funding for research.

Examples of ITC are frequently composed in novel ways the average witness will have difficulty comprehending. EVP are often very difficult to understand. Even Class A examples are likely formed with a novel arrangement of formants, distorting audio cues and making it difficult for the witness to understand. It is correctly noted that hearing EVP is often like learning a new language. To complicate this is the fact that each practitioner tends to record in what amounts to a different dialect of this novel language.

Visual examples of ITC range from something normal that is out of place to faces formed in noise. The paranormality of something normal but out of place is difficult to argue as paranormal because it is so normal. The paranormal features formed in noise are difficult to comprehend because the witness must learn to look at patterns rather than hard-edge features. It is common for a witness to say, *"Oh, I was looking for a photograph."*

There is no realistic way to police the quality of paranormal phenomena examples being presented to the public. Instead, individual practitioners must learn to self-edit and take special care to help witnesses understand what they are being shown, preferably while letting the witness independently discover the feature.

Unreliable Sensing

A single person's senses are unreliable. Anomalistic psychology [218] is based on the assumption that people too easily fool themselves into believing the mundane is paranormal. A second way of looking at this is that people's natural mental processes can produce erroneous comprehension from confusing environmental information.

One study clearly demonstrated that people tend to hear what they are told to expect (74), even if it is not present in the example. In visual ITC, people are more likely to see what the practitioner expects them to see after being told what to look for.

Experience has shown that one reliable way to assure an example represents what the practitioner thinks is to ask a number of people to examine the example and tell the practitioner what they see or hear. The most common approach to this review is the use of a witness panel.

Witness Panel Best Practice

The objective of a witness panel is not to find people who will agree with the practitioner, but to establish how the average member of the public will experience an example. If at least a majority of the panel does not report experiencing an example as the practitioner expects, then the example should be set aside and not be shared with the public.

The one consideration that the practitioner should always be aware of is that witnesses will eventually become *expert* for the practitioner's unique EVP *dialect* or ITC visual characteristic. This is unavoidable, but a periodical test that asks a new person to grade the example will help maintain the usefulness of the panel.

It is up to the practitioner to make this work.

1 It is recommended that an uneven number of people participate in a witness panel to avoid ties and simplify grading of examples. Five people are recommended as a manageable number and should provide a sufficiently large review.

2 While there are no studies of this, observations indicate that individual ability to hear follows a natural distribution determined by comprehension, hearing ability (both frequency and volume), vision and span of attention. Even after finding enough volunteers, it may be necessary to screen witnesses with known examples of known quality to find people who can be depended on to represent the average listener.

3 Examples should be saved in a file only marked as *Example (number).* The practitioner will need to track the true title. Please refer to the subsection below on *Sound File Considerations*.

4 Members of the panel should be asked to examine or listen to the example and write down what is seen or heard. It is reasonable to include alternative interpretations, but this should be limited to avoid guessing bias. Witnesses should work alone to avoid expectation bias.

5 The practitioner should compile the responses and compare them to what is thought to be in the example.

6 At this point, the practitioner must decide whether or not the example is suitable for public demonstration. It is possible that the example should be discarded. Certainly, if it is being considered for a client,

then the results of the panel may lead the practitioner not to display the example to the client.

Sound File Considerations

The Sharing EVP (162) practice should be reviewed before any example is sent to the listening panel.

It is always a good idea to include a little natural voice in sound files, such as the practitioner's voice asking a question. This provides context to help witnesses orient themselves in the recording.

Very long examples are often difficult to understand, so it may be necessary to segment the utterance into several files.

One or two syllable utterances are very often artifacts. Especially in opportunistic EVP, very brief utterances may appear to be phenomenal, but in fact be naturally occurring. It is strongly recommended that the practitioner consider context of the utterance and be prepared to discard any example that might possibly be a naturally occurring sound. Certainly, one-word EVP are not acceptable for public demonstration unless clearly relevant. For instance, a *"Bob"* utterance is suspect when asking for why a person might be in the house while *"Stuck"* might be a meaningful response.

Be leery of examples which are supposedly an answer to a question for which any response can be construed as the correct answer.

Visual Example Considerations

Transform EVP examples ultimately come down to whether or not the utterance is present and states what is reported; however, some forms of visual ITC are plagued with difficulties in distinguishing real phenomena from artifacts. Before sharing examples of visual phenomena, it is a good idea for the practitioner to become familiar with the kinds of mundane phenomena that might seem paranormal.

It is important to explain to the witness how the example was collected and the environment in which it occurred. This requires good record keeping by the practitioner. In field studies, environmental snapshots are useful for later reconstruction of the scene.

Captive Syndrome

Captive syndrome, more correctly known as Stockholm Syndrome, is roughly described as a psychological condition in which hostages develop a sympathetic point of view about their captors. As it applies to witness panels, people, especially family members, tend to want to please practitioners. It translates as a willingness to fudge a little in how they describe examples if they perceive it will make the practitioner happy.

Consideration of this tendency to error in favor of the practitioner is the reason it is very important that the practitioner does not reveal what is thought to be in the example until after the witness independently arrives at a conclusion.

Above all, the practitioner should never argue with a witness as to what is in the example. This may seem obvious, but it is very common for the practitioner to become angry when witnesses do not see what is thought to be obvious.

Study 11:
Transcommunication Research

Also see the Psi Research Commentary.

Introduction

The study of transcommunication phenomena is primarily conducted by citizen scientists. This automatically means that it is mostly self-funded and conducted without support from a community infrastructure. Studies conducted by citizen scientists often lack peer reviewed reporting. In fact, there are a lot of good, well, let us call them important studies that I am unaware of and that should be included here. Part of the reason this list of studies is limited is the lack of infrastructure and consequential lack of collaboration means I do not have the necessary resources to properly report much of the good work being done in this part of the community.

Parapsychological Research

Because I was mostly talking (complaining really) about parapsychologists in the Science Commentary, I followed it with a brief discussion of the important work some parapsychologists have done which indirectly contributes to the study of transcommunication. That is in the Psi Research Commentary. Here, I will focus on research and studies specifically applicable to the study of survival.

Authenticity of Channeled Material

We know people sometimes unconsciously color messages they think are coming from people on the other side. Most of the time, we depend on the reasonableness of the message as a measure of credibility. Some information can be verified, but we also know that information from other people in the physical can be unconsciously, psychically accessed. Keeping these factors in mind, it is clear that some people are able to provide a clear channel between a communicating personality and the medium's outward expression. What we require is a science/study-based reference that supports this belief.

Professor Paul Cunningham conducted a study intended to determine whether or not the channeled, Jane Roberts' Seth material has characteristics that might indicate a source outside of Roberts' imagination. His study of Roberts' personal writing and her channeled work, as well as her husband's writing, strongly suggests that the words attributed to Seth originated from an independent personality. (26)

This study is important because, for me, it is the only study I am aware of that provides an objective reason to think at least some channeled material may be from a second personality and not just from the author's imagination.

4Cell EVP Demonstration

In and around 2005, ATransC members participated in a series of experiments intended to assess the information gathering capability of Electronic Voice Phenomena (EVP). The studies are collectively known as the 4Cell EVP Demonstration. From the Abstract:

> *Can the etheric communicator communicate with people in different parts of the world? Can the communicator in EVP hear a question from one person, ascertain the correct answer and deliver that answer to someone else in a different part of the world?*
>
> *A series of experiments employing a specific protocol were conducted by different four-person teams. Coordination with team members who were in different parts of the USA was via email and the Idea Exchange discussion board (ATransC.org/forum/).*
>
> *One person thought of a question and told the Sender the question but not the answer. The Sender communed in some way with his or her EVP communicators, asking them to tell the Receiver the correct answer. Once notified that a question had been sent, but not the question or answer, the Receiver conducted an EVP session to ask for the answer. A fourth person (Scribe) evaluated the resulting recordings for possible EVP and made first determination of what was said, if anything. The study indicated that it is possible to use EVP to gather information, but that there are limits to the kind of information that may be accessed.* (163)

The study produced an average of 61% reasonably correct responses. An example is *"What can we, as receivers on the physical level, do to help facilitate communication from you, the transmitters in the spiritual?"* Class B+ answer: *"Just open up the portal."*

In another session, *"Who is the artist and what is the name of the painting hanging at the end of the entrance hallway in Vicki and Pete's home?"* Correct Answer: *"Marc Chagall"* and *"The Lovers."* Class B answer: *"Who painted it? ... Chagall did."*

The value of these experiments is first, an effective protocol can be designed to use EVP as a research tool, and second, the demonstration that there is some kind of interactive intelligence initiating EVP.

EVP Online Listening Trials

A major question has been whether or not EVP are objective in the sense that, without prompting, a listener will hear what the practitioner thinks is said in an EVP. A series of three online listening studies were conducted by the ATransC. From the report Abstract:

> *A common explanation for Electronic Voice Phenomena (EVP) is that the reported utterances are mundane sounds mistaken as voice forming words. This report describes three online listening trials which were conducted to determine whether or not website visitors can correctly identify words that are thought to be present in EVP examples by listening to unmarked sound files.*
>
> *A second consideration is that it is popular wisdom amongst EVP practitioners that one must learn to correctly understand EVP. A variety of approaches were tried to test this theory, including polling experienced listeners, using questions in an attempt to assess interest and predisposition to believe in EVP and asking participants to indicate experience in hearing examples. Analysis of the trials is included, along with an assessment of the reliability of the results.*
>
> *When the total number of words correctly recognized for the three trials is compared to the possible number, the overall percent of Recognized words (%Rw) is 25.2%, indicating that at least some EVP do constitute recognizable words.* (21)

A similar study conducted by Mark Leary,Ph.D. reported that *"on average, only 21% of the investigators' interpretation of particular words agreed with the consensus interpretation."* (74) Considering a few less well-organized studies, it appears that 20 to 25% agreement is about what a practitioner should expect for listeners.

Phantom Voices

It is startling how completely we can be fooled by our senses. As we learned more about transcommunication, we became increasingly alert to the problem of comprehension error. Sometimes, intelligent people with at least normal reasoning ability, and presumably normal hearing, reported hearing utterances in sound files that others could not. We have begun referring to this phenomenon as Phantom Voices.

There is a listener cueing effect in which a person will not be able to make out what is said in an EVP example, but will be able to once they are told what to expect. This effect can defeat the benefits of a witness panel and lead many, otherwise reasonable people to report hearing information that simply is not present.

To test this theory, we (ATransC) conducted an online listening test using two sound files. One contained simple brown noise (broad-spectrum noise with emphasis on voice-frequencies) and the other contained broad-spectrum noise modulated with audio pulses that simulate the cadence of speech. It was clearly stated that neither example contained voice. Possible explanations about what might cause a person to hear phantom voices were included above the hearing test and what was in the files was clearly stated. From the Abstract:

> *A frequent source of consternation for people who are asked to listen to EVP examples is their failure to hear what is reported. It is expected there will be some disagreement between listeners and practitioners. That is the nature of EVP (see Online Listening Study). However, a problem develops when listeners report hearing only noise, and doing so with example after example from the same practitioner when the practitioner insists there are paranormal voices in the examples. The question necessarily must turn to why the practitioner is hearing what others do not.*
>
> *For this study, sound files containing only noise were presented to ATransC.org online listeners who were told there was only noise and were then asked to report what they heard. The study confirmed the prevalence of people who report hearing phantom voices. The study includes a discussion as to why this may be.*

Information Gathering Using EVPmaker with Allophone

Using EVP for information gathering is something of the ultimate goal for EVP practitioners. One of the most common questions put to the unseen communicators is whether or not they can help us improve our equipment. (From my experience such help has come but has just been distracting. It appears to be true that transition does not make a person brilliant.)

The ATransC conducted a study using a computer program developed by Stefan Bion (22) which insets bits of a supplied sound file into a buffer, and then uses a relatively random process to select buffer addresses, which are concatenated into a new output sound file. EVP are thought to occur from a psi influence on the random process.

Bion also provided an audio file containing allophones generated with a speech synthesis program. This experiment required that every participant use the EVPmaker program and supplied input file containing allophones.

From the Abstract:

> *This twelve-month trial was designed to determine whether or not information not known to a participant could be requested and received via Electronic Voice Phenomena (EVP) using EVPmaker with allophones.*
>
> *A target object was left undisturbed in the same location at the beginning of each month for twelve months. Participants were asked to use only EVPmaker with allophones to produce a sound file containing the information identifying the target. The project manager did not know the target, so to qualify submissions, she screened them for reasonableness.*
>
> *Those possibly containing usable information were submitted to a listening panel of people who were familiar with EVP but who did not know the identity of the target. If a majority of the listening panel heard information as reported by the participant, the submission was considered a valid submission. Submissions were rejected if a majority did not hear what the participant heard. Valid submissions were screened for a positive response by the project manager for inclusion in the study.*
>
> *This trial did not produce positive responses based on the study's protocol. However, the abundance of non-protocol EVP captured in the study might suggest EVPmaker is not suited for the type of communication this study was designed to capture. Also, participant knowledge of the target might have to be reexamined as several targets were identified either before the target was placed or after when the participant knew what the target was. Future studies may wish to look at these non-protocol results when designing a follow up study.*

Perception of Visual ITC Images

This was a study to determine whether or not paranormal images recorded in optical-frequency noise could be consistently described by website visitors without coaching. The video-feedback technique was used for all of these examples. From the Abstract:

> *In visual Instrumental TransCommunication (Visual ITC), recognizable features are found in what should be only random optical noise. No known physical principles account for the phenomenal features and they may be found in virtually any sufficiently noisy media. The Examples and Techniques sections of this website include several such techniques for capturing the features. This report includes results of an online viewing study in which website visitors were asked to describe what they saw in unmarked visual ITC images.*
>
> *Of the seven examples, an average of 61% of respondents correctly identified the feature. Each example was presented with original, grayscale and increased contrast versions. The increased contrast version was most often correctly identified.*

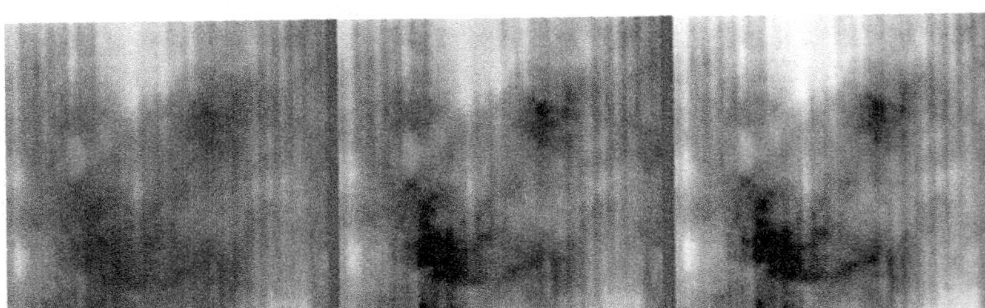

Victual Perception Study: Image 1 of the Visual Perception Study: Left is original, Center is grayscale and right is enhanced. Witnesses reported 81% correct recognition for this example.

EVPmaker with Allophones: Where Are We Now?

This was a study that was funded by the ATransC to determine if real-time, two-way transcommunication could be conducted under controlled conditions. The study came out as a *failure to replicate-style* report published in the Journal of Scientific Exploration. (88)

Because we did not manage the researchers well, the study has become more a lesson in what not to do to manage projects and how not to conduct research. For that reason, I have focused more on the lessons learned than on the organization. The full report is accessible via the Journal section of ATransC.org. (24) Some of the lessons learned are part of the Science Commentary.

Abstract

Based on a number of recent demonstrations by multiple practitioners, ATransC commissioned a study to determine the suitability of that technology for real-time, two-way communication. After three years, a failure to replicate-style report was published. This article is a discussion of procedural concerns with the study and a discussion of lessons learned which may guide future studies.

"Sfinge" Project: Instrumental Analysis of EVP Collected via a Sound-Psi Interaction

The Sfinge Project was a study conducted by members of the Italian research group, Il Laboratorio. (164) From its conclusion:

> *From the electroacoustic analyses performed on the audio samples recorded under controlled conditions, we extracted the acoustic parameters distinguishing a vocal signal. These elements allowed us to identify the phonemes constituting the words we decoded linguistically.*
>
> *By examining their structures and their anomalies, we were able to assert with documented evidence, that these events exhibit phonetic features associated with the voices. These voices have evident and absolutely original features and are structured and characterized by parameters which deviate from the typical human standards.*
>
> *The presence of formantic bands, with localized noise strengthening, confers to the voice an acoustic structure close, but not identical, to the human one. The anomalies found where the voices were recorded allow us to affirm the existence of an atypical process of formation which is still today scientifically unknown despite their objective nature.*

Abstract

The research team of Il Laboratorio from Bologna, Italy conducted a two-year long study of well-known Electronic Voice Phenomena (EVP) operator, Mrs. Lida Russo of Livorno, Italy. The microphone recording method with a commercial recorder with built-in microphone was used by the operator and professional digital devices were used to monitor the sessions. However, only a few anomalous voices of poor acoustic quality were found upon review of Mrs. Russo's audio tape. Contrary to the expected, the most interesting voices in terms of quantity and quality were recorded on the digital devices operated by Daniele Gullà.

State-of-the art software commonly used for forensic analysis was used for analyses of the resulting voices. This new approach to EVP research was introduced to the world by the researchers of Il Laboratorio and has revealed many structural electroacoustical features that prove the authenticity of these acoustic events, which now can legitimately be classified as linguistic events.

Relationship Between Physical Phenomena and Electronic Voice Phenomena in Séances with a Physical Circle

Rachel Browning conducted a multi-year study of recorded EVP in an experimental physical séance circle. (165) This is the kind of research citizen scientists are best suited to conduct. She was part of a dedicated circle and had the organizational skills to maintain a prolonged study with adequate recordkeeping to develop reasonable conclusions.

From the conclusion:

The ability to capture a voice on more than one device at the same time may indicate that the EVP is a recorded transmission of a physical sound wave and not a mechanical or electrical fault inherent to a recording device. This does not in itself provide evidence of an incorporeal source but it does suppose a source of physical generation. It also leads to the unanswerable question; how would an etherical intelligence commute thought to a sound wave that is able to be recorded at the same pitch and volume as human speech but that is not heard in the environment at the time of recording?

It would be logical having only read the study results to conclude that EVP does not provide evidence of post mortem communication. However, had you taken part in a sitting, witnessed the recording and were satisfied that no deception took place, then on playback were able to hear an anomalous voice in reply to yours, it would be logical to explore all alternative explanations, regardless of how unlikely they were. We are fortunate in continuing to experience and report physical phenomena and EVP in our sittings; the study and recordings will continue.

A Two-Year Investigation of the Allegedly Anomalous Electronic Voices or EVP

This study concerned the ability to collect EVP under controlled conditions. It is important because it constitutes a replication of other work under very controlled conditions.

Anabela Cardoso has published the ITC Journal (166) since March, 2000. This is one of the few remaining publications dedicated to the study of ITC.

Abstract

A relatively novel acoustic phenomenon has inundated the Internet and specialized literature. Several Associations, some of them with an important number of members, have formed around it in many countries. In the Anglo-Saxon world the phenomenon is called EVP (Electronic Voice Phenomenon) and is usually assumed as electronically mediated communication from or with the deceased. The first tests aimed at verifying the reality of these claims were carried out in Sweden and in Germany, in 1964 and 1970, under the direction of Professor Hans Bender from Freiburg University (Bender, 1970; 1972; 2011).

The present report describes in detail the tests designed to record the allegedly anomalous electronic voices, or EVP, under controlled acoustic conditions. Series of experiments were carried out in Vigo, Spain throughout a period of two years under conditions controlled to the highest degree achievable. Several operators were involved in the many tests conducted in Acoustic Laboratories and professional recording studios equipped with very high levels of acoustic shielding. The protocols and procedures followed in the experiments, as well as the results obtained, are herewith described. During the many experiments performed, several extra voices were recorded for which no normal explanation was found.

Observations

There have been other studies, but these are the more informative ones for this discussion. A few studies have been conducted outside of the ATransC for which we are awaiting a report.

Too few people are able to devote the time and effort to conduct studies, and a natural result is that there are usually just one or two studies for a particular question. This means that anything resembling meta-analysis is in the form of an individual's effort to make sense of what is known based on just a few empirical data points. That is one of the reasons I place so much emphasis on subject-matter specialists.

An important point for future researchers to consider is that one does not usually just decide to set up a development circle to conduct a weeklong study. From personal experience and testimony from others, it is clear that a circle may not produce meaningful phenomena at all, especially not in a short study period. With this in mind, the only realistic way to study circle phenomena is to study a willing group of sitters who are already reporting phenomena.

It must be remembered that practitioners and sitters are live subjects to whom ethical considerations must be applied. As it stands now, having seen the shabby treatment [226] some researchers have given their subjects, the atmosphere is poisoned for many practitioners when it comes to working with a would-be researcher. This is one of the issues that must be addressed for future research. (167)

So few studies and so few people conducting studies also means that what work is done needs to be directed where it will do the most good. That is one of the reasons why I maintain that research grants need to be treated like contracts.

The question of whether or not these phenomena are real has been answered. That is an important message to future researchers. The focus now should be on the nature of these phenomena. The Implicit Cosmology was modeled to provide a working hypothesis which can be tested. As such, we are very interested in supporting research that is focused on that kind of study.

A study to Test the Implicit Cosmology

The ATransC has had a study in the cue for over a year. Here is the abstract for *The Energy Profile of Transform EVP* study (133):

> *The propose of the article is to issue a call for transform Electronic Voice Phenomena (EVP) examples in the form of audio output files containing EVP and a file containing the environmental sound and/or sound used as input for voice formation. Both files must be so configured as to allow side-by-side comparison from a sound marker. The examples will be analyzed in an effort to find a relationship between EVP formation and the energy profile of the sound file which contains the example.*

An overview of the various forms of EVP and current theories for transform EVP formation are provided. A hypothesis is proposed that transform EVP are formed via stochastic amplification and that the source energy in the form of the message (that which is amplified) is made available by strategically apporting sounds already present in the physical. The study is expected to show a net increase in energy associated with EVP.

Thus far, there have been no example entries. This is a testimony to how few experienced EVP practitioners the ATransC is able to access for research. A few years ago, we would likely have had half a dozen people helping. It is not that there are no people able to produce examples. The Association no longer has the level of participation it once had.

The importance of the study is that a net increase in energy in a sound file coincident with the supposed utterance would indicate that something has been added. It may indicate support for the apport theory [286], but certainly would be another indicator of the objective nature of EVP.

Closing Comments

And so, that is the Implicit Cosmology. On the back cover, I wrote, *"Having learned, it is our lot to teach. As Hermes taught us, the three phases of a teacher are the lessons, how we live the lessons and the product of that living."*

The book is formatted as those three phases, but it is important for you to understand that you are that teacher. Everything written here is intended to help you be the best teacher possible. I have tried to convey the understanding that what I have written is probably like what will be seen as correct by future generations, but not necessarily exactly the same.

Again, you are the teacher and will probably contribute in some way as our community refines this cosmology. Acknowledging this urge to teach will help you fulfill that role with excellence.

Catch Phrases Help Describe Reality

A number of interesting phrases came to mind during the over three years required to write this book. Some are silly, all are simplistic, but they serve as a sort of short name for a much more complex thoughtform.

Having learned, it is our lot to teach.

The three phases of a student: study, comprehend and apply.

The three aspects of a teacher: study, comprehend and teach.

The Golden Rule Redo: Teach me as I teach you.

The Way of Progression: Through community comes knowledge; through teaching comes understanding; it takes a collective.

Just because you can, doesn't mean you should.

You are part of a collective: You are never alone.

Worldview represents personal reality: You create worldview with your intention and creative process. Worldview determines your happiness and unhappiness here and hereafter.

The Six Pillars of the Implicit Cosmology

> **Pillar 1, Reality:** Reality consists of life fields and expressions of life fields. (Reality Discourse)
>
> **Pillar 2, Person:** A person is an immortal etheric personality entangled with a human in an avatar relationship. (Life Fields Discourse)

Pillar 3, Purpose: We inherited the urge to gain understanding about the nature of reality from our source personality. (Prime Imperative [61] in the Organizing Principle Discourse)

Pillar 4, Mind: Unconscious psi sensing precedes conscious perception with unconscious worldview as the arbiter and conscious intention as the motivator. (Perception and Expression Discourse)

Pillar 5, Creation: The creative process is attention on an imagined outcome to produce an intended order. (Perception and Expression Discourse)

Pillar 6, Progression: Integrating acquired understanding of the nature of reality into worldview constitutes progression and expands opportunities for greater understanding. (Perceptual Agreement Discourse)

Wizard of Oz Syndrome: The misconception that people with doctorate degrees are infinitely knowledgeable and wise, that they work to help us better understand reality.

Evidence is not always evidential: To be evidential, an example must have a very well-controlled and documented *chain of custody*, and the protocol by which it was collected must be known, repeatable and observable. And of course, meaning of evidence must be based on understanding.

Wizard Complex: People with doctorates believe they are infinitely knowledgeable and wise.

Omniscient Science: Pretty much Wizard Syndrome/Complex.

Hyperlucidity: Self-delusion.

Assumed Knowledge: Often stems from hyperlucidity, but is usually a product of religious training and/or Wizard Complex. Both laypeople and academics potentially suffer from this social malady.

Academic-Layperson Partition: Separation between people without an advanced degree imposed by people with an advanced degree—an aspect of Wizard Complex.

Shit Happens: Our life lessons and future experiences are not so much in what we create as they are in how we react to what is created.

There are a few more, but you probably get my point. The message I would like you to take away is that you are immortal, your purpose is to understand your world. You are in a community with people involved in some way with things paranormal, that community is dysfunctional and that dysfunction jeopardizes your progression. Most important, you are able to manage your potential with the expression of intention based on understanding. As such, the most important point you must understand to progress is Pillar 5, Creation: The creative process is attention on an imagined outcome to produce an intended order.

For better or worse, you are a creator and thereby make your world.

Ask Questions!

The Implicit Cosmology is complex. It is likely a very different way for you to look at yourself and your world. It is important that you understand and have the opportunity to integrate at least parts of it into your life. Having spent so much time developing the Implicit Cosmology and writing this book, I have no intention of ignoring your need to ask questions. Here are a few of the ways available to you to ask questions and participate in discussion:

- Ethericstudies.org and ATransC.org websites have a contact tool at the bottom of each page.
- The *Occasional Update Email List* is a good way to stay current about what seems important at the time. There is a link to sign up for it on the Contact page of ATransC.org.
- The ATransC Idea Exchange at ATransC.org/forum/ is maintained expressly for you to ask questions and participate in discussions about things paranormal.
- The Collective at ATransC.org/bp/Collective is a wiki intended to support the development of best practices by members of the community. By becoming an editor, you can help build consensus articles about paranormal subjects.
- Keep an eye on ethericstudies.org, especially the Concepts section. I will be updating the discourses that support the Implicit Cosmology.

It is up to you.

Thank you for reading this book!

Glossary of Terms

A good citizen of any frontier subject is one who is informed about the important concepts involved in that field of study. An effective way to do this is by becoming informed about the study of trans-etheric phenomena is to learn how to speak about them with others. Doing so requires understanding of the subject which begins with understanding the terminology.

A

Academic Paywall: The Academic-layperson partition is amplified by the professional and university library systems in which paid membership is necessary to access scholarly papers, effectively assuring that most frontier subject citizen scientists [186] (laypersons) do not have access to the results of applicable research even though it is often publicly funded.

Academic-Layperson Partition: A phrase used here to describe the functional separation of academia from laypeople. The effect is that people who have a doctorate degree from an accredited institution tend to shun people who do not. For fields of study in which practical experience, ability and understanding rests with lay people or citizen scientists, this partition often produces peer reviewed studies which portray a false image of the subject.

After-Death Communication: After-Death Communication or ADC is a concept popularized by Bill and Judy Guggenheim in their book, *Hello from Heaven*. (14) From their website: ADC is defined as *"a spiritual experience, which occurs when you are contacted directly and spontaneously by a deceased family member or friend, without the use of psychics, mediums, rituals or devices of any kind."* The important points in this explanation are that ADCs are spontaneous and instruments are not required. They do include phone calls as a type, but they have been very definite that EVP is not included in the list. See Induced After-Death Communication [442].

From the ADC website, here are the twelve most frequent types of ADC people report having with their deceased loved ones:

> *Sensing A Presence; Hearing A Voice; Feeling A Touch; Smelling A Fragrance; Visual Experiences; Visions; Twilight Experiences; ADC Experiences While Asleep; Out-Of-Body ADCs; Telephone Calls; Physical Phenomena; and Symbolic ADCs*

Agnosia: Loss of the ability to interpret sensory stimuli, such as sounds or images. Agnosia was once considered a rare condition, but since the work with inattentional blindness [442], it has become clear that it is much more common than previously thought.

Anomalistic Psychology: The study of human behavior and ostensibly paranormal phenomena with the assumption that no paranormal phenomena are involved. Research in this area is often reported as an academic study, but fails to even consider a paranormal hypothesis. Essentially, a field of study intended to debunk things paranormal. See Anomalistic Psychology [218] in the Science Commentary.

Apologists: A person who speaks in defense of someone or something. This is an important form of public education if the apologist is knowledgeable about the subject and has the objective of informing and not indoctrinating. In some cases, such as in the case of skeptics [239] who brand ideas they think mainstream science does not support as pseudoscience [218].

Apophenia: The perception of patterns or connections where none exist. It may be the underlying natural response that leads a person to hear distant voices or music in noise that can *almost* be understood, but that on closer examination, do not exist and cannot be objectively recorded. It appears to be natural for our minds to seek to find order in chaotic information and that order comes from our worldview; making the unexplained familiar. See Pareidolia [452] and Gestalt [438].

Apparition: The anomalous appearance of a discarnate person; usually a spontaneous visual phenomenon. Apparitions seem more likely to manifest in moments of crisis and the message given by them is usually brief as if the contact field is unfocused. Apparitions may appear quite solid and yet objects may be seen beyond them. They also may be accompanied by bright light.

Apport: The apparent physical-to-etheric-to-physical transport of an object under the control of an etheric personality. Usually in the context of a Séance, but may occur any time. The transported object is referred to as *an apport.* Objects differ in size, may be organic or inanimate and appear not to be harmed by their movement. In the case of living objects, it is speculated that the obstructions are dematerialized to allow the life form to pass unharmed.

Apport: An apport received by the Felix Experimental Group (compared to a 2.5-inch candle). © FEG

In recent theory, it is speculated that other forms of phenomena might be explained as an apport. For instance, credible practitioners have reported examples of transform EVP which include sound known to previously exist in other media. Others have reported examples which are expressed by multiple speakers so that the phrase begins with one voice and ends with another. It is common for an utterance to clearly be a response to a question but with an unusual choice of phrasing.

It might be more efficient to find the required words in the physical and apport them to the sound track, rather than introducing the necessary influence to affect a physical process such as stochastic Amplification. See Stochastic Amplification [276] in the Foundation Principles Study and Apport [286] in the Transcommunication Study.

Appropriate Science: The population of participants in a frontier subject and the history of the subject area determine the ability of the community to document the subject. Every discipline of science has undergone an evolution as it is developed into a system of understanding or an effective tool for learning. In the beginning phase of the evolution of the community, people studying the subject are untrained and often not supported by academic institutions and literature. With this consideration, the measure of how well the discipline of science is applied to the understanding of a subject must necessarily be judged by its evolutionary progress.

A priori: A conclusion made without examination or factual study. The viewpoint that something is impossible, therefore it cannot be, is an *a priori*

conclusion if the person has made an assumption based on faith or popular wisdom and not on evidence or actual study of the subject. The skeptic community is especially guilty of following an *a priori* assumption about anything they think is not explicitly supported by mainstream science.

Aspectation: The creative process of intention acting on an imagined result produces aspects of reality which are a subset of personality's personal reality [88]. For instance, when a person imagines driving a new car, the car and its environment are formed out of a subset of how the person imagines reality (personal reality). The *little me* personality in the car is an aspect of the person as the person imagines him or herself to be. See the Principles of Aspectation [63] in the Organizing Principles Discourse.

Assumed Knowledge: *Everyone is an expert!* is a common complaint of people who have studied transcommunication for many years. As people come into this field and learn a little from fellow enthusiasts, some too quickly decide they are more knowledgeable about things paranormal than others. Even without doing the work of studying existing material, they assume their expertise is greater than anyone who has been in the field for many years.

Assumed knowledge is also prevalent amongst academically trained researchers. Many holding doctorates tend to automatically disregard anything said by people who do not hold doctorates even though they may have considerably more practical experience in the subject. In the academic-layperson partition, the person appears to assume that college training, and in most cases for transcommunication, a degree in psychology sufficiently prepares the person to study any subject without input from subject matter specialists [462].

Attention Matrix: In the Implicit Cosmology the Functional Areas of a Life Field Diagram [xxiii] illustrates the functional areas of a life field. Attention functions as an attractor matrix in which intention, emotion and curiosity are influences and represent the etheric equivalent of physical energy. Visualization and expression are fundamental processes which operate with these influences.

Throughout the Implicit Cosmology, functional areas exist only so long as, and to the extent that attention is on them. For instance, a visualized aspect of reality is created within the matrix of the personality's attention and ceases to exist, other than as memory, when attention is withdrawn.

Attractor: This is a term associated with chaos theory. (72) An attractor is an element of a set that acts as a locus of influence for the other elements in the set. In effect, they tend to favor the attractor element in some way.

An attractor for an etheric field is probably best described as purpose. In a life field, personality is the attractor. In the reality field, Source is the attractor. In a collective field, the local source for that collective is the attractor. See Field [438].

Aura: An energy field or field of radiation thought to surround a physical, living organism. Those who can see an auric field (clairvoyance [430]), say it consist of bands of color. The auric field may be a combination of an electromagnetic byproduct of biological processes and the psi influence associated with the morphic field [102]. See Biofield [428].

Automatic Writing: Text produced while in a disassociated state, presumably originating from a discarnate personality. Also known as autoscript, this is probably best thought of as a means of more directly accessing the person's etheric personality or as a more lucid awareness of the Perceptual Loop [112].

Similar abilities are automatic drawing, painting and psychic art. These are artistic expression without the control of the conscious self. Sometimes this is done in complete darkness and at great speed. See Trance [463].

Avatar: In Hinduism, an avatar (from Sanskrit for *descent*) is a deliberate descent of a deity to Earth. The term can be translated into English as *incarnation*. A person is an etheric personality entangled with a human body in an avatar relationship. There is a discussion of the avatar concept in the Personality-Centric Perspective Discourse.

B

Best Practices: Methods and procedures found to be most effective. Best practices are not rules, laws or standards which people are expected to follow. In principle, best practices are to be used as a guide to help a person be aware of that which is known to work and that which has inherent pitfalls. Best practices are an excellent means of maintaining actionable aspects of the public memory. It is reasonable to ask someone if they followed best practices as a means of gaging their work.

Big Circle: The Big Circle consists of our loved ones in the etheric along with their loved ones in the physical. The realization that this group existed in the etheric, and was trying to connect to their loved ones here in the physical world, prompted members of the Association TransCommunication to form a special group called *Big Circle – Bridge to the Afterlife*, which is now referred to simply as the *Big Circle*. This group is composed of bereaved parents, spouses, siblings and friends whose mission is to build a channel between the physical and etheric aspects of reality—a Bridge to the Afterlife.

The people who formed the circle have mostly moved on to other interests, but it is thought to remain a viable way of forming a contact field [100] as a cooperative community [64]. The ATransC remains available to assist in this endeavor by way of the Idea Exchange at ATransC.org/forum/.

Billet Reading: Billet reading involves reading and commenting on questions written on folded and sealed paper—without breaking the seal.

Biofield: Life-based emanation sometimes referred to as a *human energy field.* From the physical person perspective (clairvoyance), the biofield may be a combination of bioelectric byproduct and psi field influence perceived as a visible field of energy. The net effect is that the biofield presents useful information about the living organism. See Aura [427].

Black Box Analysis: Engineers design models for systems they are trying to understand. One way to develop a model is to figuratively put the subject in a *black box* with the known input and output clearly defined. Not knowing for sure what is inside the *box*, engineers must design a set of functional areas *within the box* that will produce the observed outputs. This is an effective technique for organizations to reverse engineer copyrighted or patented technology in an effort to develop alternative designs.

It is important to note that, when black box modeling is used for a metaphysical cosmology, a plausible explanation may be developed that does not correctly describe reality. Even so, the resulting model may be useful as a guide for further development. Also see Black Box Analysis [14] in the Theory Introduction Discourse.

C

Cabinet: An enclosed space in which mediums are thought to *condense* the psi field that is necessary for physical phenomena. A cabinet may be as simple as a curtain hung across the corner of a room. Some mediums sit in the cabinet while others sit outside of it. Many great mediums who produced physical phenomena found a cabinet unnecessary. See Contact Field [100].

Chakras: Also known as energy centers or spiritual centers. According to Hindu and occult teaching, there are usually seven major centers associated with ductless glands in the physical body. As with the aura, these centers, which are a mixture of colors of varying brightness, are seen clairvoyantly [430]. Some practitioners say that they work on the chakras to stimulate a healing of the physical body. See Biofield [428] and Aura [427].

Channel: A person who is able to mentally *step aside* with deep trance to allow an etheric communicator to speak through his or her physical body. The distinction between a medium and a channel is usually made by who is communicating and for what purpose. Generally, a medium communicates messages from discarnate people. A channel is usually thought to be bringing through more profound information from *higher* beings, usually an entity who has probably not experienced a physical lifetime, and frequently, a personality representing a group of personalities. See the Mediumship Study.

Citizen Scientist: Ideally, people who study nature are trained to do so by an academically certified organization, but in many cases, the magnitude of a research task may be such that a few academically trained scientists must depend on volunteer assistants for routine tasks such as data collection and field support. These layperson research supporters are sometimes referred to as citizen scientists. In the early days of radio, many ham radio operators could have been referred to as citizen scientists.

There are relatively few academically trained scientists actively conducting research in frontier fields, in which case, most of the organized study is conducted by enthusiasts. Hauntings investigation groups are a good example, and of course, the study of ITC has historically been shunned by academics, making it necessary for citizen scientists to fill the resulting void. See Appropriate Science [425] and the Science Commentary.

Clairaudience: This is a French word that means clear hearing. It is the ability to sense presumably psi information as sounds rather than impressions or images. To the person who experiences this, it is as if hearing someone speak; however, not with the physical ear because it is not an acoustical phenomenon. See Clairvoyance [430] for more detail.

Claircognizance: This is a French word that means clear knowing. It is the ability to sense presumably psi information as a knowing, sometimes described as a gestalt understanding. See Clairvoyance [430] for more detail.

Clairsentience: This is a French word that means clear sensing. A person senses characteristics of the etheric communicator or circumstance but does not hear or see the information. See Clairvoyance [430] for more detail.

Clairvoyance: This is a French word that means clear seeing. It is the ability to sense presumably psi information as mental images. The perception may be of objects, people and/or scenes, and can be from the present, past or future. The clairvoyant experience may be spontaneous or induced as through meditation, scrying or other methods of divination.

The term clairvoyance has become a catchall term which includes all forms of sensing information in the psi field. Since this term is so commonly used in Spiritualism [461] and by the New Age culture [450], parapsychologists coined *Extra Sensory Perception* or ESP, which is apparently being replaced with psi functioning [356]. In mediumship, anomalous information reception is a currently popular term. See the Perception and Expression Discourse.

Collective: In the Implicit Cosmology, reality is seen as being hierarchical [61] with Source [55] as the top life field and the reality field. Aspects of Source inhabiting sub-realities as imagined by those aspects. Altogether, the life fields represent a collective that might be called the Source Collective. Each aspect is seen as the source of many sub-aspects so that each aspect would represent a collective. The Principle of Aspectation Diagram [57] illustrates the relationship amongst rounds of aspectation.

Coloring: A message purported to come from the etheric that has been influenced by the medium's own thoughts and ideas. Mediums are trained to give what they receive even if they do not understand the message. They are trained to guard against trying to interpret or translate a message to assure that their own thoughts and views are not placed into the message. Nevertheless, it is arguable that all forms of transcommunication are colored in some way because they must traverse the medium's perceptual processes. An exception to this may be trans-influences that involve a contact field [100] without the need for an entangled avatar.

Concept: A fundamental idea; a root thoughtform from which systems of thought can be derived. For instance, the concept of something to protect a foot is the fundamental concept for shoes, sandals and such. In Etheric Studies [204], the intention of *concept* is to refer to the underlying thoughtforms [99] from which reality is formed.

Conceptual: Reality is experienced in a conceptual form, as opposed to an objective one. Information can only be treated in an etheric form. As the expression of the information emerges into consciousness, it is assigned objectivity so that the physical world is conceptual, but experienced based on habit as objective. See the Perception and Expression Discourses.

Consciousness: The *I think I am this* aspect of the *I am this* personality. Conscious self is the experiential aspect of a life field. By comparison, the personality, as the core intelligence of a life field, is the purposeful aspect. Personality and all of the perception and expression processes are mostly unconscious, as compared to the conscious self.

Contact Field: The concept is that the rapport [68] amongst mutual cooperating personalities and persons results in an intentionality field [100] useful for trans-etheric influences. See the Contact Field [100].

Contact Healing: Physical contact is made between the sitter and the healer-practitioner. Also called *laying on of hands* or spiritual *healing*. Absent, distant and non-contact healing is healing done without contact and/or at a distance. As a means of managing expectations, *healing intention* may be a more appropriate term for healing. See the Healing Intention Study.

Control: The etheric personality thought to be controlling the entranced medium and the flow of a séance.

Cooperative Community: An effort to express understanding is necessary for progression. Collectives are inherently cooperative communities. A person is attracted to communities of like-minded people cooperating to facilitate progression. This is an Organizing Principle [54].

Corollary: A proposition that follows with little or no proof required from a previously proven statement.

Cosmology: In metaphysics [447], a conceptual model representing the relationship of the component parts of reality. In Etheric Studies, the cosmology relating such concepts as personality, the physical verses the greater aspects of reality, transition and transcommunication are modeled as the Implicit Cosmology. Not to be mistaken for an astronomical cosmology addressing only the physical universe.

Tree of Life	Influence	Self	Plane
1 Crown	Cosmic Self		
2 Wisdom	Life Force	Causal Body	
3 Understanding	Intuition	Buddic Body	Divine Soul
0 Knowledge			Gnosis
4 Mercy	Memory		Higher Mental Plane Retrospective
5 Severity	Volition		Higher Mental Plane Prospective
6 Beauty	The Ego	Egoic Body	Ego Plane
7 Victory	Desire	Desire Body	Lower Mental Plane Retrospective
8 Splendor	Intellect	Mental Body	Lower Mental Plane Prospective
9 Foundation	Automatic Consciousness	Astral Body	Astral Plane Vital Soul
10 Kingdom	Sensation	Physical Body	Physical Plane
	Fire		Man
	Water		Animal Kingdom
	Air		Plant Kingdom
	Earth		Mineral Kingdom

Hermetic Qabalah: The hierarchical cosmology as described in the Hermetic Qabalah (as if a layer cake) is useful in conversation as a means of indicating here and relatively not here as one moves into other aspects of reality.

Traditional metaphysical cosmologies portray reality as what might be thought of as a layer cake arrangements of component parts. See the Hermetic Qabalah Cosmology Diagram above. The Implicit Cosmology is modeled as a distributed array of fields related to the influence of a single source. See The Cosmology of Imaginary Space Discourse.

Cross-Correspondence: One or more forms of phenomena are experienced as separate events and found to produce information that agrees. For instance, a medium might say that the loved one is standing behind a person, and at the same time, a photograph of the person contains a disturbance that *seems* to be the shape of a person.

Cross-Reference: Different mediums simultaneously receive the same message with a request to forward the message to the same person.

Crystal Gazing or Scrying: A form of divination done by gazing at a surface that appears to have great depth such as a crystal ball, pool of water or tilted mirror in a featureless enclosure. After a time of apparent cloudiness in the depth of the material, pictures and scenes are said to appear. This is the form of divination said to have been used at the time of the Greek Oracles and also by Nostradamus. See Divination [434].

Cultural Contamination: Worldview is populated with cultural influences. Worldview is also the functional area of mind representing memory, beliefs and understanding that produces personal reality for conscious self. If a concept is in the cultural domain, it should be expected to be an influence in worldview.

For instance, there was little agreement in the cultural view of what an extraterrestrial looks like until publication of the cover of Whitley Strieber's 1987 book *Communion*, which portrayed a *typical* gray extraterrestrial. Today, close encounters with aliens, can be expected to include the now standard alien description which agrees with the *Communion* book cover. That kind of cultural contamination must be considered when evaluating the usefulness of any report based on perception.

D

Death-Bed Visions: In typical death-bed visions, a person near the moment of transition reports seeing the appearance of deceased loved ones and relatives. The dying brain explanation does not explain why the person reports seeing only people on the other side and not just anyone from memory. There have been instances in which the person reported seeing a friend who had recently transitioned without the person or person's family knowing of the transition.

Dematerialization: The disappearance of matter or a materialized form. Parts of the medium's body have been recorded as temporarily disappearing during séances in which physical phenomena has taken place.

Development Circle: A group of people working together over an extended period of time to develop their individual mediumistic abilities and/or to establish objective contact with cooperating personalities. The circle may be formatted as a class which includes meditation and learning how to work with psi and communicate across the veil. This is an excellent format for developing mediumship and healing intention, as many techniques can be used to familiarize the student with psi functioning.

In some cases, the circle works in total darkness with the intention of using ectoplasm [435] or psi for materialization (ideoplastic structures [104]). A classroom circle may meet for only a few months, but one seeking trans-etheric phenomena in the séance room may sit for many years before establishing contact. This is one reason why there are so few materialization mediums still practicing today.

Differentiation: The mechanism by which aspects of reality are formed. See the Principle of Aspectation [63] in the Organizing Principles Discourse.

Direct Drawing, Painting and Writing: This is accomplished without human contact with the materials used. The materials combine directly to produce the artistic result. The process is believed to be under the control of a discarnate personality. It is probable that the effect is accomplished via the person acting as a practitioner. See Physical Mediumship [32] in the Trans-Survival Hypothesis Discourse.

Direct Voice: A voice heard in the air without any visible source. The voice often emanates from a trumpet, which serves as a condenser and amplifier for the voice. But with a sufficiently strong contact field [100], the voice can be heard from any part of the room. The theory of how this is done is that ectoplasm is withdrawn from the medium's or sitter's body to produce a larynx or ectoplasmic voice box.

Direct voice phenomena have been reported outside of the séance room as well. It is one of the possible explanations for EVP occurring on more than one recorder even though it was known that no one was speaking at the time. In such a case, the voice should be heard in real time by observers. See Physical Mediumship [32] in the Trans-Survival Hypothesis Discourse.

Discarnate: A person who has transitioned out of a physical lifetime. A personality that is not associated with a physical body.

Divination: Discovery of things that are hidden using various practices such as dowsing, astrology, numerology, palmistry, crystal gazing and various cards used for divination such as the Tarot. This is a form of psi functioning [356] as information acquisition using a device or object for attention entrainment.

Doorkeeper: See Control [431].

Dowsing: Using a rod, twig or pendulum to discover information such as locating water for a well or finding a lost object. Dowsing is often used as a means to help focus the mind for clairvoyance.

E

Earthbound Entity: A term often used to describe entities believed to have remained perceptually close to the physical aspect of reality because of habits, attachments or ignorance of their true state. Earthbound *spirits* are often thought to be responsible for haunting activity. Rescue circles seek to assist such personalities to complete their transition.

As directors of the ATransC, we have found no support for this concept and advise that it is inappropriate to think discarnate personalities require rescue. The concept of perceptual agreement appears to assure that a personality will be attracted to the aspect of reality with which it is in most agreement. A personality might perceptually associate with the physical aspect of reality, but as a purposeful act and not as an involuntarily *stuck* personality. We wonder how many loved ones come to say hello but have been rejected by a supposed clearing of the house.

Ectoplasm: A column of ectoplasm in front of the medium in the Felix Experimental Group circle. (184) ©

Ectoplasm: Typically, a white substance that is sometimes produced by the medium during séances. It is usually produced via the mouth but reportedly might come from any orifice. In red light, we have witnessed the FEG medium produce it in copious amounts, often pulling the white substance from his mouth with both hands.

Skeptics typically claim that ectoplasm is just cloth the medium put in his mouth before the séance. The material does tend to hold shape as cloth does, but it never smells of bile or any other bodily odor. The times I witnessed the FEG medium's preparation for a séance, he drank nearly a liter of orange-colored juice. Doing so guaranteed that any cloth he regurgitated to fake ectoplasm would have come out orange. It did not.

Once in good red light, the medium walked about the séance room letting each sitter feel a handful of ectoplasm by way of the medium touching the person's head with it. It was very wet, and water could be seen dripping from it, even though other times it only had a slight white, self-luminous quality and left no water mark where it was on the unpainted plywood floor.

The ectoplasm is produced as a tool. It can become semi-rigid and move on its own so as to lift a spirit light or trumpet to swing about the room. That is usually in darkness, but we have seen a hand rise from a pile of the substance as if it was a white glove [32] with no hand in it. We could see it wave about as if greeting the sitters; this, without the attention or attendant movement of the medium. Inspection of the cabinet before and after each séance revealed no mechanism to make the ectoplasm move. Certainly, he carried nothing with him.

Ectoplasm is able to assume the shape of a person, even to the extent of flesh tones. We have seen cameo-like faces [32] embedded in it, and sometimes those have been recognized as loved ones. I have seen the face of Hans Bender, the medium's etheric control.

Electronic Voice Phenomena or EVP: EVP are anomalous, intelligible speech produced in electronic devices. They may be heard as a real-time output from the recording device but are more generally heard on review of a subsequent recording. The voices are not acoustical so that they cannot be heard in the air. No currently understood physical processes account for the existence of EVP. They are considered a subset of Instrumental TransCommunication (ITC). Also see the Best Practices: Classification Study.

Energy Healing: See Healing Intention [440].

Entanglement: The concept of entanglement is used to indicate the binding rapport between aspects of reality which are linked by attention. The attention of the Core Intelligence of a life field on the life field's component parts established a link of integrating influence or entanglement. In the same way, the attention of an etheric personality on its avatar (physical body) during a lifetime creates a link of entanglement. An important example of entanglement is the link of influence binding aspect of a shared personalities into a collective as a cooperative community [64].

The nature of this connection is not clear but the most obvious indication of its nature is the difficulty a person has in disassociating from the body. Entanglement is binding, because once established, the resulting rapport is a living link of attention that is maintained by personality's intention to gain understanding.

Entanglement is thought to be a common characteristic of nested formative fields [98]. In this concept, the top organizing field of an organism is entangled with the related subfields composing the organism.

Etheric: Etheric is an old term referring to heaven. From the body-centric perspective, the etheric is that aspect of the reality field that is not physical. From the personality-centric perspective, the physical is an aspect of the etheric. The term has been traditionally used to describe the subtle, underlying media supporting reality.

As a generic term for the conceptual space of reality, the *Etheric* concept will probably be subdivided as more is understood about the greater reality. For instance, today, the parapsychologist's psi field is considered in the implicit Cosmology a subset of the etheric.

Etheric Communicator: All trans-etheric influences are thought to be under the intelligent and mindful direction of a personality. In the Trans-Survival Hypothesis, personality is native to the etheric, and is seen as a communicator. The personality initiating the influence may be that of a physical person or that of a discarnate person.

Etheric Field: Conceptually, a field represents a set of elements with common characteristics and bound together by a common influence. Etheric fields are regions of reality that can be modeled as a node or nexus (attractor [427]) associated with a set of related elements. It is useful to model reality as a single etheric field (reality field [98]) expressed by a first cause or Source. In this view, the reality field is the top etheric field, from which all other fields have been derived via aspectation [63]. In this way, the reality field has been differentiated as a hierarchy of nested etheric fields. See the Etheric Fields Discourse for more on etheric fields.

Etheric Studies: The study of all things etheric. This is not an individual science, such as biology or physics. It is a field of study concerned with the nature of the etheric aspects of reality. By comparison, physics and biology are sciences concerned with the physical aspect of reality and would be part of *Physical Studies*. The field of Etheric Studies is intended to complement physical science, as it is thought that discoveries in Etheric Studies will often extend the definition of physical principles while occasionally requiring the definition of new principles. Caution here, this is a term I have adopted. It is not widely accepted. See the Etheric Studies Commentary.

Experiencer: A person who witnesses or experiences phenomena.

F

Field: A set of elements with related characteristics which are bound into a system by a common influence (the attractor [427]). In metaphysics, a field is bound by the extent of influence the attractor is able to exert. Also in metaphysics, a field is not located by coordinates. Instead, they are located by association with other fields. See the Etheric Fields Discourse.

First Sight Theory: First Sight theory is a term used by James Carpenter. (13)

Fractals: See Life Field [444].

Frontier Subject: Any field of study that has not been embraced by mainstream society, especially mainstream science. A frontier subject may appear to be pure fantasy or it may have some supporting evidence. A frontier subject is one that has not benefited from having a well-trained and well-funded community vet the subject for reasonableness and supporting evidence. It is assumed that frontier subjects will eventually merge with mainstream understanding or be shown as unsupported belief. A similar concept is emergent science.

G

Gestalt: A system or set of elements that can be experienced as a whole; something that is understood as a whole, such as a gestalt realization resulting in sudden overall comprehension. The idea is that information in the etheric is managed as concepts which are apparently conveyed by gestalt bundles of related concepts. Our minds, which are oriented toward a building-block perspective of reality, then converts this *thought ball* [463] into understanding based on our worldview. See Thoughtform [99] in the Etheric Fields Discourse.

An interesting explanation for how practitioners and listeners might find EVP where there are none is seen in the Gestalt Laws of Perceptual Organization, which includes (168):

The Law of Proximity: Stimulus elements that are closed together tend to be perceived as a group.

The Law of Similarity: Similar stimuli tend to be grouped, this tendency can even dominate grouping due to proximity.

The Law of Closure: Stimuli tend to be grouped into complete figures.

The Law of Good Continuation: Stimuli tend to be grouped as to minimize change or discontinuity.

The Law of Symmetry: Regions bound by symmetrical boarders tend to be perceived as coherent figures.

The Law Simplicity: Ambiguous stimuli tend to be resolved in favor of the simplest.

Ghost: Any visitation from a discarnate personality. The presence of the personality might be sensed as a feeling or as the result of some psychokinetic influence such as lights turning on and off. In extreme cases, the personality manifests as an apparition which may be seen by more than one person and even photographed. It is the apparition one usually thinks of as a ghost.

In ghost hunting lore, a personality is said to be earthbound or *stuck* because it has unfinished business, does not know it is dead or has some emotional tie to a place or person. Ghost *recue* is a common practice in which a person attempts to convince the personality to move on.

There is speculation that some apparitions are really just the residual body consciousness or body image behaving as a recorder ghost by repeating more or less the same habitual actions. See the Life Fields Discourse.

Glossolalia: Speaking in a language unknown to the medium or person doing the speaking.

God: God, Source, First Cause, Infinite Intelligence and Prime Creator are terms referring to an intelligence which is considered the source of all of reality. *God* is a term most commonly thought of in the religious sense as a divine being. In the Implicit Cosmology, *Source* is used in lieu of *God* to avoid religious implications. See the Source Discourse.

Guide: A continual, benevolent, protective influence thought of as an advanced entity that voluntarily returns to the physical locale with a purpose in mind. Some guides say that they are only the mediums for even higher or more advanced entities. They may also be loved ones who remain perceptually close to help loved ones still in the physical.

H

Haunting: Paranormal disturbances usually attributed to the personality of a discarnate person. Various phenomena are reported as haunting events including moving objects, strange noises, lights, cold spots, unpleasant smells and the appearance of ghostly figures. See Trans-Etheric Influence [464].

Hauntings Investigation: An organized study of a reportedly haunted location following a protocol to use instruments and sensitives to sense the local psi field. Tests may include measurements of temperature, magnetism, broadcast electromagnetic radiation, light and sound. Often a mental medium or psychic will participate. Gathered data is compared with historical records and a report is made intended to show whether or not trans-etheric influences have been detected, and if so, how they relate to the known history of the location. This is known as ghost hunting when such investigations are more in the form of a club outing with no intention of publishing a paper.

Healing Intention: When a person thinks of another person, a link of influence is established between the two. In the context of the Implicit Cosmology, this is referred to as rapport. The nature of this rapport (the link) is determined by the person's attention, visualization and intention. When intention is focused on the wellbeing of the other person, the resulting expression of intended order is seen as a healing influence.

The various healing modalities are modeled as healing intention, but with unique points of view as to how healing is expected to occur. Some modalities are more about setting expectations and conditioning the sitter's worldview to accept healing. Others depend on a conditioned expectation that there will be some tactile/sensed effect such as a difference in heat, stiffness or vitality near the surface of the body, depending on the system.

Healing intention is thought to apply more to the avatar body mind (body image) but certainly has the potential of influencing the etheric personality. See the Healing Intention Study.

Home Circle: A group of friends who meet on a regular basis, usually in a person's home, to develop their mediumistic abilities. Some circles sit for specific purposes such as absent healing, meditation, psychic development, mediumship and physical phenomena. See Séance [459].

Hyperlucidity: A short-term change in behavior marked by the tendency to find phenomena *everywhere* despite considerable testimony to the contrary by peers. The concept comes from occasional reports of odd behavior exhibited by some people when they visit an emotionally charged place. This condition is known by such names as Paris syndrome and Jerusalem syndrome (65). In regard to paranormal phenomena, the effect is the experiencer's mistaken belief in experiencing paranormal phenomena such as reporting voices in recordings when there are none, seeing features in otherwise featureless visual media and misattributing thoughts as transcommunication. See Lucidity [151] and hyperlucidity [152] in the Mindfulness Discourse.

Hypothesis: A hypothesis is a statement that is assumed to be true for the sake of discussion. Validity of a hypothesis is dependent on verification of predictions derived from the hypothetical statement. A hypothesis is typically based on one or more assumptions that imply supporting concepts.

The Trans-Survival Hypothesis is designed to unify current understanding of personality, survival of personality after transition and the nature of reality in which personality continues after transition. It is not proposed as a truth, but is a speculation presented as a unified view of current understanding (specifically Tom Butler's understanding).

Testable predictions should be able to be derived from the hypothetical statement which, if shown to be true, should be used to evolve the hypothetical statement toward a well-tested representation of reality.

I

Ideoplastic Structure: The *Ideo-* prefix means idea or image. *Ideoplastic* is used here to mean the nature of objects formed in the creative process as mind-to-object expression. *Ideoplasy* means the process of formation. The term was coined by Max Vorworm, but is used here in the sense intended by the German parapsychology icon Dr. Freiherr von Schrenck-Notzing as *"Everything that can be remembered, also things the person has read or observed or is interested in, all these memories can find a materializing reflection in the influence field"* [of the séance room]. (169) See the Perception and Expression Discourse.

Imagined Reality: A special case of personal reality is involved in the creative process. The perceptual process involves comparing environmental information with worldview. As it is modeled here, part of the comparison process involves forming an imagined version of how the information may be understood and then comparing that with worldview. If it agrees, then the imagined version becomes the expression of the understanding to conscious self or as a psychokinetic influence to the environment. The person's personal reality (worldview) is changed to accommodate the expression. See the Perception and Expression Discourse.

Inattentional Blindness: The failure to notice a fully visible, but unexpected object because attention was engaged on another task, event or object. (170)

Incredulity Blindness: A category of agnosia [424], in which an audio or visual example of a phenomenon is not experienced because it is foreign to a person's worldview. There are at least two forms of the experimenter effect. One is the difference in experimental results *collected* by *believers* and *skeptics*. The second is due to the difference in reported results between a *believer* and a *skeptic*.

Induced After-Death Communication: The Guggenheims maintain After-Death Communications (ADC) are spontaneous, which is why Allan Botkin (171) refers to his EMDR (172) induced ADCs as induced or IADC™. Eye Movement Desensitization and Reprocessing (EMDR) is a therapeutic technique in which the practitioner guides a client to move his or her eyes back and forth while dwelling on an emotional memory. This is over simplification. In practice, the technique helps remove the emotional charge of a memory. Based on his work with war veterans, Botkin developed a technique in which clients have impressively vivid communication with transitioned loved ones, thus the *Induced* part of ADC.

Infinite Intelligence: In Spiritualism, Source, Prime Creator, First Cause, God, whatever you think of when you think of the influence that started reality in the first place, is known as Infinite Intelligence. Yes, it is correct to say that Infinite Intelligence is God. Spiritualists do believe in God. But there are a few important differences in the way Spiritualists understand God. Most importantly, Infinite Intelligence is not a father figure that offers rewards and punishment to his children. See the Source Discourse.

Informed: *Informed* is used here to mean *based on* or *influenced by*. Perception is informed by worldview. Informed visualization [133] is visualization based on what is known about the objective. In fact, all visualization is informed by the past and current understanding by way of worldview. See the Hypothesis of Formative Causation [103].

Instrumental TransCommunication (ITC): Communication across the veil with the aid of technology. Trans-etheric influence is a more global term meaning communication between the physical and the greater reality with or without devices. ITC occurs via technology as images, voices and text messages initiated by communicating etheric personalities.

To be ITC, phenomena needs to be objectively experienced, meaning that they are sufficiently substantial to be recorded in physical media that can be witnessed by others. And, technology of some form is employed to enable the phenomena.

Paranormal experiences that are witnessed by one or more people as mental perception, such as a shared deathbed vision, are not considered ITC because they cannot be recorded. While probably real, in a pragmatic sense, they are not considered objective. Poltergeist and other psychokinetic activity, direct voice and materializations that are able to be recorded satisfy the ITC requirement, but unless they involve technology to manifest, they would not be considered ITC. See the Audio ITC and Video ITC Studies.

Intended Order: This is one of the foundation concepts of the Implicit Cosmology. A common factor in ITC is the transformation of broad-spectrum, chaotic energy into intelligently intended features such as the voice in EVP and faces in visual ITC. Current research is showing that output of random event generators increase in order when in the presence of strongly expressed intention. There is evidence in precipitation phenomena, such as spirit art, that the precipitation material undergoes a transformation from chaotic energy to intended order as raw material is brought via the etheric (apported [425]) and organized as informed by the imagined outcome. This research suggests that the basic element of the creative process is the emergence of intended order. See Intended Order [136].

Intention Field: Expression of intention accumulates as an intention field which represents all of the imagery, thoughts, understanding and beliefs associated with the expression. Intention fields persist as long as the personality maintains attention on the intended outcome. This is a form of etheric field. See Intention Field [100] in the Etheric Fields Discourse.

Interested Observer: A person is considered an interested observer or witness if the person is present for an experimental session, knows about it or in some way has attention on the session. In the study of trans-etheric influences, it is not necessary for the person to be physically involved to have an influence on the process. At this time, there is no way to shield from attention, even if it is only subconscious.

J

(Empty)

K

Karma: A principle in Theosophy, Buddhism and Hinduism maintaining that we are the cause of our present conditions. These conditions are said to be governed by the Law of Action which holds that for every action there is a reaction with our deeds, acts and attitudes creating our future and making us ultimately responsible for ourselves.

From the perspective of the Implicit Cosmology, karma is probably more correctly thought of as understanding of the operation of reality, with *bad* karma being imperfect understanding.

Knowing: The current best practice for measuring accuracy of understanding is an iterative test. That is, if something is presumably understood, but that understanding changes with the next relevant experience, then it must be only belief. If, however, understanding stands up to repeated tests, then a degree of confidence about its *correctness* can be assumed. Here, that is referred to as knowing. Knowing must always be tempered with the certainty that the nature of reality is evolving and what is known today may be different tomorrow. Thus is the nature of living systems.

L

Levitation: The movement of a person or object, usually up or down. Not always paranormal, but as paranormal phenomena, levitation is presumably caused by the expression of intention as psychokinesis (or the older term, telekinesis). Levitation is a commonly observed phenomenon in physical mediumship, especially during séances. See Psychokinesis [457].

Life: In the Implicit Cosmology, reality is modeled as life fields and expressions of life fields. Life is modeled with the characteristics shown in the Functional Areas of a Life Field Diagram [xxiii]. See the Life Field Discourse.

Life Field: In the Implicit Cosmology, life is considered the fundamental component of reality. If modeled as a fractal, then Source is the top fractal.

Fractals occur as a top fractal, subsequent to which is a hierarchy of fractals of diminishing scale.

A fractal is structure, shape or form that is expressed in a similar way at all scales. The Sierpinski Triangle for instance, below, is formed by connecting the midpoints of each side of a triangle to form three triangles inside the first, and repeating that as scale allows. The triangle is a fractal which is repeated at many different scales. The veining of a river drainage system or branching of plants can also be seen as fractal patterns. See the Life Fields Discourse.

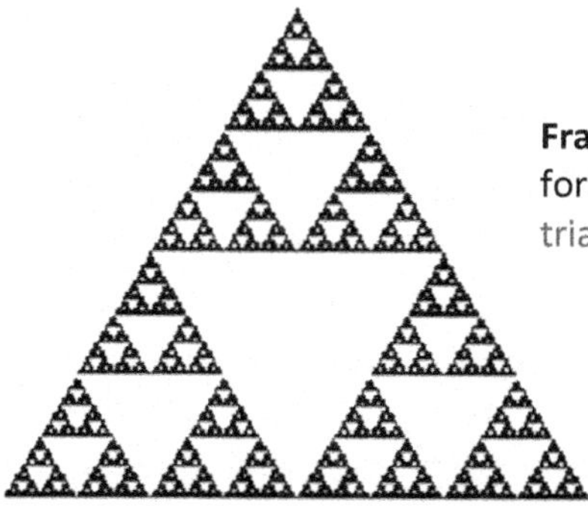

Fractal: The Sierpinski Triangle is formed as nested equilateral triangles. The triangle is a fractal.

Lucidity: The concept of consciousness has little meaning if talking about the person as an etheric personality. A more appropriate term is lucidity in the sense of a clear and open channel of awareness between the *I think I am this* (conscious self) and the true *I am this* intelligent core (personality).

It is becoming increasingly evident that a clear channel of communication between physical consciousness and the etheric personality is a factor for management of psi information. For instance, if people are informed about their environment on an unconscious level by way of psi sensing, then lucidity would be a measure of how aware the person is of that connection. More about lucidity in the Mindfulness Discourse.

Luminous Phenomena: These frequently occur in association with physical mediumship; however, luminous discharges have been reported during healing intention sessions, as well. A technique for detecting the influence of intentionality at a distance is to monitor luminosity of cucumbers. (173)

M

Materialization: In metaphysics, materialization is generally the appearance of any formation of physical material, apparently *out of thin air*. It is considered one of the phenomena of physical mediumship, and the objects are traditionally thought to originate first as ectoplasm [435] which then takes the shape of the materialized object. Such objects usually retain their physical shape for a short period of time, depending on intention. There might be many different objects or discarnate people materialized during a physical séance with no remaining evidence that they were present.

Materialization might be in the form of precipitation [454], in which physical material acts as the raw stuff acted on by intention to produce an image or object. In this case, the object may be relatively permanent. Lore has it that such precipitated forms sometimes require a period of time; say twenty-four hours, to stabilize before becoming permanent.

An object might be apported [425] to the present location, in which case, it would presumably have existed elsewhere in the physical, but then dematerialized, moved in some way in the etheric and then materialized in the present location. An apported object is thought to be a permanent form.

In the Implicit Cosmology, ectoplasm is considered just another form of materialization. The mechanism for this expression is the Attention Complex producing intended order. In this cosmology, all objects of reality are modeled as ideoplastic structures.

Matrixing: This is a term that has apparently been coined by one of the ghost hunting TV programs. It is used instead of the more commonly accepted term, pareidolia, to indicate that a person is mistaking an ordinary object or stimulus as paranormal because the person is not seeing it as it really is. In the interest of maintaining communication with people outside of the hauntings investigation community, pareidolia is the recommended term. See Pareidolia [452].

Matter-Through-Matter: In the work of the Society for Research in Rapport and Telekinesis (SORRAT) (174), this phenomenon has been produced on many occasions with a variety of items being removed or introduced into sealed jars and locked containers.

In physical séances, clothing that has been stitched on has been removed from mediums. Rings of wood, leather, iron and other matter have been linked together. Items have been reported passing through tables and in and out of closed séance rooms. See Apports [286] in the Nature of Transcommunication Study.

Medium: A person who is able to sense etheric personalities and convey

SORRAT: Film canister passing through the glass wall of an aquarium used as a *mini-lab* used by the Society for Research in Rapport and Telekinesis. (149)

information from them to people in the physical. The task of a medium is to be as clear and open a channel as possible to avoid coloring [430] the message. This is usually accomplished by the medium entering into trance. Inspirational speaking, for instance, is usually accomplished while the person is in a light trance. Mediums giving *spirit greetings* in a Spiritualist meeting and mental mediums giving full readings also typically work in a light trance. A physical medium producing materialization phenomena is thought to be in a more or less complete or deep trance in which the medium's personality has *stepped aside* so that the personality of the communicating personality may directly communicate using the medium's physical body.

Mediumship: Functioning as a medium. See the Mediumship Study.

Metaphysics: In philosophy, the examination of the nature of reality, including the relationship between personality and the greater reality. Models used to describe a set of metaphysical concepts are referred to as a cosmology [432]. In Etheric Studies, metaphysics is the study of an ordered set of concepts [430] which are described in a cosmology.

Mind: The Implicit Cosmology is a model that has been developed around functional areas of a person as a nonphysical personality entangled with a human in an avatar relationship. In this context, mind is the collection of functional areas described in the Life Field Complex Diagram [xxv].

The term, *consciousness* is used as a continuum from fully consciousness awareness to unconscious so that *conscious self* represents the physical point of view of a personality that is entangled with a human in an avatar relationship (body-centric).

The mental processes associated with perception and expression are modeled in the Perception and Expression Diagram [xxiv]. In that, the rules of perception are thought to agree with the proposed corollaries in First Sight Theory. (13)

A person in a light trance might be fully aware but a person in a deep trance would be unconscious. Trance-like states presumably allow the person greater awareness along the continuum between conscious and unconscious. A sleeping person is considered unconscious; however, if the person has achieved some degree of lucidity, he or she may have greater access to normally unconscious processes.

Mind Tensor: In the context of etheric space, the term can be used to signify how a life-field navigates amongst etheric fields. The Perceptual Loop is thought to be incapable of processing information that is very different from that which is in worldview. This is the foundation of the Principle of Perceptual Agreement [67].

According to the Random House Kernerman Webster's College Dictionary. (2010), a tensor is a mathematical entity with components that change in a particular way in a transformation from one coordinate system to another. Tensors are becoming an important tool in the study of large-scale quantum fields. The term is not used in this book because it overly complicates the description of this model.

Modes of Transcommunication: There are two fundamental forms of transcommunication. A person can initiate communication across the veil via an expression of intention or instrumentally as in EVP [292]. This form of transcommunication is considered **induced**. When communication is induced via the expression of intention and without the use of devices, it is considered **direct**. When instruments are used such as for EVP, the communication is considered to be **instrumental**.

For most people, communication across the veil is **spontaneous**, in that the etheric communicator might approach the person or there might be contact with a thoughtform as a result of an environmental influence that stimulates the creative process. Spontaneous transcommunication can be either direct as in mental impressions or instrumental as in unexpected voices in recording media.

Morphic Field: According to Rupert Sheldrake (9), this is *"a field within and around a morphic unit which organizes its characteristic structure and pattern of activity. Morphic fields underlie the form and behavior of holons or morphic* *units at all levels of complexity. The term morphic field includes morphogenetic, behavioral, social, cultural, and mental fields. Morphic fields are shaped and stabilized by morphic resonance from previous similar morphic units, which were under the influence of fields of the same kind. They consequently contain a kind of cumulative memory and tend to become increasingly habitual."* See the Etheric Fields Discourse.

Morphic Resonance: According to Rupert Sheldrake (9), this refers to *"the influence of previous structures of activity on subsequent similar structures of activity organized by morphic fields. Through morphic resonance, formative causal influences pass through or across both space and time, and these influences are assumed not to fall off with distance in space or time, but they come only from the past. The greater the degree of similarity, the greater is the influence of morphic resonance. In general, morphic units closely resemble themselves in the past and are subject to self-resonance from their own past states."* See the Etheric Fields Discourse.

Morphogenetic Field: See Morphic Fields, above.

N

Natural Law: Natural Law is defined as an *"ascertained working sequence or constant order among the phenomena of nature"* (NSAC Spiritualism (2)) In concept, Natural Laws include all of the principles governing the operation of reality. They are thought to be the same everywhere. Understanding the principles of Natural Law permits the person to live in accordance with the operation of reality, which is thought to be the objective of living. The degree to which a person understands and is able to live in accordance with the principles is a measure of the person's progression [73] which is the degree to which a person understands the *true* nature of reality.

There is little agreement about the principles of Natural Law. For instance, from a metaphysical view, Organizing Principles are described in fundamental concepts such as aspectation and perceptual agreement. In sociological views, the Golden Rule *Do unto others as you would have them do unto you* and *like attracts like* are usually included in lists of Natural Law. There is also considerable confusion about the scale or context to which principles apply. So to be specific, in Etheric Studies, Organizing Principles are described in terms of fundamental concepts within the context of Reality, Formation and Personality.

Some principles such as the Principles of Aspectation, Life Field and Perceptual Agreement appear to be universal. However, it is important to note that the Implicit Cosmology assumes Source is learning about itself via aspect personalities. As such, Source's nature is expected to evolve as understanding is gained, meaning that the Organizing Principles also evolve. This is a fundamental concept of the Hypothesis of Formative Causation. (9)

New Ager: The idea of a new age begins with the idea that, in terms of astrology, we are entering into the utopian Age of Aquarius. There is no single organization that can be identified as *The New Age Movement.* It has become a worldwide community identified by a common point of view. New Agers are typically interested in developing personal potential, understanding nature and living as a good citizen of the universe. Beliefs include a universal god of which everyone and everything is part, personal responsibility, respect for the life in all things and the connectedness of all things. Being seen as a New Ager is perhaps a badge of honor because it signifies a personal desire to be the best citizen possible in Mother Nature's creation. A person who is seen as a New Ager is not necessarily a person who believes in transcommunication, survival of personality or a greater reality.

Nonlocal: Ubiquitous or everywhere at once. It may be correct to say that *everywhere is here*. This is a physical space perspective. In the etheric, movement is accomplished by changing one's mind. See NonLocal and Nontemporal [94] in the Etheric Fields Discourse.

Nonphysical: See Etheric [437].

Nontemporal: The concept that all time is now in etheric space. There is some indication that presentiment, the physiological reaction to an event before it actually occurs, is best explained with First Sight Theory [126]. Presentiment is a short-term phenomenon while most reports of precognition indicate a long-term effect. In that, the concept of emerging thoughtforms representing potential futures might better explain the effect.

A second factor is the concept of potential. Rather than precognition as sensing the future, there is better support for the idea of sensing the potential of a particular future. As such, a nontemporal characteristic of etheric space is probably incorrect. See the NonLocal and Nontemporal [94] in the Etheric Fields Discourse.

Normalize: Normalization is the process of reducing two or more different sets of information to a common denominator that can be considered the root for both sets. For instance, conversation about car races and a different conversation about a Sunday drive. The common denominator is car. We automatically find car in our memory and use it as a reference to make sense of the conversations.

O

Object of Reality: In the Implicit Cosmology, reality exists in two forms: life fields which are morphic fields for physical organisms, and expressions of life which are ideoplastic structures. Expressions are described in terms of thoughtforms that have been given substance in the creative process, where *substance* only has meaning in the venue in which it is imagined. See the Perception and Expression Discourse.

Objective Phenomena: Phenomena that have external reality and that can be perceived and observed by others. Objective phenomena are considered the physical manifestation of a subjective influence.

Ouija Board: A board with letters of the alphabet used as an aid for transcommunication. A pointer attached to a small, easily moved platform is placed on the board. One or more people acting as sitters in the circle lightly place their fingers on the platform. The platform is expected to move, presumably under the control of discarnate entities, to spell out messages one letter at a time. See Planchette [454].

Outlier: In statistics, an outlier is a data point that falls *very far* outside of the norm. The logic is that an outlier may be due to faulty measurement or some otherwise unidentified experimental error. Alternatively, one extreme measurement might give a false impression of what is average. As a general rule, outliers are discarded to avoid inappropriately biasing a dataset. It is noteworthy that Class A EVP are, by definition, outliers.

P

Paranormal: Subjects that are concerned with mind-body duality (nonphysical personality), the psi field, continuation of consciousness beyond bodily death (survival), the natural environment of mind in which personality survives (the etheric) and communication across the mind-body interface (transcommunication) are considered paranormal.

Such subjects as reincarnation, out-of-body experiences and near-death experiences are not addressed in this book. They are indirectly related to trans-etheric influences, but probably because of their subjective nature, they have little bearing on the study of things paranormal. The same is true of the many people who experience spontaneous After-Death Communication if the experience is not objectively shared by others.

Phenomena thought to be physical but not ordinary, such as UFOs, cryptozoological creatures and parallel realities would likely be considered paranormal, but they are not usually included in Etheric Studies, unless a psi field explanation is involved.

Paranormalist: This term has become useful for identifying people who are more than a little interested in paranormal phenomena, or who are studying or practicing some technique related to the paranormal. Examples of paranormalist interests include the study of psi phenomena (telepathy, remote viewing, telekinesis), healing intention (biofield healing, distant healing, healing prayer), hauntings investigation, transcommunication (mediumship, ITC, channeling) and survival (the etheric, personality, cosmology).

Paranormalist Community: The paranormalist community consists of anyone in some way interested in paranormal phenomena. The community is not homogenous, in that it is composed of a number of sub-communities of interest related only by shared interest in the phenomena.

Parapsychologist: Specifically, a person who is academically trained in the subject of parapsychology. Anyone with an advanced degree can be accepted into the parapsychological organizations as a full member. However, there is a problem with this, in that too many people claiming to be parapsychologists do not have the necessary training.

It is common for academically trained people to cross over from a different specialty, usually psychology, to conduct parapsychological studies. These crossover scientists should be considered citizen scientists, depending on the extent to which they inform themselves in the subject.

Parapsychology: A frontier subject; a field of study about the evidence for psi phenomena, such as telepathy, clairvoyance and psychokinesis, which are not explained by mainstream science. These phenomena are forms of human potential described as psi functioning. In the Implicit Cosmology, the study of trans-etheric phenomena is referred to as Etheric Studies, while it can be said that parapsychology is Psi Studies. See the Etheric Studies Commentary.

Pareidolia: A form of apophenia [424] (seeing patterns where there are none) in which the person finds meaning in images and sounds when there in none. Pareidolia and apophenia are terms used to describe a very real human tendency to mistake the mundane as paranormal. An example might be the deer hunter who mistakes tree limbs for antlers (buck fever) or seeing faces in clouds (Rorschach test). See Gestalt [438].

Peer: Frontier subjects seldom have a large community of subject matter specialists who are academically trained. In many cases, it is practitioners who have come to the subject from another field who may be the only qualified people. Yet mainstream culture demands that an academically trained person, even if the qualifications are in an unrelated field, should be given authority over the possibly more qualified practitioners who may not hold a degree. Of course, this is an untenable approach to understanding the field of study. In order to distinguish between different kinds of experience and expertise in Etheric Studies, a peer is identified as either an *academic peer* (trained in an accredited educational institution) or a *practical peer* (may be academically trained outside of the field or self-trained in the field).

Peer Review: The examination of a research report by others who are trained in the same field. There are no, or very few people working in Etheric Studies who are academically trained in a related subject. Consequently, review of articles for reasonableness of hypothesis, protocol and conclusions is necessarily accomplished with probably not academically trained subject matter specialists (practical peers). Because of this necessary departure from academic standards, the process of judging articles is better described as vetting rather than peer review. See Vetted [465].

Perception: In the Implicit Cosmology, perception is the result of the decision made in the Perceptual Loop as illustrated in the Perception and Expression Diagram [xxiv]. If the *Agree?* decision produces a *Yes*, the person will experience (perceive) the stimulus, probably more or less as it is. If the decision is *Maybe*, the person more likely experiences a version of what is intended based on worldview. A *No* decision will likely stop conscious perception of the information. See the Perception and Expression Discourse.

Perceptual Agreement: This is an often-used term in the Implicit Cosmology. The Principle of Perceptual Agreement [67] holds that *Personality must be in perceptual agreement with the aspect of reality with which it will associate.* The objective of progression toward greater understanding is to bring worldview (personal reality) into agreement with nature as it is and not as the person is taught. See the Perception and Expression Discourse.

Personality: Personality is defined here as the etheric essence of a person. From the perspective that a person is a physical body that has evolved in the physical and mind which has evolved in the etheric, personality is that part of a person that has evolved in the etheric. Personality existed before this lifetime and will continue after this lifetime. It can be thought of as an etheric field containing the essence of who the person is including a measure of the person's understanding which is realized as progression and an inherited urge to acquire understanding by way of experience.

Phenomena: In the context of Etheric Studies, any expression of intention into physical form; a trans-etheric influence. Examples include mediumship, clairvoyance, healing intention, ITC and psychokinesis.

Planchette: A small, table-like device that can be easily moved over a surface containing letters, words or other information. In application, one or more participants lightly place fingers on the device. It is thought to move under the control of an etheric personality. A planchette might also have a pencil attached so as to mark a paper when it is moved. See Ouija Board [451].

Point of view: Used to describe the perspective or viewpoint in literature and art. In terms of living a life, people also have a point of view, and just as the author intends the reader to come away with a specific perspective of the story, so does a person's point of view direct us toward understanding a specific perspective of experiences. See the Point of View Commentary.

Potential Field: Events are accompanied by a probability signature referred to here as a potential field. In this concept, an event might exist in the etheric as a potential to occur. Each potential field represents the possible characteristics of an event, so that an event that has occurred may have been preceded by many potential fields representing different or optional potential outcomes.

Practitioner: In Etheric Studies, a person conducting a recording session for EVP or ITC; a person who applies principles intended to cause phenomena; may be instrumental or it may be direct as in mental mediumship. See Modes of Transcommunication [448].

Precipitation: By definition, a precipitate is something that condenses out of a solution and falls as a deposit. Rain or snow is one example. A precipitation medium is one who is able to work with etheric helpers to produce physical objects, via precipitation. A good example of this is the Reverend Hoyt Z. Robinette as described in the article, "Sitting with Hoyt Robinette." (31) Currently available evidence indicates that the precipitate originates in the physical and is in effect, dematerialized, transported via the etheric and rematerialized according to the intended order expressed by the medium (may be under the direction of communicating entities). In effect, this is the same process thought to be involved with apportation of object, except that the precipitate becomes objective in a new form. See Materialization [445].

Precognition: Knowledge of something in advance of its occurrence, especially by extrasensory perception; clairvoyance.

Presentiment: The sense that something is about to happen. In parapsychology, studies indicate that a person will unconsciously react to a stimulus just before it is delivered. When the response is in reference to an event that occurs in the near future (seconds, minutes), it is referred to as presentiment; however, if the event is relatively more in the future (hours, years), this is usually referred to as precognition.

There is growing understanding that every event has a psi signature, and that the body unconsciously senses this signature. Research now suggests that this signature may also relate to a growing potential for the event to occur. Presentiment and precognition seem to support the concept of a nontemporal nature of reality, meaning that all time is now. However, sensing that something is about to happen may be sensing the potential for the event to occur, rather than clairvoyantly seeing into the future to experience the event. There seems to be more support for the idea that time exists and is not nontemporal.

Prime Imperative: This is a rather flamboyant way of saying that the one purpose everyone seems to share is to gain understanding from experience. This concept is fundamental to the Implicit Cosmology. That is, Source exists and reality is an expression of Source. Curious to understand the nature of its own expression (implications, the *so what*), it is speculated that Source expressed aspects of itself to explore reality and eventually return understanding of its nature. As aspects of Source, we share this curiosity and the urge to understand...everything. In effect, this understanding gained during a lifetime is our *payload* as we transition out of this lifetime and become more in agreement with Source. See Prime Imperative [61] and Progression [73] in the Ordering Principles Discourse.

Process: A process is an ordered sequence of events. In metaphysics, creation through process is an important alternative to the idea that a fully mature reality might spring into existence. The creative process is described as attention on the expression of an imagined outcome with the intention of making it so. See the Perception and Expression Discourse.

Progression: Progression represent the degree to which a personality satisfies its inherited urge to gain understanding. Progression can be described as spiritual maturity from a body-centric perspective. See Prime Imperative [61] and Progression [73] in the Ordering Principles Discourse.

Pseudoscience: Pseudoscience means *false science*. It is a derogatory term coined by skeptics (175) to label subjects with which they disagree. This disagreement is seldom based on known instances of bad science, but rather because, in the Skeptical view, the subject is not supported by orthodox science. This term is virtually always used in conjunction with efforts to convince an audience to dislike, mistrust or even fear the subject.

Pseudoskeptic: Pseudoskeptics or *false skeptic* is a term that refers to people who reject ideas without examination if they do not seem to be supported by mainstream science. Since *pseudoskeptics* proudly refer to themselves as *skeptics* and it is under that title that they are found on the Internet, the ATransC refers to them as *skeptics*. See the Skeptics Commentary

Psi: Term coined by J. B. Rhine to represent the unknown force involved in Extrasensory Perception (ESP) and psychokinesis. It is popularly used today to mean *psychic*. Psi represents processes and elements in nature or in the human personality, which appear to transcend the accepted limiting principles of science.

Psi Field: Etheric fields are described in parapsychology as a psi field. While physical energy is local and time dependent, the psi field is experienced as being everywhere at the same time (nonlocal). The term is used in the Implicit Cosmology as it is intended in parapsychology, but here, psi fields are modeled as interconnected etheric fields, each instance being associated with a life field's perception. See the Etheric Fields Discourse.

Psi Functioning: The expression of intention on the subtle energy referred to as *psi*. See psi functioning [356] in the Mediumship Study.

Psych-: This prefix is generally seen in psychology for *mind* or *mental*, which in paranormal phenomena, assumes the person is the direct causative agent when it has not been determined that such influence is not accomplished in cooperation with or by an etheric personality.

Psychic: Ability to access information by sensing the nature of the psi field. In the past, this ability has been referred to as Extra Sensory Perception or ESP, but more recently it is known as psi functioning. It is correct to say that a person is psychic or to refer to a psychic ability. A psychic may not necessarily function as a medium, meaning the person may not be intending to access information from a discarnate personality. A psychic may be able to foretell the future, excel in psychometry and be clairvoyant but would not say that this information was coming from an etheric personality. See the Mediumship Study for a discussion about the distinction between psychic and medium.

Psychokinesis (P.K.): Movement of an object via psi functioning. This is a term used in parapsychology. Telekinesis is an alternative term meaning to have an influence from a distance. See Telekinesis [463].

Psychometry: In psychometry, information about a place, object or the owner of the object is accessed by touching an object, an auric field or the local environment such as near a person or in a building. This is a form of psi functioning in which the practitioner senses information carried by objects or places in which their history is thought to be imprinted.

Q

(Blank)

R

Raps: A common feature of physical séances. Noises are heard as knocks, pops, cracks and bangs. Coded communication is possible. Current research indicates that the raps originate within the material, suggesting they are psychokinetically induced.

Rapport: Relationship, especially one of mutual trust or emotional affinity. Transcommunication messages have emphasized the importance of rapport within a circle wishing to develop physical phenomena.

The attention/intention relationship amongst life fields can be characterized as rapport. Likewise, the link of entanglement between etheric personality and its avatar and between a top organizing field and its subfields might be characterized in terms of rapport. See the Rapport [68] in the Ordering Principles Discourse and the SORRAT (174) and (149) references.

Reality: Reality is seen in the Implicit Cosmology as a fundamental characteristic of all there is. And here, *all there is* is defined as the life field referred to in this cosmology as Source. That would make Source's life field the reality field. While there is one reality, experienced reality is relative because it is perceived by individuals as personal reality.

If Source is just an aspect of an even greater reality, then that greater reality is well beyond the scope of this cosmology. Here, all things begin with Source as a basic assumption. See the The Nature of Reality Discourse.

Reincarnation: The rebirth of the personality in another physical body. Based on the Implicit Cosmology, it appears that some people may choose to reincarnate into this physical venue for learning; however, it is as likely that a person will decide to move on to a different venue for learning, depending on the need for experience. It should be noted that a past life regression is capable of calling up a thoughtform representing someone's lifetime that might help answer the sitter's needs to learn, even though it is not necessarily that person's past lifetime. This is a common explanation for why so many people have been important historical figure in their past.

Researcher: While an EVP practitioner conducts routine recordings for the collection of information, a researcher conducts experiments following a protocol designed to provide verifiable data. In frontier subjects, it is more often the practitioners who must conduct research.

Residual Energy: Living organisms are thought to emanate a form of energy sometimes referred to as the biofield. This is usually the heat, light and electromagnetic signature of biological processes. Life forms are also thought to have a psi field which is associated with their organizing field and mental expressions.

From the body-centric perspective, some characteristics of these fields are believed to accumulate in a location or in association with an object such as a ring. This residual signature is thought to be expressed in greater *abundance* when the person expresses extreme emotions. Thus, locations with long-time human occupancy, especially hospitals and theaters, are thought to have a strong presence of this residual field.

From the personality-centric perspective, a contact field is developed in relationship with a point of attention. This is not an energy-based explanation, but a focus of attention and rapport-based model. See Contact Field [100] in the Etheric Fields Discourse.

Retrocognition: Awareness or knowledge of facts of past events or persons acquired by means of psychic ability.

Retro Familiar Storytelling: Modification of an initial memory or sense based on feedback from the environment. In mediumship, this is seen when the medium elaborates on the message based on the sitter's feedback. For instance, the Medium might say "I see a big tree." The sitter might respond with "I like to read under a big tree." In Retro Familiar Storytelling, the feedback from the sitter might produce a response from the medium as "Yes, I see that. You like the warmth from the sun." This is coloring based on information that was not in the originally sensed message.

Any modification of how a person senses memory, a psi signal or visualization based on feedback from the environment should be considered a form of storytelling that possibly colors the original sense beyond its initial meaning. This effect is probably best seen as a sloppy mental habit that can be managed through mindfulness. See Retro Familiar Storytelling [352] in the Artifacts and Perceptual Error Study.

Revelation: Something revealed especially a dramatic disclosure of something not previously known or realized.

S

Science: Science is based on the point of view that reality is knowable. It is practiced as the scientific method. The objective is to produce understanding in a form that can be tested and applied as engineering solutions.

Scientism: The belief that science, the scientific method and work product is the only way to validate reality. In practical terms, scientism holds that, if something is not recognized by mainstream science, then it is not real and is, therefore, impossible. When people under the cloak of authority of science advise the public about any subject without analysis of available information about that subject, they are effectively practicing scientism. (106) Skeptics are, by definition, believers in scientism.

Scientific Method: A systematic approach to gaining understanding about a subject. There are a number of techniques commonly used for research. Statistical analysis, for instance, is used as a means of determining whether or not there is a significant difference between control and experimental results. Also common is meta-analysis which involves reviewing many research reports to identify trends.

It is becoming evident that techniques which are useful for physical phenomena may not be appropriate for the study of trans-etheric influences. As such, it should be cautioned that one size does not fit all. (176)

Séance: A group of people sitting for the purpose of obtaining psychic manifestations or for establishing transcommunication. The presence of an established physical medium is helpful, but in development circles, the group may seek to have one of their members emerge as a new physical medium.

Séances are typically conducted in a completely dark room; however, some choose to work in dim light or red light. See Home Circle [440].

Sitter: The attendees of a séance or sitting; a person who is the point of attention for a medium, healing intention or for psi functioning.

Skeptic: Perhaps the most important trait for anyone involved in the study of frontier subjects is the ability to maintain an open mind while practicing discernment. This attitude is sometimes referred to as skepticism; however, when a person is skeptical of something without reason or to favor a belief over objectivity, that person is popularly known as a skeptic.

Rather than practicing discernment, skeptics actively campaign to teach the public to see new thought as a hazard to society. As is already occurring in some governments, including the USA, this vilification of frontier subjects has the potential to cause social and governmental reaction that could at the least prevent further study and possibly provoke action harmful to people studying these subjects. (7)

A potentially useful alternative term for the kind of skepticism practiced by skeptics is scientism [459] which means the ideological belief that science (mainstream science) is the only authority on the nature of reality. This is where the idea comes from that *"If it is impossible according to (main-stream) science, then it cannot be."* The rest of this ideology is that anything not accepted by mainstream science must be pseudoscience [456].

Skepticism in itself is a healthy attitude so long as it is accompanied by open-minded investigation. Perhaps a better term for open-minded skepticism is *discernment*. See the Skeptics Commentary.

Skeptic's Syndrome: Assumed knowledge which is usually based on faith, belief or popular wisdom but seldom based on empirical evidence. The syndrome manifests as the assumption that one's own understanding of a subject is more correct than that held by others without regard to evidence to the contrary. This attitude is maintained in the face of empirical evidence by making unsupported assumptions, belittling those who do not agree and attempting to associate the opposing view with previously discredited concepts. See Assumed Knowledge [426].

Source: Source is used in the Implicit Cosmology as the name for the initial cause of reality. This is the same concept as God, First Cause, Infinite Intelligence and Creator Personality. In this hypothesis, Source is credited with the same attributes identified with life fields [107] including self-determination, point of view, purpose and an urge to understand local reality. In fractals, Source is the top fractal. See the Source Discourse.

Spirit: As common usage, spirit refers to the vital principle or animating force of life; the indestructible essence of self-conscious life. The word *spirit* is often used to describe a discarnate person. The more correct term for a person who has transitioned from a physical lifetime is a person *in spirit* or a discarnate person. The term, *spirit*, is closely related to religious belief, and so is sometimes avoided when discussing metaphysics.

When the healing influence is said to be derived directly from Source (God) and undifferentiated by any other personality, it is said to be coming from Spirit. Here *Spirit* is written with a capital *S*. In common usage, *spirit*, written with a lower case *s* is a reference to the psi field as it is influenced by intentionality. See Etheric [437].

Spiritual: Anything relating to Source (Infinite Intelligence, God, Prime Creator or First Cause). It is reasonable and proper to refer to the purpose and origination of reality as spiritual. A person expressing the high ideals of citizenship in the greater reality might be referred to as spiritual or being spiritual. (Yes, the ideals themselves are an abstraction, but the point should be clear that a conscious effort to be a good citizen is paramount.)

Spiritual Healing: See the Healing Intention Study.

Spiritualism: A system of belief based on the acceptance of the survival hypothesis, transcommunication and healing intention. It assumes that all trans-etheric influences including acquisition of information via mediumship and healing intention exists as the cooperation between a Medium and a discarnate helper. The National Spiritualist Association of Churches (NSAC) (2) defines Spiritualism as the *"Science, Philosophy and Religion of continuous life, based upon the demonstrated fact of communication, by means of mediumship, with those who live in the Spirit World."*

Spiritualism, especially NSAC Spiritualism, is listed here because it does not require members of the community to believe anything beyond the Declaration of Principles (10), and does offer a community of likeminded seekers. It maintains nine principles of the Declaration of Principles constitute the NSAC tenet, which are consistent with the Implicit Cosmology. The only belief is the existence of Infinite Intelligence. The rest is acceptance of survival, an assumption of personal responsibility and the expectation that a person will seek to understand the nature of reality and live in accordance with that understanding.

Stochastic Resonance: Stochastic resonance (131) is an effect that results in a weak signal being amplified in a nonlinear system when a large noise background is applied. A common factor in many forms of trans-etheric influence is the presence of noise. It is speculated that a weak telekinetic influence (signal) is made stronger via stochastic amplification. In electronic circuits, the active regions of tubes and transistors provide the necessary nonlinear condition for amplification.

Storytelling: Claiming what was experienced is phenomenal by telling a reasonable story that seems to make the claim sensible. This is a way of leading a witness into believing an event or example is paranormal even though it is mundane. In EVP, something to the effect of *"I asked where my book was and recorded, 'ten ... other ...yes.' The entity was telling me I was only ten feet from it and that it was somewhere other than where I was looking and that 'yes,' I will find it."* In fact, the possible EVP had nothing to do with the question, but the story seemed to make it sound like a good answer. See the Artifacts and Perceptual Error Study.

Subjective Phenomena: See Conceptual [430].

Subject Matter Specialist: A person who has more than average experience as a practitioner; may or may not be a researcher, experience consistent with appropriate science [425]; participates in peer review; sometimes a citizen scientist [429]. (177)

Subliminal: Below the threshold of conscious perception; not sufficient to produce conscious awareness but able to cause a response by way of the Attention Complex. See the Perception and Expression Discourse.

Subtle Energy: See Etheric [437], Etheric Field [437] and Psi Field [456].

Super-Psi Hypothesis: A theory designed to explain observed psychic abilities, but from the perspective of psychology. Rather than allowing for survived personality, it depends on the belief that, once created, information exists forever. The hypothesis proposes a vast reservoir of residual conscious energy (fossil memory) which is accessible by people who exhibit the ability to psychically access information.

This hypothesis depends on mind as a product of brain, but which is then independent of brain and continues after the brain dies as a *survived* memory. It also depends on the existence of a psi field. See the Trans-Survival Hypothesis Discourse.

Survival: In the context of metaphysics, the continuance of the personality after the change called death. It is the fundamental doctrine of Spiritualism [461]. See the Transition Experience [166] in the Progression, Teaching and the Community Discourse.

Survival Hypothesis: See the Trans-Survival Hypothesis Discourse.

T

Telekinesis: The movement of objects with the influence of intention; have an influence from a distance; psi functioning. Also known as Psychokinesis [457].

Thought Ball: Robert Monroe (53) described the method of communication between people he encountered during his out-of-body travels as an exchange of thought balls he called *rotes*. Thought balls may be thought of as thoughtforms, or more specifically, intention fields. The term is useful in describing an exchange of information that is sensed as a gestalt [438] impression that includes intention, feelings and imagery as an instantaneous experience. See Intention Fields [100] in the Etheric Fields Discourse.

Thoughtform: Thoughtforms are treated as etheric fields that function as a container for information. In their simplest form, they are a collection of related information bound by or attracted to a central theme. The intended outcome (expression) of the creative process is generally thought of as a thoughtform which acts as a formative influence for physical action (motion, voice), conscious thought or an ideoplastic structure. See the Etheric Fields Discourse.

Trance: A spontaneous or induced sleep-like condition or altered state of consciousness that permits the subject's body to be used by a discarnate personality for communication. Trance may be very light with little difference than consciousness, either outward or inward. It may be very deep with the medium totally unaware of the surroundings or the messages being spoken via the body. Upon coming back to consciousness, the person so entranced may remember nothing of what has happened while in the trance state.

Trans-: Across, from there to here. Used here as across the interface from the greater reality to the physical aspect of reality. *Etheric* is used as shorthand to indicate the greater reality.

Transcommunication: A trans-etheric influence that is intended to be communication; trans-etheric communication. Not all trans-etheric influences are considered communication. Some, such as residual influence phenomena may simply happen without regard to who is experiencing the event.

Ernst Senkowski (129) coined the term Instrumental TransCommunication or ITC to describe meaningful anomalies including extraordinary voices, computer text, recorded images and images found on video displays which are suggestive of a continuance of life after death. (Because of common usage in this field, terms with a *trans-* prefix can be used without the hyphen.) Note that after-death communication, EVP and mediumship should be considered forms of transcommunication.

Trans-etheric: From the etheric aspect of reality to the physical aspect of reality. The *trans-* prefix is always used here to mean *across the veil*.

Trans-etheric Communication: See Transcommunication above.

Trans-etheric Influence: The expression of intended order from the etheric to the physical when an etheric personality communicates with a physical person or when a hauntings event occurs. All trans-etheric influences are considered conceptual influences producing objective effects. See Transcommunication above.

Transfiguration: Recognizable portrayals of discarnate persons are formed over the medium's features. This phenomenon usually takes place in dim light as ectoplasm is thought to be used to produce the transformation.

Transform EVP: In ITC, available physical energy is transformed into words or images. That is, the resulting voice in EVP has much the same frequency profile as the background noise. The physical process thought to be involved is stochastic amplification [276] of a weak signal imposed via the psychokinetic expression of intention; active component such as a transistor.

Transition: A term used to describe the process of *dying* out of this physical aspect of reality (this lifetime) and the return of the conscious focus of attention to the more natural nonphysical environment. Transition is a more accurate term for what is currently called *death*.

Trans-Survival Hypothesis: A theoretical model designed to put current understanding about the continuation of personality after transition (physical death) into a form that can be used to predict outcomes of future research. This is the ATransC version of the survival hypothesis used in parapsychology. It is rather different from the parapsychological version because it incorporates what has been learned in the study of transcommunication.

This hypothesis is an attempt to identify objective studies which support the Survival Hypothesis over the Super-Psi Hypothesis, and, of course, the Physical Universe Hypothesis. The Implicit Cosmology is designed to provide a model of reality that is based on the implications of the Trans-Survival Hypothesis. See the Trans-Survival Hypothesis Discourse for a full explanation

U

Understanding: The presumed objective of progression is returning to Source an understanding of its true nature. Understanding is resident in worldview. It informs perception by way of the Perceptual Loop, and is therefore a controlling factor for a person's personal reality. In practice, understanding decides what of reality a person is able to experience (Perceptual Agreement). See the Progression, Teaching and the Community Discourse.

V

Vetted: In Etheric Studies, *peer review* is used in the sense of vetting a manuscript, report or article with emphasis on reasonableness of assumptions, hypothesis and conclusions. A properly vetted article will have been commented on by people with relevant experience and training and those comments will have been incorporated into the document. The comments, reviewers and their relevant qualifications should be noted as part of the public record. See Peer Review [214] in the Science Commentary.

W

Worldview: This term is central to explanations about the nature of personality. Worldview is considered a functional area which is much like a database. It represents personal reality and is the controlling influence for perception. See the Life Fields Discourse.

X

Xenoglossy: Writing in tongues unknown to the medium.

Y-Z

(Blank)

References

1. Butler, Tom and Butler, Lisa. *There is No Death and There are No Dead.* Reno, NV: AA-EVP Publishing, 2003. ISBN: 0-9727493-0-6, atransc.org/resources/books.htm.
2. *National Spiritualist Association of Churches* (NSAC), nsac.org.
3. *Academy for Spiritual and Consciousness Studies, Inc.* (ASCSI), ascsi.org/.
4. Butler, Tom (et al). Best Practices and References. *The Collective,* atransc.org/bp/Collective.
5. Butler, Tom. Concerns with Wikipedia. *Etheric Studies,* ethericstudies.org/viewpoint/concerns_with_wikipedia.htm.
6. Butler, Tom. *Handbook of Metaphysics: A plain English Discussion of New Age Concepts.* The Christopher Publishing House, Hanover. ISBN: 0815804857, 1994, p. 385. ethericstudies.org/writing/books.htm#Handbook_of_Metaphysics
7. Biography of Wilhelm Reich. *The Wilhelm Reich Infant Trust,* wilhelmreichtrust.org/biography.html.
8. Stanford, Rex G. "Psychological Concepts of psi Function: A Review and Constructive Critique." Etzel Edited by Cardeña, John Palmer and David Marcusson-Clavertz. *Parapsychology: A Handbook for the 21st Century.* McFarland, 2015, Chapter 8, pp. 94-109.
9. Sheldrake, Rupert PhD. "Morphic Resonance and Morphic Fields." *Rupert Sheldrake.* sheldrake.org/research/morphic-resonance/introduction?.
10. About Spiritualism. *Spiritualist Society of Reno.* spiritualistsocietyofreno.org/about_spiritualism.htm.
11. Decision-making May Be Surprisingly Unconscious Activity. *Science Daily.* Max-Planck-Gesellschaft, www.mpg.de, 2008. sciencedaily.com/releases/2008/04/080414145705.htm.
12. Forman, Robert K. C. "An Emerging New Model for Consciousness: The Consciousness Field Model." *Studies in Neuroscience, Consciousness and Spirituality,* 2011, Vol. Volume 1, pp. pp 279-288. rd.springer.com/chapter/10.1007/978-94-007-2079-4_17
13. Carpenter, James C, Ph.D. *First Sight: ESP and Parapsychology in Everyday Life.* Rowman & Littlefield Publishers, 2012. ISBN 978-1-4422-1392-0 (ebook).
14. Guggenheim, Bill and Judy. *Hello From Heaven.* Bantam Books, 1996. after-death.com.
15. Butler, Tom and Lisa. "From Our Viewpoint—Proof of Survival." *Association TransCommunication.* 2003. atransc.org/articles/butler-survival.htm..
16. Rousseau, David. "The Implications of Near-Death Experiences for Research into the Survival of Consciousnes." *Journal of Scientific Exploration,* Vol. 26, No. 1, 2012, 2012, Vol. 26.
17. Butler, Tom and Lisa. "Faces in Reflected Light." *Association TransCommunication.* 2011. atransc.org/examples/faces_in_reflected_light.htm.

18. Butler, Tom and Lisa. "Faces in Light Reflected from Crystals." *Association TransCommunication.* 2006. atransc.org/examples/light_from_crystals.htm.

19. Downey, Margaret. "ITC experiments using Light Reflected from Water." *Association transCommunication.* 2007. atransc.org/techniques/techniques_downey_itc.htm.

20. Butler, Tom. "Perception of Visual ITC Images." *Association TransCommunication.* 2010. atransc.org/journal/visual_perception_study.htm.

21. Butler, Tom. "EVP Online Listening Trials." *Association TransCommunication.* 2008. atransc.org/journal/online_listening_trials.htm.

22. Bion, Stephan. EVPmaker. tonbandstimmen.de/evpmaker/index_e.htm.

23. Downey, Margaret. "Randomly Processed Speech Synthesis EVP." *Association TransCommunication.* 2008. atransc.org/examples/examples_voice_synth_1.htm.

24. Butler, Tom. "EVPmaker with Allophones: Where are We Now?" *Association TransCommunication.* 2011. atransc.org/journal/evpmaker_study_response.htm.

25. Emily Williams, Kelly. "Some Directions for Mediumship Research." *Journal of Scientifi c Exploration,* Vol. 24, No. 2, pp. 247–282, 2010, 2010.

26. Cunningham, Paul F. "The Content-Source Problem in Modern Mediumship Research." *The Journal of Parapsychology,* Rivier University, Department of Psychology. 76(2), 295-319., 2012. rivier.edu/faculty/pcunningham/Publications/CunninghamJP_Fall-2012-Vol-76-(2)-295-319.pdf.

27. Butler, Tom and Lisa. "Stewart Alexander. *Association TransCommunication.* 2011. atransc.org/circle/stewart_alexander.htm.

28. Butler, Tom and Lisa. "A Visit to the Felix Experimental Group." *Association TransCommunication*, 2010, ATransC NewsJournal, Vol. 29.

29. Butler, Lisa. "David Thompson and The Circle of the Silver Cord Séance." *Association TransCommunication.* 2009. atransc.org/circle/david_thompson_seance1.htm.

30. Butler, Lisa. The Silver Fox Circle. *Association TransCommunication.* 2012. atransc.org/circle/silver_fox_circle.htm.

31. Butler, Tom and Lisa. "Sitting with Hoyt Robinette." *Association TransCommunication.* 2010. atransc.org/circle/hoyt_robinette.htm.

32. Butler, tom and Lisa. "Viewpoint: David Thompson." *Association TransCommunication.* 2012. atransc.org/circle/thompson_seance_viewpoint.htm.

33. Copeland, Martha. *I'm Still Here.* AA-EVP Publishing, 2005. ISBN: 0-9727493-1-4. atransc.org/resources/evp_books.htm

34. "EVP Examples by Martha Copeland." *ATransC.* atransc.org/examples/martha_copeland_evp.htm.

35. Radin, Dean and Utts, Jessica. "Experiments Investigating the Influence of Intention Experiments Investigating the Influence of Intention." *Journal of Scientific Exploration,* Vol. Vol. 3, pp. pp. 65-79, 1989.

36. Mason, Lynne, Patterson, Robert and Radin, Dean. "Exploratory Study: The Random Number Generator and Group Meditation." *Journal of Scientific Exploration,* Vol. Vol. 21, pp. pp. 295–317, 2007.

37. Butler, Tom. "Understanding Balance by Understanding Personality Types." *Etheric Studies.* 2010. ethericstudies.org/concepts/(cc)tom_butler2010-balance_and_personality_types.pdf.

38. Bargh, John A. "Our Unconscious Mind." *Scientific American*, Scientific American, Vol. Volume 310, 2015.

39. Bargh, John A. and Morsella, Ezequiel. "The Unconscious Mind." *US National Library of Medicine, National Institutes of Health.* 2008. ncbi.nlm.nih.gov/pmc/articles/PMC2440575/.

40. Carpenter, James. "First Sight: A Model and A Theory of Psi." *James Carpenter.* drjimcarpenter.com/about/documents/FirstSightformindfield.pdf

41. Dean Radin, Leena Michel, Karla Galdamez, Paul Wendland, Robert Rickenbach. "Consciousness and the double-slit interference pattern: Six experiments." *DeanRadin.com.* 2012. deanradin.com/papers/Physics%20Essays%20Radin%20final.pdf.

42. Butler, Tom. "Examples of Other ITC Techniques: SORRAT." *Association TransCommunication.* 2005. atransc.org/examples/examples_other_techniques.htm.

43. Butler, Tom and Lisa. "Hans Bender's Message at Reno Séance's." *Association TransCommunication.* 2013. atransc.org/circle/hans_bender_speaks.htm.

44. Morgana, Aimee. "The N'Kisi Project." *Rupert Sheldrake.* 2002. sheldrake.org/nkisi/.

45. Holman, Peggy. "Chapter 1. What Is Emergence?" *Peggy Holman, Cultivating leadership for complex times.* 2014. peggyholman.com/papers/engaging-emergence/.

46. Harris, Thomas A. M.D. *I'm OK - You're OK.* July 6, 2004. Harper Perennial , 1969.

47. Butler, Tom. "The Emerald Tablet." *Association TransCommunication.* 2000. ethericstudies.org/writing/emerald_tablet.htm.

48. Littin, Shelley. "UA Study: Your Brain Sees Things You Don't." *University of Arizona,* 2013. uanews.org/story/ua-study-your-brain-sees-things-you-don-t.

49. Joshi, Sheila. James Carpenter's First Sight model and neurological damage-induced psi openings. *Blog: Neuroscience and Psi.* August 11, 2012. neuroscienceandpsi.blogspot.com/2012/08/james-carpenters-first-sight-model-and.html.

50. "Zero-Point Energy." *Calphysics Institute.* calphysics.org/zpe.html.

51. Global Consciousness Project: Meaningful Correlations in Random Data. noosphere.princeton.edu/.

52. "Glossary," *Rupert Sheldrake.* sheldrake.org/Resources/glossary/index.html.

53. Monroe, Robert. *The Monroe Institute.* monroeinstitute.org/.

54. Merrill, David W. and Reid, Roger H. *Personal Styles and Effective Performance.* Chilton Book Company, 1981.

55. Kinser, Patricia Anne. "Brain Structures and their Functions." *Serendip Studios.* 2000. serendip.brynmawr.edu/bb/kinser/Structure1.html.

56. Allen, James. *As A Man Thinketh.* Cornerstone Books, 1902.

57. *Psyleron Consciousness Technologies and Research.* psyleron.com/.

58. Butler, Tom. "Hermes Mercurius Trismegiatus, The Emerald Tablet." *Etheric Studies.* ethericstudies.org/writing/emerald_tablet.htm.

59. Builders of the Adytum. bota.org/.

60. Butler, Tom. "Relating John 14 in the Bible to Metaphysical Concepts." *Etheric Studies.* 2000. ethericstudies.org/writing/john14.htm.

61. Waite, Arthur Edward. *The Pictorial Key to the Tarot.* 1911.

62. Grimes, Roberta A. "How Gospel Analysis Can Be Combined with Afterlife Evidence and Traditional Science to Help Us Better Understand Consciousness". *Proceedings, The Academy for Spiritual And Consciousness Studies, Inc.* 2013.

63. Jane Roberts Learning Center. sethlearningcenter.org/.

64. Butler, Tom. "Ethics as a Personal Code for Mindfulness." *Etheric Studies.* 2016. ethericstudies.org/essays/code-of-ethics/.

65. Durst, Dr Rimona. "Jerusalem syndrome." *The British Journal of Psychiatry, 176: 86-90.* 2000. bjp.rcpsych.org/content/176/1/86.full.

66. Butler, Tom. "Phantom Voices." *Association TransCommunication.* 2012. atransc.org/journal/phantom-voices.htm.

67. Butler, Tom (et al). "Best Practices: EVP Witness Panel." *ATransC Collective.* 2013. atransc.org/bp/Witness_Panel.

68. Butler, Tom. "On Being a Good Witness." *Etheric Studies.* 2011. ethericstudies.org/culture/witness.htm.

69. Chopra, Deepak. "Only Spirituality Can Solve The Problems Of The World." *Huffpost Healthy Living.* 2010. huffingtonpost.com/deepak-chopra/only-spirituality-can-sol_b_474221.html.

70. Olivia, Olivia. "The Selfless Gene." *Manchester.edu.* 2007. users.manchester.edu/Facstaff/SSNaragon/Online/texts/201/Judson-SelflessGene.pdf.

71. "What is Chaos Theory?" *Fractal Foundation.* fractalfoundation.org/resources/what-is-chaos-theory/.

72. Milnor, Dr. John W. "Attractor." *Scholarpedia.* 2011. scholarpedia.org/article/Attractor.

73. Butler, Tom. "Radio-Sweep: A Case Study." *Association TransCommunication.* 2009. atransc.org/journal/radiosweep_study.htm.

74. Leary, Mark. "A Research Study into the Interpretation of EVP. *Association TransCommunication.* 2013. atransc.org/journal/radio-sweep_study2.htm.

75. *Forever Family Foubndation.* foreverfamilyfoundation.org/.

76. "Beatnik." *Wikipedia.* 2015. en.wikipedia.org/wiki/Beatnik.

77. Moore, David W. "Three in Four Americans Believe in Paranormal." *Gallup.* June 16, 2005. gallup.com/poll/16915/three-four-americans-believe-paranormal.aspx#1.

78. C.J. Efthimiou, R. Llewellyn. "Is Pseudoscience the Solution to Science Literacy?" *Cornell Univerity Library,* 2006. arxiv.org/pdf/physics/0608061v1.pdf.

79. Moster, Todd. "The Afterlife Files." *Kickstarter.com.* 2015. kickstarter.com/projects/afterlifefiles/the-afterlife-files.

80. Butler, Tom. "Comparing Mind-Body Models." *Etheric Studies.* 2014. ethericstudies.org/concepts/unfinished_hypothesis.htm.

81. Varvoglis, Mario Ph.D. "The Sheep-Goat Effect." *Parapsychological Association.* archived.parapsych.org/sheep_goat_effect.htm.

82. "Nothingness." *Stanford Encyclopedia of Philosophy.* 2007. plato.stanford.edu/entries/nothingness/.

83. Declan J. O'Donnell, JD and Sherry E. Bell, PhD. "Plato and Aristotle's Ether: Revisited and Redefined." *Bob Krone's Publications.* 2013. bobkrone.com/sites/default/files/Plato and Aristotle's Ether By Declan J. O'Donnell, JD and Sherry E. Bell, PhD.pdf

84. Einstein, Albert. "Ether and the Theory of Relativity," University of Leydon. *Technische Universität Hamburg-Harburg (TUHH).* 1920. tu-harburg.de/rzt/rzt/it/Ether.html.

85. MacRae, Alexander. *The Mystery of the Voices.* Self-published CD, Scotland, 2000.

86. "Relative Permeability." *HyperPhysics, Georgia State University.* 2014. hyperphysics.phy-astr.gsu.edu/hbase/solids/ferro.html#c5.

87. *Society for Psychical Research.* spr.ac.uk.

88. *Society for Scientific Exploration.* scientificexploration.org/.

89. *Parapsychological Association.* parapsych.org/home.aspx.

90. Butler, Tom and Lisa. "Welcome to the Association TransCommunication." *Association TransCommunication (ATransC).* atransc.org/.

91. Edited by Cardeña, Etzel, Palmer, John and Marcusson-Clavertz, David. *Parapsychology: A Handbook for the 21st Century.* McFarland, 2015. ISBN 978-0-7864-7916-0, Ebook ISBN 978-1-4766-2105-0.

92. "What is the Dark Night of Soul?" *Got Questions Ministry.* gotquestions.org/dark-night-soul.html.

93. McLeod, Saul. "Kolb - Learning Styles." *Simple Psychology.* 2013. simplypsychology.org/learning-kolb.html.

94. "About NSFNET." *NSFNET.* nsfnet-legacy.org/about.php.

95. "What is Anomalistic Psychology?" *Goldsmiths, University of London.* 2015. gold.ac.uk/apru/what/.

96. Butler, Tom. "Critiquing ITC Articles written by Imants Barušs." *Etheric Studies.* 2010. ethericstudies.org/viewpoint/failure_to_replicate.htm.

97. Braude, Stephen. "Investigations of the Felix Experimental Group: 2010–2013. 2014." *Journal of Scienti c Exploration, Vol. 28, No. 2, 2014.* academia.edu/7593753/Investigations_of_the_Felix_Experimental_Group_2010-2013.

98. Nahm, Michael. "The Development and Phenomena of a Circle for Physical Mediumship." *Journal of Scientific Exploration,* Vol. 28, No. 2, 2014 2014. anomalistik.de/images/pdf/sdm/sdm-2014-08-nahm.pdf.

99. *Paranormal Review.* Society for Psychical Research, Spring 2015.

100. Butler, Tom. "The Arrogance of Scientific Authority." *Etheric Studies Essays.* 2015. ethericstudies.org/essays/the-arrogance-of-scientific-authority/.

101. "The Belmont Report: Office of the Secretary, Ethical Principles and Guidelines for the Protection of Human Subjects of Research." *The National Commission for the Protection of Human Subjects of Biomedical and Behavioral Research.* 1979. hhs.gov/ohrp/humansubjects/guidance/belmont.html.

102. Irwin, Harvey J. "The Views of Parapsychologists: A survey of members of the Parapsychological Association." *Journal of the Society for Psychical Research*, 2014, Vol. 78.2, pp. 85-101. ISSN: 0037-1475.

103. Edwards, Chris and McCluskey, Neal. "Downsizing the Federal Government: Higher Education Subsidies." *CATO Institute.* 2009. downsizinggovernment.org/education/higher-education-subsidies.

104. "Jean-Baptiste Lamarck (1744-1829)". *University of California Museum of Paleontology.* ucmp.berkeley.edu/history/lamarck.html.

105. "IONS Directory Profile: Dean Radin." *Institute of Noetic Sciences (IONS).* noetic.org/directory/person/dean-radin/.

106. "Scientism Explained." *Everything Explained Today.* everything.explained.today/Scientism/.

107. "Pseudoscience." *Wikipedia.* en.wikipedia.org/wiki/Pseudoscience.

108. "Science and Engineering Indicator 2006: Chapter 7: Science and Technology: Public Attitudes and Understanding." *National Science Foundation.* 2006. nsf.gov/statistics/seind06/c7/c7s2.htm#c7s2l3.

109. Stemman, Roy. "Skepticism: The New Religion." *Spiritualist Society of Reno.* 2010. spiritualistsocietyofreno.org/articles/stemman_new_religion.htm.

110. "Wikipedia:Requests for arbitration/Pseudoscience." *Wikipedia.* 2009. en.wikipedia.org/wiki/Wikipedia:Requests_for_arbitration/Pseudoscience.

111. "Science and Engineering Indicator 2012: Chapter 5: Academic Research and Development: Financial Resources for Academic R&D." *National Science Foundation.* 2012. nsf.gov/statistics/seind12/c7/c7s2.htm.

112. "Wikipedia:Requests for arbitration/Fringe science." *Wikipedia.* 2009. en.wikipedia.org/wiki/Wikipedia:Requests_for_arbitration/Fringe_science.

113. Kelley, Charles R. Ph.D. "What is Orgone Energy?" *Kelley-Radix.* 1999. kelley-radix.org/downloads/what_is_orgone_energy.pdf.

114. Charman, Robert A. "An Unusual Form of Radiation has a Reproducible Effect in the Laboratory." *Association TransCommunication.* 2010. atransc.org/psi/unusual_energy.htm.

115. Weiler, Craig. *PSI WARS: TED, Wikipedia and the Battle for the Internet.* Amazon Digital Services, Inc., 2013. p. 270. ISBN: 1494389002, EAN13: 9781494389000.

116. "Rupert Sheldrake." *Wikipedia.* en.wikipedia.org/wiki/Rupert_Sheldrake.

117. Weiler, Craig. *The Weiler Psi.* weilerpsiblog.wordpress.com/the-big-ted-controversy-of-2013-part-1/.

118. "Wikipedia:Expert retention." *Wikipedia.* en.wikipedia.org/wiki/Wikipedia:Expert_retention.

119. "Wikipedia:Expert editors." *Wikipedia.* en.wikipedia.org/wiki/Wikipedia:Expert_editors.

120. "Learning styles. *Wikipedia."* en.wikipedia.org/wiki/Learning_styles.

121. "Wikipedia:Five pillars." *Wikipedia.* en.wikipedia.org/wiki/Wikipedia:Five_pillars.

122. "Wikipedia:Requests for arbitration/Paranormal." *Wikipedia.* 2007. en.wikipedia.org/wiki/Wikipedia:Requests_for_arbitration/Paranormal.

123. Butler, Tom. "ATransC White Paper on Transcommunication with emphasis on Electronic Voice Phenomena." *Association TransCommunication.* 2014. atransc.org/theory/white_paper-evp.htm.

124. "Welcome to MediaWiki.org." *Wikimedia Foundation.* mediawiki.org/wiki/MediaWiki.

125. *"Attribution-ShareAlike 3.0 Unported (CC BY-SA 3.0)." Creative Commons.* creativecommons.org/licenses/by-sa/3.0/.

126. "CZ:Paranormal Subgroup." *Citizendium.* en.citizendium.org/wiki/CZ:Paranormal_Subgroup.

127. "Search Paranormal Community." *Association TransCommunication.* atransc.org/links_community.htm.

128. Butler, Tom. "Clarity of Communication." *Etheric Studies Essays.* 2015. ethericstudies.org/essays/clarity-of-communication/.

129. Senkowski, Ernst. "Instrumental TransCommunication." *World ITC.* 1995. worlditc.org/c_07_senki_00_content.htm.

130. Woodward, Scott. "The Veil of Forgetfulness." *scottwoodward.org.* scottwoodward.org/veil.html.

131. Abbott, Mark D. McDonnell and Derek. "What Is Stochastic Resonance? Definitions, Misconceptions, Debates, and Its Relevance to Biology." *PLOS: Computational Biology*, May 29, 2009. ncbi.nlm.nih.gov/pmc/articles/PMC2660436/

132. Butler, Tom. "Transcommunication Via Telephone Calls." *Association TransCommunication.* 2012. atransc.org/examples/telephone_calls.htm.

133. Butler, Tom. "The Energy Profile of Transform EVP." *Association TransCommunication.* 2013. atransc.org/research/transform_evp_study.htm.

134. Wagner, Stephen. "Edison and the Ghost Machine." *About.com Paranormal Phenomena.* paranormal.about.com/od/ghostaudiovideo/a/edison-ghost-machine.htm.

135. Rogo, D Scott and Bayless, Raymond. *Phone Calls From The Dead.* Prentice-Hall, Inc., New Jersey, 1979.

136. Jürgenson, Friedrich. *Voice Transmissions with the Deceased* (German to English Translation, T. Wingert & G. Wynne, 2001). *Friedrich Jürgenson Foundation, Sweden.* 1964. http://atransc.org/resources/evp_books.htm.

137. Raudive, Konstantin. *Breakthrough: An Amazing Experiment in Electronic Communication with the Dead.* New York: Taplinger, Colin Smythe, Ltd., 1971.

138. Gullà, Daniele. "Computer–Based Analysis of Supposed Paranormal Voice: The Question of Anomalies Detected and Speaker Identification." *Association TransCommunication.* 2004. atransc.org/journal/gulla-voice_analysis.htm.

139. Butler, Lisa. "Precursor Sounds in Physical Phenomena." *Association TransCommunication.* 2002. atransc.org/articles/precursor_sound.htm.

140. *Report of an Anomalous Speech Products Experiment Inside a Double Screened Room.* MacRae, Alexander. Journal of the Society for Psychical Research, 2003.

141. Weisensale, Bill. "Eliminating Radio Frequency Contamination for EVP." *Association TransCommunication.* Spirit Voices, Issue 3, 1981. atransc.org/articles/weisensale_rf.htm.

142. Butler, Lisa. "EVP Examples from Lisa and Tom Butler." *Association TransCommunication.* atransc.org/examples/butler_evp_cal-neva.htm.

143. Butler, Lisa. Recording Thoughts of the Living. *Association TransCommunication.* 2001. atransc.org/articles/recording_thoughts_of_living.htm.

144. Butler, Tom (et al). "Best Practices: Control Recorder for EVP." *Assoociation TransCommunication Collective.* 2007. atransc.org/bp/Control_recorder_for_EVP

145. Braude, Stephen E. "Survival or Super-psi?" *Journal of Scientific Exploration,* Vol. 6, No. 2, 1992.

146. Talbot, Michael. *The Holographic Universe.* NY, HarperPerennial, 1991.

147. Estep, Sarah. *Voices of Eternity.* New York: Fawcett Gold Medal Book, Ballantine Books, 1988. ISBN 0-449-13424-5.

148. Audacity Audio Management Program. *sourceforge.net/download/windows.* audacity.sourceforge.net/.

149. Grattan-Guinness, I. "Real Communication? Report on a SORRAT Letter-Writing Experiment." *Journal of Scientific Exploration*, 1999, Vol. 13. citeseerx.ist.psu.edu/viewdoc/download?doi=10.1.1.518.5976&rep=rep1&type=pdf.

150. Rouvas-Nicolis, Catherine and Nicolis, Gregoire. "Butterfly Effect." *Scholarpedia.* 2009. scholarpedia.org/article/Butterfly_effect.

151. Radin, Dean, PhD. "Selected Peer-Reviewed Publications on Psi Research." *Dean Radin.com.* deanradin.com/evidence/evidence.htm.

152. "Facts About Telomeres and Telomerase?". *UT Southwestern Medical Center.* utsouthwestern.edu/labs/shay-wright/research/facts-about-telomeres-telomerase.html

153. "Darwin's Theory Of Evolution - A Theory in Crisis". *About Science.* darwins-theory-of-evolution.com/.

154. Monroe, Robert. *Far Journeys.* Main Street Books, 1985. ISBN 0-385-23182-2.

155. "Far Journeys: The Mystery of Loosh." *Nexus Magazine.* March 02, 2012. nexusmagazine.com/articles/doc_view/197-far-journeys-the-mystery-of-loosh.

156. *Franz Anton Mesmer.* anton-mesmer.com.

157. Bengston, William F., Ph.D. "The Energy Cure." *Bengston Research.* bengstonresearch.com/.

158. Abbott, Christopher J. "Breath Photography." *Association TransCommunivation ATransC News Journal,* 2013, Vols. Vol. 23, Num. 3.

159. Butler, Lisa. "Did Seth 3 Appear in the Bennetts' Crystal Experiments?" *Association TransCommunication NewsJournal,* 2006, Vols. Vol. 25, Num. 3.

160. Butler, Tom. "Locating EVP Formation and Detecting False Positives." *Association TransCommunication.* 2010. atransc.org/journal/false_positives.htm.

161. Butler, Tom. "The Formation of EVP." *Association TransCommunication.* 2009. atransc.org/journal/evp_formation.htm.

162. Butler, Tom (et al). "Best Practice: Sharing EVP." *ATransC Collective.* 2013. atransc.org/bp/Sharing_EVP.

163. Butler, Tom. "4Cell EVP Demonstration." *Association TransCommunication.* 2005. atransc.org/journal/4cell_experiment.htm.

164. P. Presi, D. Gullà, G. Gagliardi, G. Lenzi. "SFINGE Project: Instrumental Analysis of EVP Collected via a Sound-Psi Interaction." *Association TransCommunication.* 2006. atransc.org/journal/presi-analysis-of-evp.htm.

165. Browning, Rachel. "Relationship Between Physical Phenomena and Electronic Voice Phenomena in Séances with a Physical Circle." *Association TransCommunication.* 2013. atransc.org/journal/browning-evp-in-seances.htm.

166. Cardoso, Anabel. "ITC Journal." *Instrumental Transcommunication.* itcjournal.org/index.php.

167. Butler, Tom. "Debunking Survival Under Cover of False Academic Authority" *Etheric Studies Essays.* 2014. ethericstudies.org/essays/scientists-attack-medium/.

168. "Perceptual Organization." *ACM SIGGRAPH.* Association for Computing Machinery's Special Interest Group on Computer Graphics and Interactive Techniques. siggraph.org/education/materials/HyperVis/vision/percorg.htm.

169. Schrenck-Notzing, Freiherr von. "Phenomena of materialisation: a contribution to the investigation of mediumistic teleplastics." *Archive.org.* 1923. archive.org/details/phenomenaofmater00schr. 39999056769407.

170. *Scholarpedia.* scholarpedia.org.

171. Botkin, Allan. *Induced After-Death Communication.* induced-adc.com/.

172. EMDR. Eye Movement Desensitizatoion and Reprocessing Therapy. *EMDR Institute, Inc.* emdr.com/.

173. Kokubo, Hideyuki, Takagi, Osamu and Nemoto, Yasuyuki. "Spatial Distribution of Healing Power: Spatial Distribution of Healing Power." *Society for Anomalistics.* 2011. anomalistik.de/images/pdf/sdm/sdm-2011-10-kokubo.pdf.

174. Butler, Tom and Lisa. "Examples of Other ITC Techniques: SORRAT." *Association TransCommunication.* atransc.org/examples/examples_other_techniques.htm.

175. Butler, Tom. "Skeptic: On Practicing Healthy Skepticism and Being a Skeptic." *Etheric Studies.* 2011. ethericstudies.org/culture/skeptic.htm.

176. Butler, Tom. "The Scientific Method: As it Applies to The Study of Transcommunication." *Etheric Studies.* 2014. ethericstudies.org/culture/scientific_method.htm.

177. Jones, Andrew Zimmerman. "Decoherence and the Measurement Problem." *About.com Physics.* physics.about.com/od/quantumphysics/fl/Decoherence-and-the-Measurement-Problem.htm.

178. Myers, Steve. "Myers Briggs Personality Types." *Team Technology.* teamtechnology.co.uk/tt/t-articl/mb-simpl.htm.

179. David W. Merrill, Roger H. Reid. *Personal Styles and Effective Performance.* Chilton Book Company, 1981.

180. Weitz, Edmund Prof. Dr. "Mandelbrot set explorer." *Edi Weitz.* 2013. weitz.de/mandelbrot/.

181. Tymn, Michael E. "Imperator and Rector." *Survival After Death.* survivalafterdeath.info/spiritcontrols/imperator-rector.htm.

182. Tippit, Sarah. "Study: Brains Function in Clinically Dead." *NBC News, About,* 2003. abcnews.go.com/Technology/story?id=98447

183. Russell, Ronald. "*Far Journeys* by Robert Monroe: An Excerpt from the new book." *Intuitive Connections Network.* 1992. intuitive-connections.net/2007/book-monroe.htm. ISBN-10: 0385231822.

184. Mügge, Kai. "Felix Experimental Group." *Felix Experimental Group.* felixcircle.blogspot.com/.

185. McTaggart, Lynne. *The Intention Experiment: Using Your Thoughts to Change Your Life and the World.* Free Press, 2007. ISBN-10: 0743276957, ISBN-13: 978-0743276955.

186. *Near-Death Experience, Consciousness and the Brain.* Lommel, Pim Van. Taylor & Francis Group, LLC, 2006, World Futures, The Journal of New Paradigm Research, Vol. 62, pp. 134–151. ISSN 0260-4027 print / 1556-1844 online.

187. Kardec, Allan. *The Spirits' Book.* White Crow Books (Revised edition) 2010, 1857. ISBN-13: 978-1907355981, ISBN-10: 1907355987.

188. Dean Radin, Nancy Lund, Masaru Emoto, and Takashige Kizu. "The Effects of Distant Intentions on Water Crystal Formation: A Triple Blind Replication." *Journal of Scientific Exploration,* 2008. deanradin.com/papers/emotollproof.pdf

189. Butler, Tom. "Trans-survival Hypothesis." *Association TransCommunication.* 2015. ethericstudies.org/concepts/survival_hypothesis.htm.

190. Butler, Lisa. "Media Watch." *The National Spiritualist Summit.* 2001-2012.

191. Butler, Tom and Lisa. "Friedrich Jürgenson." *Association TransCommunication.* atransc.org/articles/jurgenson.htm.

192. Bion, Stefan. "EVPmaker." *www.Tonbandstimmen.de.* tonbandstimmen.de/evpmaker/index_e.htm.

193. Hefner, Alan G. "Akashic Records." *The Mystica.* themystica.com/mystica/articles/a/akashic_records.htm.

194. Butler, Lisa. "Voice in Death Same as that in Life." *Association TransCommunication.* 2005. atransc.org/journal/voice_analysis-sonia.htm.

195. Gale, Thomson. *Encyclopedia of Occultism & Parapsychology.* ISBN-13: 978-0810385702, 2000.

196. Blank

197. McTaggart, Lynne. *The Field.* Harper Perennial, 2003. ISBN: 978-0-06-093117-9.

198. Radin, Dean. *Entangled Minds.* Paraview Pocket, 2006. ISBN: 10-1-4165-1677-8.

199. Takagi, Osamu, et al. "Meditator's non-contact effect on cucumbers." *Academic Journals.* 2013. academicjournals.org/article/article1380558158_Takagi%20et%20al.pdf.

200. Butler, Tom. "Peer Reviewed Online Journals." *Etheric Studies.* 2011. ethericstudies.org/culture/online_journal.htm.

201. Moga, Margaret M., Ph.D, Bengston, William F. Ph.D. "Anomalous Magnetic Field Activity During a Bioenergy Healing Experiment." *Journal of the Society for Scientific Investigation,* 2010, Vol. 24. bengstonresearch.com/content_assets/docs/JSE_Geomagnetic_Moga_Bengston.pdf

202. "Conan Doyle's Return." *The American Museum of Photography.* Winnipeg, Manitoba, Canada, 1873-1935. photographymuseum.com/doylefalg.html.

203. *SpeakJet™.* speakjet.com. Magnevation LLC.

204. Kelly, Theresa M., MsD. *Models of Psi Mediation: A Classical and Quantum Approach.* 2011. scireprints.lu.lv/202/1/Models-of-Psi-Mediation.pdf.

205. Butler, Tom and Lisa, Editors. "ATransC Online Journal." *Association TransCommunication.* atransc.org/journal.htm.

206. Butler, Tom. "Pseudoscience." *Etheric Studies.* 2012. ethericstudies.org/culture/pseudoscience.htm.

207. Pell, Rich. "Audio illusions that will fool your ear (and brain)." *EE Times.* 2007. eetimes.com/author.asp?section_id=1&doc_id=1283413.

208. Gurstelle EB, de Oliveira JL. "Daytime parahypnagogia: a state of consciousness that occurs when we almost fall asleep." *US National Library of Medicine, National Institute of Health.* 2004. ncbi.nlm.nih.gov/pubmed/14962619.

209. Saw, James T. "Gestalt." *Design Notes: Art 104 Sesign and Composition.* Palomar College, 2000. daphne.palomar.edu/design/gestalt.html.

210, 211 Blank

212. "David Chalmers - the hard problem." *Conscious Entities.* consciousentities.com/chalmers.htm.

213. Nahmias, Eddy. "Why we have free will." *Scientific American.* January 2015. scientificamerican.com/article/why-we-have-free-will/.

214. Singer, Emily. "A Comeback for Lamarckian Evolution?" *MIT Technology Review.* 2009. technologyreview.com/news/411880/a-comeback-for-lamarckian-evolution/.

215. *American Society of Standards in Mediumship and Psychical Investigation.* assmpi.org/.

216. Nashawaty, Chris. "The Jerusalem Syndrome: Why Some Religious Tourists Believe They Are the Messiah." *Wired Magazine.* 2012. wired.com/2012/02/ff_jerusalemsyndrome.

217. Miller, Glenn M.D. "Voyager Syndrome." *Glenn Miller M.D.* glennmillermd.com/voyager-syndrome.

218. Blank

219. "Talk:Pseudoscience." *Wikipedia.* January 15, 2015. en.wikipedia.org/wiki/Talk:Pseudoscience#Learning_styles_as_pseudoscience.

220. Butler, Tom and Lisa. "The EVP of Tom and Lisa Butler 2." *Association TransCommunication.* atransc.org/examples/evpexamples_butler2.htm.

221. Butler, Tom and Lisa. "Faces on the Wall." *Spiritualist Society of Reno.* 1998. spiritualistsocietyofreno.org/articles/face_on_wall.htm.

222. *Interdisciplinary Laboratory For Biopsychocybernetics Research (Il Laboratorio).*

223. Butler, Tom. "The Monroe Way." *Association TransCommunication.* 2008. atransc.org/articles/monroe_way.htm.

224. Blank

225. Hogan, Craig Ph.D. *Self-guided Afterlife Connections.* selfguided.spiritualunderstanding.org/.

226. Zahradnik, Walter von Lucadou & Frauke. "Predictions of The Model of Pragmatic Information About RSPK." *Parapsych.org.* 2004. archived.parapsych.org/papers/09.pdf.

227. "Kaleidoscope." *LunaPic.* lunapic.com/editor/?action=kaleidoscope.

228. Butler, Tom (et al). "Best Practices: Classification." *ATransC Collective.* 2013. atransc.org/bp/Classification.

Index